To Mike Ca

compli

N. Sharma

May 9, 2002

Sawai Jai Singh and
His Astronomy

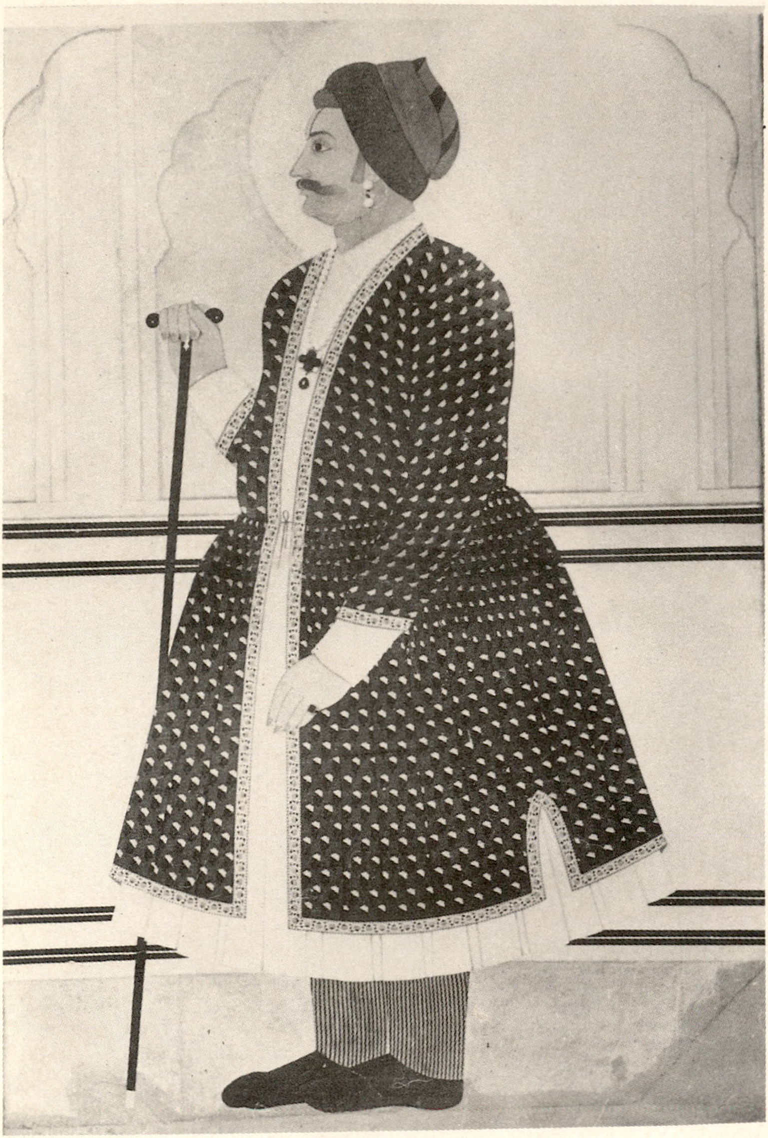

Sawai Jai Singh (1688-1743)
(Courtesy Sawai Man Singh II Museum, Jaipur)

Sawai Jai Singh and His Astronomy

VIRENDRA NATH SHARMA

MOTILAL BANARSIDASS PUBLISHERS
PRIVATE LIMITED • DELHI

First Edition: Delhi, 1995

ISBN: 81-208-1256-x

Also available at:

MOTILAL BANARSIDASS

41 U.A. Bungalow Road, Jawahar Nagar, Delhi 110 007
120 Royapettah High Road, Mylapore, Madras 600 004
16 St. Mark's Road, Bangalore 560 001
Ashok Rajpath, Patna 800 004
Chowk, Varanasi 221 001

PRINTED IN INDIA

BY JAINENDRA PRAKASH JAIN AT SHRI JAINENDRA PRESS,
A-45 NARAINA, PHASE I, NEW DELHI 110 028
AND PUBLISHED BY NARENDRA PRAKASH JAIN FOR
MOTILAL BANARSIDASS PUBLISHERS PVT. LTD.,
BUNGALOW ROAD, DELHI 110 007

Dedicated
To
My Parents

PREFACE

Around 1978, I first became interested in Jai Singh's instruments. Jai Singh built his masonry instruments in the day and age of the telescope. The question, why someone would build instruments of masonry and stone in the age of the telescope, intrigued me. I found no convincing answers in the existing literature. I also found that literature on Jai Singh's astronomical endeavors was scanty. In fact, the last authentic book on Jai Singh's instruments had been written almost 60 years ago. Trying to seek an answer to the question of why Jai Singh built his instruments the way he did, and also to learn more about the instruments, I researched primary sources and investigated the instruments in detail. This book is an outcome of this effort.

The resource materials on Sawai Jai Singh's astronomy are scattered in many languages, such as Rajasthani-Hindi, English, French, German, Latin, Portuguese, Persian, and Sanskrit. I have relied on my friends and colleagues to translate the materials for me. The translations were done by Anne Hintz (French), Lila Huberty (French), Rudolph J. Schlueter (German); Yaqub Oomar and Mahboob A. Siddiqi (Persian); late Fr. Leo Bourque (Latin), and Eugene Gibas (German, French, Portuguese). I am grateful to these friends.

I am indebted to S.M.R. Sarma for reviewing the draft chapters of this book and for making useful suggestions and comments. I am also thankful to S.M.R. Ansari, Anjani Mehra, Gopal Narain Bahura, and David Pingree, whose comments have benefitted the book.

Sincere appreciation is expressed to A.K. Das, Yaduvendra Sahai, the authorities of the Sawai Man Singh II Museum, Jaipur, for making the museum holdings readily available to me for my study. Om Prakash Sharma, the supervisor of Jantar Mantar, Jaipur, saw to it that my investigations at the Jantar Mantar should proceed smoothly. I appreciate his help.

The photographic developments and enlargements were done by Thomas Frantz. J. Asha Sharma painstakingly edited the manuscript. I am thankful to her. General thanks are offered to many people, particularly, Eugene Gibas, who from time to time aided me in this project. The funds for the research were provided by the Smithsonian Institution, National Science Foundation of USA (Grant No. INT-8018996), UW-Fox Cities Foundation, and the UW-Centers.

The project of research and writing the book required me to spend many hours, which often came at the expense of my family. I express my appreciation for the patience and understanding displayed by my children and by my wife, Manorama Sharma.

August 1, 1994

Virendra Nath Sharma

PREFACE

CONTENTS

ILLUSTRATIONS

Chapter 2

Chapter 3

Chapter 4

Chapter 5

Chapter 6

Chapter 7

Chapter 8

Chapter 9

Chapter 10

Chapter 11

Chapter 12

ABBREVIATIONS

Bahura I	Gopal N. Bahura, *Catalogue of Manuscripts in the Maharaja of Jaipur Museum*, Jaipur, 1971
Bahura II	Gopal N. Bahura, *Literary Heritage of the Rulers of Amber and Jaipur*, Jaipur, 1976
Bhatnagar	V.S. Bhatnagar, *Life and Times of Sawai Jai Singh*, Delhi, 1974
Bhavan	Gokulchandra Bhavan, *A Guide to the Observatories in India*, (in Hindi), Varanasi, 1911
Bibliography	A. Rahman, *Science and Technology in Medieval India—A Bibliography of Source materials in Sanskrit, Arabic and Persian*, New Delhi, 1982
BORI	Bhandarkar Oriental Research Institute, Pune
CESS	David Pingree, *Census of the Exact Sciences in Sanskrit*, 4 Vols., Philadelphia, 1970-1981
DK	Dastūr Kaumvār records of the Jaipur state, Rajasthan State Archives, Bikaner
Garrett	A. ff. Garrett, *The Jaipur Observatory and Its Builder*, Allahabad, 1902
Hunter	William Hunter, "Some Account of the Astronomical Labours of Jayasinha, Rajah of Ambhere, of Jayanagar," *Asiatic Researches*, No. 5, pp. 177-211, (1799)
Kaye	George R. Kaye, *The Astronomical Observatories of Jai Singh*, Calcutta, 1918, reprint, New Delhi, 1982
Lettres	*Lettres Édifiantes et Curieses, concernant L'Asie, L'Afrique et L'Amérique, avec Quelques Relations Nouvelles des Missions et des Notes Géographiques et Historiques*, Tome Deuxiéme, Paris, 1843
Museum	The Sawai Man Singh II Museum, City Palace, Jaipur
Pannekoek	A. Pannekoek, *A History of Astronomy*, New York, 1961
Ptolemy	*Almagest* or *Suntaxis Matematikè* of Claudius Ptolemy, published as *Ptolemy's Almagest*, tr., G. J. Toomer, New York, 1984
R.S.A.	Rajasthan State Archives, Bikaner
RORI	Rajasthan Oriental Research Institute, Jodhpur
Sayili	Aydin Sayili, *The Observatory in Islam*, Ankara, 1960
SSMC	*Siddhānta-samrāṭ*, ed., Murlidhar Chaturveda, Sagar Univ., 1976
SSRS	*Samrāṭ-siddhānta*, ed., Ram Swarup Sharma, New Delhi, 1967

Tieffenthaler	*Description Historique et Geographique de l'Inde*, ed., Jean Bernoulli, Vol. I, Berlin, 1786
Yantraprakāra (J)	*Yantraprakāra* of Sawai Jai Singh, Ms. 261/31-MJM or No. 31, Sawai Man Singh II Museum, Jaipur
Yantraprakāra (S)	*Yantraprakāra of Sawai Jai Singh*, ed. and tr., Sreeramula Rajeswara Sarma, *Supplement to Studies in History of Medicine and Science*, Vols. X and XI, New Delhi, 1986, 1987
ZMS	*Zīj-i Muḥammad Shāhī*, Ms. Add. 14373; Dept. of Oriental Mss., The British Lib., London, England

CHAPTER I

SAWAI JAI SINGH AND HIS TIMES

The Rulers of Amber

Sawai Jai Singh (Savā'ī Jaya Siṁha), the statesman-astronomer of India, was born to a royal family on November 3, 1688 in the town of Amber in the state of Rajasthan, India.[1] His ancestors belonged to the Kachavāhā dynasty of the Rajputs founded in 967 by Dūlaha Rāya (d. 1006) after he conquered small towns in the countryside surrounding the present city of Jaipur.[2,3] The hill town of Amber, which became the capital of the Kachavāhā rulers for the next 700 years, was conquered by Dūlaha Rāya's son, Kākila Deva (d. 1036). The rulers of the Kachavāhā clan remained petty chieftains for a long time until Bhāramala, or Bihārīmala (d. 1573), recognizing the ascendancy of Mughal power, accepted the suzerainty of the emperor Akbar in 1562, and thereby made the house of Amber the most influential of all the Rajput houses serving the Mughals. Bhāramala's descendants, Māna Singh (1543-1614) and Mirzā Raja Jai Singh (1611-1667), were highly regarded nobles at the Mughal court and were entrusted with important missions.

By the time Jai Singh was born, the house of Amber had lost much of its influence at the imperial court, such that Jai Singh's father Raja Biśana Singh or Viṣṇu Singh (ruled 1689-1700) occupied a relatively minor post in the imperial service. The emperor deployed him to suppress an insurrection around Mathura, a holy city about 150 km south of Delhi, and later posted him to Kabul, Afghanistan, where he passed away at the age of 27.

Education

The Kachavāhā rulers were patrons of learning and paid careful attention to the education of their offspring.[4] They hired competent teachers and sent their sons to study at Varanasi, several hundred kilometers away. Mirzā Raja Jai Singh, the great-grandfather of Jai Singh, sent his two sons there to study at a Sanskrit college.[5] Viṣṇu Singh, Jai Singh's father, also paid close attention to the education of his sons although he had to be away from his home for extended

---

[1] For a biography of Sawai Jai Singh, see Bhatnagar.

[2] Harish Chandra Tikkiwal, *Jaipur and the Later Mughals*, pp. 1-5, Jaipur, 1974.

[3] Harnath Singh Rawal, *Genealogical Table of the Kachavāhās*, f. 3, Jaipur.

[4] Mirzā Raja Jai Singh patronized the famous Hindi poet Bihārī.

[5] A French traveller Jean Baptiste Tavernier confirms in his memoirs that the princes studied at the college. *Travels in India by Jean Baptiste Tavernier*, tr., V. Ball, Vol. II, pp. 234-235, London, 1889.

periods of time. He received progress reports of his sons' education from trusted servants. Jai Singh's early education was similar to that of any other prince of his time. He learned Hindi, Sanskrit, Persian and possibly some Arabic. Along with these languages he also learned mathematics and, of course, the martial arts to prepare him for a Rajput's career in the army.

Jai Singh displayed an interest in mathematics and astronomy at an early age. He had two manuscripts on astronomy copied for him when he was only thirteen. The Sawai Man Singh II Museum of Jaipur still preserves these manuscripts.[1] The *Zīj-i Muḥammad Shāhī*, an astronomical text which Jai Singh compiled later, reads:[2]

> " . . . Sawai Jai Singh from the first dawning of reason in his mind, and during its progress toward maturity, was entirely devoted to the study of mathematical science (astronomy), and the bent of his mind was constantly directed to the solution of its most difficult problems; by the aid of the supreme artificer he obtained a thorough knowledge of its principles and rules."

Jai Singh's formal education was cut short at the age of eleven when his father died, and he had to take charge of the administration. He ascended the throne of Amber on January 25, 1700. However, he continued his studies along with the discharge of his princely duties and soon acquired mastery over the subjects of astronomy and mathematics.

Jai Singh the Ruler

When Jai Singh inherited the rule of his state, the Mughal empire was on the decline. Shortsighted policies of the aging monarch, Aurangzeb (ruled 1658-1707), had alienated the majority of the population, and rebellions were breaking out in all parts of the empire. Immediately after the death of Jai Singh's father, the emperor summoned Jai Singh to the imperial court in the South, where the emperor was making futile attempts to suppress a guerrilla war waged by the Marathas. In the South, Jai Singh met Jagannātha, a young man well versed in mathematics and astronomy. The contact with Jagannātha apparently further stimulated the budding interest of the young prince in astronomy and mathematics.[3,4]

While Jai Singh was still in the South, the aging Emperor passed away (1707), and a war of succession broke out between his sons. In this war Jai Singh

---

[1] These manuscripts are *Bhāsvatī* (No. 5572), and *Vārāhī Saṁhitā* (No. 5569), Bahura II, p. 54.

[2] This translation is based on Hunter.

[3] Bahura II, p. 54.

[4] For further details about Jagannātha see Ch. 12.

fought on the losing side. At the critical moment of a crucial battle, he switched over to the winning side. But the victor, Aurangzeb's eldest son, the future Emperor Bahādur Shāh, did not look favorably upon Jai Singh's act of switching sides. He annexed Jai Singh's ancestral state to the imperial domain. However, Jai Singh and the other Rajput chieftains, who had also lost their ancestral lands to the imperial domain, united themselves. They drove the Mughals out of their territories and forced the newly crowned Emperor to concede his claim to their hereditary lands. In exchange, Jai Singh, along with the other Rajputs, accepted the Emperor as his suzerain and promised to serve the empire as his forefathers had done before.

During the decade following this episode, the political situation in Delhi became very uncertain. Several emperors ascended the "peacock throne" in succession, only to be deposed one after the other by natural or unnatural means. Finally, the "king makers" at the Imperial Court installed a great-grandson of Aurangzeb as the emperor and a semblance of stability returned to the country. The newly installed emperor, Muḥammad Shāh, ruled for a long time, and Jai Singh remained on reasonable terms with him. Jai Singh served the empire in various capacities, such as governor of the provinces of Agra and Malwa. Jai Singh's acceptance of Muḥammad Shāh as his overlord was more an act of convenience than anything else, as he and the Emperor never trusted each other.

Jai Singh spent his life during one of the most troubled, uncertain, and critical periods of Indian history, and he was involved directly or indirectly in just about every conflict of his time. When Jai Singh ascended the throne of Amber, his state had shrunk considerably in prestige and influence. The total area under direct control of the house of Amber was no more than 3000 square miles,[1] and that too was taken away from his possession by the Emperor Bahādur Shāh. However, through his Machiavellian policies in which the end justified the means, Jai Singh extended his power and influence far beyond the three districts he had inherited. At one time, he and another Rajput raja, Ajīta Singh of Jodhpur, "held all the country from thirty *kosa* (about 100 km) of Delhi, where the native land of Jai Singh began, to the shores of the sea at Surat (a prosperous port on the west coast of India)."[2] In the 1730's, from the gates of Delhi to the banks of the river Narmada in central India, Jai Singh was the supreme authority.[3] A French Jesuit traveller, Claude Boudier, confirms that Jai Singh indeed possessed a great deal of authority throughout the Mughal

---

[1] Bhatnagar, p. 270.

[2] Khāfī Khān's contemporary chronicles translated by Charles Elliot and published as *The Later Mughals*, p. 100, 3rd ed., Susil Gupta, Calcutta, 1960.

[3] Muḥammad Shafī Wārid, *Tārīkh-i Hindī*, pp. 119-120, as quoted by William Irvine in the *Later Mughals*, Vol. 2, p. 278, reprint, New Delhi, 1971.

empire. "The passports issued in the name of the Raja," he writes, "were more respected than those issued in the name of the Emperor."[1]

With great wealth and resources available to him as a powerful ruler, Jai Singh embarked upon an ambitious program of reviving astronomy in his country. To this effect, he designed instruments, built observatories, compiled an excellent library assembled competent astronomers of different scientific backgrounds, and sent a fact-finding scientific mission to Europe. His scientific career lasted for more than 20 years until his death in 1743 at age 54.

Survey of Astronomy in India

When Jai Singh embarked upon his quest of infusing new life into Indian astronomy, there were primarily two different traditions of astronomy, namely, the Hindu and the Islamic, and these traditions had become, by and large, stagnant.

The Hindu School of Astronomy

Hindu astronomy has its roots in the pre-Christian era. Around 500 B.C., Lagadha wrote *Vedāṅga Jyotiṣa*, the only text available to us on the subject from that period. The text contains only 43 verses.[2] When Indian civilization came in contact with the Greeks in the third century B.C., elements of Greek astronomy began to permeate Hindu astronomy. From about 400 A.D. onward, Hindu astronomy developed rapidly, and the astronomers composed a series of astronomical texts or *Siddhāntas*. These *Siddhāntas*, or canons of Hindu astronomy, contained some Greek ideas and also original concepts regarding mathematics and astronomical computations. Varāhamihira (d. 587), in his *Pañcasiddhāntikā*, summarizes five *Siddhāntas* popular in his time. The most important of these *Siddhāntas* was the *Sūryasiddhānta*. The famous astronomers of the period between 400 A.D. and 1200 A.D. are Āryabhaṭa (b. 476), Bhāskarācārya I (*fl.* 629), Brahmagupta (b. 598), Lalla (*fl.* 8th c.), Vateśvara (*fl.* 900), Śrīpati (*fl.* 1000) and Bhāskarācārya II (b. 1114). These astronomers wrote commentaries on earlier works, determined better parameters of planetary motions, and devised instruments of observation.

---

[1] Extract from a letter of Claude Boudier to Etienne Souciet dated 17 January 1736. Archives des Jesuits de la Province de Paris, Fonds Brotier, Vol. 68, ff. 143-45, as quoted by William A. Blanpied in "The Astronomical Program of Raja Sawai Jai Singh II and Its Historical Context," *Japanese Studies in the History of Science*, No. 13, p. 94, (1974).

[2] The text of Vedāṅga Jyotiṣa has been published by K. V. Sarma. Sarma believes that Lagadha wrote his text about 1180 B.C. See *Vedāṅga Jyotiṣa of Lagadha in its Ṛk and Yajus Recensions*, p. 7, New Delhi, 1985.

Theoretical progress in Hindu astronomy came to a standstill after Bhāskarācārya II.[1] The astronomers after Bhāskarācārya II were content to write commentaries on early works or by composing *karaṇas*, or practical manuals, to simplify the task of calculations which were based on other *siddhāntas*.[2,3] No doubt some of these texts were well written and became famous in time; nonetheless, the fact remains that no new substantial knowledge was added to the existing stock. In fact, by the time of Kamalākara (*fl.* 1658), the Hindu astronomers, including Kamalākara himself, thought any new knowledge to be impossible. They advocated the supremacy of the ancient texts, such as the *Sūryasiddhānta*. They went a step further and advocated that any concepts not based on the *Sūryasiddhānta* must be presumed false.[4] For instance, according to Kamalākara, an *amāvasyā*, or new moon, is not the time of the conjunction of the moon and the sun, as it ought to be, but a calculated result based on the *Sūryasiddhānta*.[5] This tendency became so pervasive with time that astronomers such as Nityānanda (*fl.* 1628/1639) found it necessary to cloak the facts borrowed from the Islamic school of astronomy with a fable that the facts were not new but had already been revealed to Bhāskara (Romaka) by the sun god, *Sūrya*, when he was born in the *Yavana* country under the curse of the Lord Brahmā.[6] The emergence of such attitudes greatly hampered the progress of astronomy in India.

The noteworthy astronomers of the period between Bhāskarācārya II and Sawai Jai Singh included Mahādeva (*fl.* 1316), Padmanābha (*fl.* 1398), Makaranda (*fl.* 1478), Keśava II (*fl.* 1496), Gaṇeśa Daivajña (*fl.* 1520), and Nityānanda (*fl.* 1628/1639). Mahādeva (*fl.* 1316) compiled planetary tables, known as the *Mahādevī Sāraṇī* in 1316 for calculating the apparent and mean positions of the planets. Makaranda wrote a text for preparing a *pañcāṅga*, or almanac, around 1478. Keśava was a good observer. He wrote *Grahakautuka* in which he used his own observations. Keśava's son Gaṇeśa Daivajña was an astronomer of high

---

[1] For Hindu astronomy after Bhāskarācārya see Shankar Balkrishna Dikshit, *History of Indian Astronomy*, tr., R. V. Vaidya, Vol. 2, pp. 114 ff., Delhi, 1969. Also Gorakh Prasad, *Bhāratīya Jyotiṣa kā Itihāsa*, (in Hindi), pp. 186 ff, Hindi Samiti, Govt. of Uttar Pradesh, Lucknow, 1974.

[2] In a *karaṇa*, a contemporary date is chosen as the epoch by its author on which a conjunction of the moon with its higher apsis occurs. The author then gives the longitudes of the planets for the epoch chosen. In a *karaṇa*, corrections to the astronomical parameters are also applied. See *History of Astronomy in India*, ed., S. N. Sen and K. S. Shukla, p. 11, New Delhi, 1985.

[3] For the *Karaṇa* and *Siddhānta* literature, see David Pingree, *Jyotiḥśāstra, A History of Indian Literature*, ed., Jan Gonda, Vol. 6, pp. 17-40, Otto Harrassowitz, Wiesbaden, 1981. Also Sen and Shukla, *Op. Cit.*, pp. 7-15.

[4] Dikshit, *Op. Cit.*, pp. 162-163.

[5] Gorakh Prasad, *Op. Cit.*, p. 196.

[6] For a discussion of the *Zīj-i Nityānandī Shāhjahānī*, see David Pingree, "Islamic Astronomy in Sanskrit," *J. for the History of Arabic Science*, No. 2, Vol. 2, pp. 315-330, (1978). Pingree writes, "The most enlightened Indian astronomer (Nityānanda) of the seventeenth century did not dare to base his claims on the evidence of the senses." Nityānanda completed his text in 1628.

caliber and wrote many texts on the subjects of astronomy. His *Grahalāghava* became very popular throughout the country. Nityānanda wrote *Siddhāntasindhu* in 1639. As the inventory of Jai Singh's library suggests, he had access to most of this literature.[1]

Texts on Instrumentation

During the period between Bhāskarācārya II and Jai Singh, astronomers recognized the need for observations. They either devised a number of observation instruments of their own or borrowed designs from outside India. At least one such instrument whose origin we can trace from outside India was the astrolabe. Mahendra Sūri wrote a text on the astrolabe in 1370, which was based on an Islamic text. Padmanābha (*fl.* 1398) wrote *Yantraratnāvali*, a treatise on instruments, in which he describes the Dhruvabhrama yantra, an instrument for telling time at night.[2] Cakradhara wrote *Yantracintāmaṇi*, a treatise describing the quadrant, on which Rāma Daivajña wrote a *ṭīkā*, or commentary, in 1625.[3] The instruments described in these texts were portable and were not necessarily erected in masonry and stone. The importance of observation and periodic revision of the astronomical parameters was recognized by astronomers such as Nīlakaṇṭha. In his *Jyotirmīmāṁsā* he emphasizes this fact.[4]

Instruments

The early instruments of Hindu astronomy, as described in the *Sūryasiddhānta*, were rather simple. However, by the early 18th century, astronomers had constructed a large variety of more complex instruments. These instruments ranged from a simple *Śaṅku*, an upright rod, to highly sophisticated astrolabes.[5] In Table 1-1 we give a brief survey of the instruments of the pre-Jai Singh era.

A comparatively large number of time-measuring instruments in Table 1-1 suggests that Hindu astronomers paid more attention to measuring time than to measuring the coordinates of stars and planets. Further, descriptions of their instruments suggest that the instruments were portable, small in size, and that the least count of those measuring coordinates had been no better than a fraction of a degree.

---

[1] R.S.A., Jaipur State, Pothikhana records.

[2] CESS, Series A, Vol. 4, pp. 170-172.

[3] CESS, Series A, Vol. 3, pp. 36b-37b; also, Vol. 4, p. 88. Also see Pingree, Jyotiḥśāstra, *Op. Cit.*, p. 53.

[4] K. V. Sarma, "A Survey of Source Material," in Sen and Shukla, *Op. Cit.*, p. 16.

[5] R. N. Rai, "Astronomical Instruments," in Sen and Shukla, *Op. Cit.*, pp. 308-336.

Time Measuring Instruments

	Instrument	Author/Text	Type
1.	Naḍikā yantra	Vedāṅga Jyotiṣa	Clepsydra,[1] outflow water clock
2.	Ghaṭikā yantra	Sūryasiddhānta	Sinking bowl clepsydra[2]
3.	Nāḍīvalaya	Bhāskarācārya II	Equinoctial sundial
4.	Phalaka yantra	Bhāskarācārya II	A kind of sundial
5.	Kartarī yantra	Lalla and Śrīpati	Equinoctial sundial
6.	Dhruvabhrama yantra	Padmanābha	Measuring time by the rotation of the Big Dipper
7.	Kapāla yantra	Varāhamihira	Hemispherical sundial
8.	Pratoda yantra	Gaṇeśa Daivajña	Whip-shaped gnomonic device[3]
9.	Yantrarāja	Mahendra Sūri	Astrolabe

Coordinate Measuring Devices

	Instrument	Author	Type
1.	Turīya yantra	Cakradhara	Quadrant[4]
2.	Golānanda	Cintāmaṇi Dīkṣita	Armillary sphere[5]
3.	Yantrarāja	Mahendra Sūri etc.	Astrolabe

Table 1-1 Instruments in Hindu astronomy before Jai Singh.

---

[1] Sreeramula Rajeswara Sarma, "Water Clocks and Time Measurements in India," to appear in *Aligarh J. Oriental Studies*.

[2] *Sūryasiddhānta*, 13.23.

[3] *Pratoda Yantra by Sh. Gaṇeśa Daivajña*, ed. by Shakti Dhar Sharma, Kurli, 1982.

[4] *Yantracintāmaṇi* of Cakradhara, *ṭīkā* by Rāma Daivajña. CESS, *Op. Cit.*, Ref. 23.

[5] Dikshit, *Op. Cit.*, pp. 232-233, CESS, Series A, Vol. 3, pp. 50a-50b; and Vol. 4, p. 94.

Observatories in India

Information on observatories in India is meager. However, we do know that a number of prominent astronomers, patronized by kings, carried out their own observations. These observations are mentioned in *karaṇas*, or practical manuals. The places of these observations, if operated for a reasonable period of time, technically could be called observatories. A court astronomer, Śaṅkaranārāyaṇa (*fl.* 869), mentions a place with instruments in the capital city of king Ravi Varmā of Kerala.[1] ᶜAbd al-Rashīd al-Yāqūtī (15th c.) reports an observatory in the city of Jājilī.[2] Similarly, ᶜAbdu'llāh Shukrī al-Qunawī (16th c.) mentions observatories at the cities of al-Kankadaz or al-Dharkanak.[3] The two cities remain unidentified, however. According to certain authors, one of these cities may be identified with Ujjain because a *Kanak-dar* bearing resemblance with al-Dharkanak is described in some Persian-Arabic literature as a place east of Arabia with the zero degree longitude.[4,5] Ujjain did indeed have the zero degree longitude designation in the astronomy of India. The Emperor Humayun (d. 1556) is reported as having a personal observatory at Kotah, near Delhi, where he himself took observations.[6]

Nadvi claims that the Emperor Shahjahan (ruled 1627-1658) contemplated building an observatory at Jaunpur, a place about 300 km east of Delhi, but did not go ahead with the plans.[7] Although there were similar plans during Shahjahan's reign, no observatories were built during the Mughal rule before Jai Singh. During the Mughal period, observations were conducted by individual astronomers on a disorganized basis.

The Islamic School of Astronomy in India

With the invasion of India from the northwest, beginning in about 1000 A.D., came the art and science of the Muslim world. Later, with the ascendancy of the Mughals in the mid-16th century, a large number of scholars migrated to

---

[1] Unfortunately, Śaṅkaranārāyaṇa does not give any details about the place or the instruments the observatory had. Sen and Shukla, *Op. Cit.*, p. 16.

[2] Sayili, p. 359.

[3] *Ibid.*

[4] For instance, see Ansari in *History of Indian Astronomy*, ed., Sen and Shukla, *Op. Cit.*, pp. 363-364.

[5] O'Leary mentions, "Brahmagupta (*circ.* 628) ... worked in Ujjain where there was an observatory." However, O'Leary does not give any reference for his information. Brahmagupta did carried out his own observations, but it is open to question if there was an observatory at Ujjain. See De Lacy O'Leary, *How Greek Science Passed to the Arabs*, p. 105, London, 1979.

[6] *Tauzak-i Jahāngīrī* (Memoirs of Jahangir), edited by Sayyid Aḥmad, Aligarh, p. 267, cited by Blanpied, *Op. Cit.*, p. 112.

[7] Sayyid Sulaimān Nadvi, "Muslim Observatories," *Islamic Culture*, Vol. 20, pp. 280-281, (July 1946).

India from the Islamic countries to seek fame and fortune. The scholars brought with them astronomical literature, such as the *Zījes*. "A *Zīj* consists essentially of the numerical tables and accompanying explanations sufficient to enable a practicing astronomer, or astrologer, to solve all the standard problems of his profession, i.e., to measure time and to compute planetary and stellar positions, appearance of the moon, and eclipses."[1] The *Zījes* brought from abroad were copied in India, and some indigenous ones were also produced. According to Khan Ghori, Maḥmūd bin Umar prepared the first such *Zīj* on Indian soil sometime during the reign of Iltutmish (1210-1236). This *Zīj* was written a little before the Maragha observatory in Central Asia came into operation around 1259.[2] Another text, *Zīj-i Jāmi'-i Maḥmūd Shāh Khiljī*, was prepared around 1448 and then revised in 1661-62. The author of this *Zīj* was apparently familiar with al-Kāshī's *Zīj-i Khāqānī* and with the *Ilkhānic* tables.

A large amount of *Zīj* literature of different levels of sophistication was available to a scholar of Persian and Arabic in India during the Mughal period. Abu'l Fadl, a well respected courtier of Akbar, counted 86 different *Zījes* in India.[3] During the reign of Akbar, Mullā Chānd wrote a *Tas'hīl*, or gloss on the *Zīj-i Ulugh Begī*, a copy of which is preserved at the Sawai Man Singh II Museum of Jaipur.[4] In 1597, during the reign of Shahjahan, Farīd al-Dīn wrote *Zīj-i Shāhjahānī*, which was also based on *Zīj-i Ulugh Begī*. These *Zījes* did not incorporate any new observations but were updated versions of the *Ulugh Begī Zīj*. Jai Singh had a copy of the *Ulugh Begī Zīj* in his library.[5]

The Islamic Elements in Hindu Astronomy

The assimilation of the Islamic school of astronomy into Hindu astronomy was slow.[6] The first noticeable assimilation took place during the reign of Fīrūz Shāh Tughlak (ruled 1351-1388), when Mahendra Sūri wrote a treatise on the astrolabe. Mahendra Sūri spent some time at the court of Fīrūz Shāh Tughlak, and there he composed his work about 1370, after learning the construction and use of the instrument from a Muslim astronomer.[7] A commentary on the works was written by Malayendu Sūri more than ten years later in 1382.

---

[1] E. S. Kennedy, "A survey of Islamic Astronomical Tables," *Transactions of the American Philosophical Society*, New Series, Vol. 46, Part 2, p. 123, May 1956.

[2] S.A. Khan Ghori, "Development of *Zīj* Literature in India," in Sen and Shukla, *Op. Cit.*, p. 30.

[3] Khan Ghori, *Op. Cit.*, pp. 45-47, has reproduced the list of the *Zījes* from *A'in-i Akbrī*.

[4] Bahura I, p. 72.

[5] R.S.A., Jaipur State, Pothikhana records.

[6] For the assimilation of the Islamic elements in Hindu astronomy, see David Pingree (1978), *Op. Cit.*

[7] *Yantrarāja* of Mahendra Sūri, 2. Nirnaya Sagar Press, Bombay, 1936. Also CESS, Series A, Vol. 4, pp. 393-395.

Mālājīta Vedāṅgarāya wrote *Pārasīprakāśa* in 1643, in which he gave the equivalent Sanskrit words for the astrological terms in Arabic and Persian.[1] A similar work was written by Vṛjabhūṣaṇa in 1660. Pingree has traced elements of Islamic *Zījes* in several Sanskrit texts. For example, he has found such elements in Viśrāma's *Yantraśiromaṇi* composed in 1615, in Haradatta's *Jagadbhūśaṇa Sāraṇī*, written in 1638, and later in the Trivikrama's tables, compiled in 1704.[2]

Kamalākara, who has been mentioned in another context above, was well aware of certain elements of Islamic astronomy. For example, in his *Siddhāntatatvaviveka* he included a number of topics, such as clouds, hail, earthquakes, meteors, and geometrical optics that are not found in Hindu books.[3] Jai Singh, because of his unique position as a man of means, had easy access to all this literature.

SURVEY OF EUROPEAN ASTRONOMY

The Revival of the Heliocentric Theory

By the time Jai Singh embarked upon his pursuit of astronomy, European astronomy had made remarkable strides. Great advances had been made in the theoretical and the observational aspects of this science. Copernicus (1473-1543) had published his *De Revolutionibus Orbium Coelestium*, or *On the Revolutions of the Heavenly Spheres*, in 1543. In this text, Copernicus revived the Greek idea of a heliocentric cosmos. Then 66 years later, in 1609, Kepler (1571-1630) published *The New Astronomy*, in which he expounded the first two laws of planetary motion. Ten years later in 1619, Kepler published his third law of planetary motion. Finally, in 1687, a year before Jai Singh's birth, Newton published the *Principia*, in which he gave celestial mechanics a solid theoretical footing. In his book, Newton described the laws of mechanics and those of universal gravitation, with which orbital motions of celestial bodies could be calculated with great certainty.

Mathematics

Europe had made notable progress in mathematics also. For trigonometric functions of angles, Rheticus (1514-76) developed tables in 15 figures with angles in increments of 10″ of arc. A reduced version of these tables was

---

[1] Sen and Shukla, *Op. Cit.*, p. 18

[2] Pingree believes that the *Ulugh Begī Zīj* had been available to Hindu astronomers at Delhi and Varanasi in Sanskrit translation.The copy of the *Ulugh Begī Zīj* brought from Surat and now preserved at Jaipur does not seem to be the one available to the astronomers at Delhi and Varanasi. For the details of this *Zīj* see Ch. 11.

[3] Dikshit, *Op. Cit.*, p. 164, and Pingree (1978), *Op. Cit.*, Ref. 19, p. 323.

published in 1596.[1] Advances were also made in developing useful formulae in spherical trigonometry. The logarithm, which made calculations easier, was invented by John Nepier of Scotland, and in 1628 the logarithmic tables in the decimal system were published in Holland. The tables were extended soon afterwards to 10 decimal places, making the task of astronomical calculations much easier. Jai Singh acquired a book, *Description and Use of the Sector and Other Instruments*, published in 1637, which contained the logarithm of numbers, in nine digits.[2]

In the 17th century, Blaise Pascal (1623-62) and others invented geared calculating machines.[3] The calculating devices were important, because in that age mathematical techniques were still rather crude. The most important instrument for calculating is said to have been the sector, a hinged and graduated ruler.[4] The instrument remained in use in one form or other until the 18th century.[5]

Instrumentation and Observatories

The beginning of a great renaissance in instrument making took place in the second half of the 15th century in Germany.[6] Nuremberg and Augsburg became the centers of instrument making, and they produced instruments of considerable ingenuity and of exquisite craftsmanship. This continued until the Thirty Years War (1618-1648). From Germany the interest in instrument making spread to Western Europe. By the last quarter of the 16th century, the craft of instrument making found fertile ground in England, France, Italy, Denmark, and Holland. Tycho Brahe's (1546-1601) instruments were built in the workshops of Augsburg, Germany.

While measuring an angle in the sky, an astronomer encounters several problems such as proper alignment of his sighting device and accurate reading of the graduations. Tycho Brahe obtained his alignment by using multiple slits arranged in a square configuration.[7] Later astronomers solved the problem of alignment with the telescope. Because the large-size instruments are difficult to handle, the astronomers of this period opted for comparatively small instruments

---

[1] Pannekoek, p. 200.

[2] Edmond Gunter, *Description and Use of the Sector and Other Instruments*, London, 1636. The Sawai Man Singh II Museum has a copy of this book. Although the Museum does not give the date of the acquisition of this work, it is certain that it was Jai Singh who acquired it sometime after 1730.

[3] Charles Singer, E. J. Holmyard, A. R. Hall and Trevor I. Williams, editors; *A History of Technology*, Vol. III, p. 627, London, 1957.

[4] *Ibid*, p. 627.

[5] *Ibid*, p. 628.

[6] *Ibid*, p. 585.

[7] Victor E. Thoren, "New Light on Tycho's Instruments," *J. for Hist. Ast.*, Vol. 4, pp. 27-29, (1973).

with imaginative division schemes so that fractions of a degree could be measured. Tycho used transversals or diagonal lines on the degree divisions.[1] With this method, he artificially lengthened the interval between divisions without going to a larger arc.[2] The Vernier scale, which made this artificial lengthening of divisions unnecessary, came into use about 1630.[3] The micrometer, which greatly facilitated fine measurements, was invented by William Gascoigne around 1640.[4]

An excellent book on the subject of instrumentation was written in 1709 by Nicolos Bion (d. 1733), the chief instrument maker of the king of France. The book's English translation, with a supplement written by Edmund Stone (1700-1768), appeared in 1758.[5] Bion describes just about every type of instrument in use in the 17th and early 18th centuries.

The Telescope

The true history of the invention of the telescope is unknown. The telescope was certainly known in Europe by 1608.[6] Galileo (1546-1642) heard about the telescope in 1609 and built one for himself. After improving it, Galileo directed the instrument toward the sky and observed lunar mountains, phases of the planet Venus, and the moons of Jupiter. After Galileo's observations, the course of astronomy began to be dominated by this instrument. Astronomers were slow in accepting the telescope as a measuring apparatus. Its general acceptance had to await the invention of the cross-wire at the prime focus. The credit for using a cross-wire for astronomical purposes goes to Jean Picard at the Royal Observatory of Paris in 1667. In the beginning, because of the relatively crude optics of the instrument, improvements in the technique of observing were not impressive. Observers such as Hevelius of Danzig could match telescopic measurements with their non-telescopic sights. However, when Flamsteed, using a refined instrument, recorded positions of stars with a precision of ± 10 arc sec, naked-eye observations became things of the past, and no quadrant or sextant was used without a telescopic sight after that. A missionary, Jean Richaud (1633-1693), brought a telescope to India from Siam

---

[1] The transversal method of increasing the effective width of a division was invented much before Tycho. The author has seen it on the drawing of a quadrant in *Kitāb Taᶜlīm Ālāt Zīj* of ᶜAbd al-Munᶜim al-ᶜĀmilī (16th century). Ms., British Museum, PS 8320580, Add. 7702.

[2] *Tycho Brahe's Description of His Instruments and Scientific Works*, tr., Hans Raeder, Elis Stronmgren and Bengt Stromgren, pp. 141-144, Copenhagen, 1946.

[3] Pierre Vernier, *Encyclopedia Britannica, Micropedia*, Vol. 12, p. 325, (1987).

[4] Pannekoek, p. 258.

[5] *Construction and Principal Uses of Mathematical Instruments* of Nicolos Bion, tr. by Edmund Stone, London, 1758, reprint, London, 1972.

[6] For a history of the telescope, see Pannekoek, pp. 253-260.

around 1689 and conducted some telescopic observations.[1] Jai Singh had at least one telescope in his possession, and this we will discuss in a later chapter.

Clocks

The measurement of time is an important aspect of astronomical observations. In the late 14th century, mechanical clocks began to replace sundials and other time-reckoning devices. The first mechanical clock is said to have been made by Giovanni de Dondi between 1348 and 1362. This clock indicated the motions of the sun, the moon, and the five planets with a series of gears. The use of the pendulum as a controlling device to achieve precision was suggested by Galileo and independently discovered by Huygens (1629-95). The pendulum clocks fulfilled the need for the accurate measurement of time demanded by astronomers. To achieve a high degree of accuracy, Thomas Tompion in 1676, built two spring winding clocks with 13-foot pendulums, beating two seconds, for the Greenwich observatory. These clocks could go without winding for almost one year. Flamsteed (1646-1719) used these clocks in his work on astronomical tables.[2]

Observatories

By the time Jai Singh decided to build his observatories, a number of important observatories had been built in Europe. The most famous of these was Tycho Brahe's observatory, built in 1576, which remained in operation for 20 years.[3] Tycho's instruments were bigger in size than those of his predecessors. His largest instrument was a vertical quadrant 2.1 m in radius. However, his most famous instrument was a mural quadrant of radius 1.94 m, constructed in 1582. The transversals of this instrument were divided to measure one-sixth of a minute of arc. Wesley, who has compared Tycho's results with computer-generated values, finds that the individual reading-error of some of these instruments, such as the large equatorial armillary, was as much as 0;1,23 of arc.[4] With these instruments, Tycho conducted observations

---

[1] Kameswara N. Rao, A. Vagiswari, and Christina Louis, "Father J. Richaud and Early Telescope Observations in India," *Bull. Ast. Soc. India*, Vol. 12, pp. 81-85, (1984).

[2] Singer et al, *Op. Cit.*, p. 667.

[3] Tycho Brahe's observatory included a mural quadrant, revolving wooden azimuthal quadrant, revolving steel quadrant, portable quadrant, small brass quadrant, astronomical sextant, large equatorial armillary, northern equatorial armillary, and southern equatorial armillary. See Raeder et al, *Op. Cit.*

[4] The average error of these instruments ranged anywhere from 32″ to 49″ of arc. See Walter G. Wesley, "The Accuracy of Tycho Brahe's Instruments," *J. for Hist. Ast.*, pp. 42-53, Vol. 9, (1978).

for almost 20 years. His observations later became the basis of the famous laws of planetary motions discovered by Kepler.[1]

The Royal Observatory of Paris

The French king Louis XIV built the first national observatory in Europe at Paris between 1667 and 1672. An engraving of the observatory shows a number of instruments such as a mounted semicircular arc almost 2 m in diameter, a telescope more than 3 m long, an armillary sphere, and some other equipment. In the early days of the observatory, famous names, such as Cassini, Jean Picard, Ole Römer, and Christian Huygens were associated with it. The observatory had more than one tubeless refractor telescope with which Cassini discovered the four satellites of Saturn and the division of its rings known as the Cassini division. Cassini published in 1693 a table of Jupiter's moons, with which the longitude of any place could be determined in relation to the meridian of Paris.[2] At this observatory, Philip de La Hire prepared his astronomical tables with a telescope-fitted quadrant of 1.5 m radius.[3] Soon after the founding of the Royal Observatory, a number of private citizens, such as Charles Messier (1730-1817), built their own observatories in Paris.[4] The Royal Observatory of Paris was in operation when the scientific delegation dispatched by Jai Singh visited Europe.[5]

In addition to the observatories at Paris, there were also observatories in the outlying provinces of France and in countries such as Germany and Italy, although they were not as famous as those of Paris and Greenwich.

The Royal Observatory of England at Greenwich

King Charles II of England founded the Royal Observatory at Greenwich in 1675.[6] The observatory's purpose was to predict accurately the apparent motion of the moon and to fix the position of the stars for the benefit of sailors who wanted to determine their longitude at sea. John Flamsteed was its first director, or the Astronomer Royal. Flamsteed, using a 2.1 m radius sextant of steel, fitted with a micrometer eyepiece, prepared star charts with an accuracy of ± 10 sec of arc. He published in 1729 his catalog of the coordinates of 3000 stars. In addition to the sextant, the Royal Observatory included a mural quadrant and some long-focus telescopes as it is evident from an engraving

---

[1] For details of Tycho's observations see Pannekoek, pp. 204-216.

[2] Nicolos Bion, *Op. Cit.*, p. 167.

[3] Pannekoek, p. 279.

[4] Messier is famous for his catalog of nebulae and star clusters.

[5] See Ch. 13.

[6] For a description of the Royal Observatory of Greenwich, see E. Walter Maunder, *The Royal Observatory, Greenwich*, Religious Tract Society, London, 1900.

printed in the star-catalog. Jai Singh acquired a copy of this work, which may still be seen at the Sawai Man Singh II Museum of Jaipur. Flamsteed died in 1719. After his death, his widow removed the instruments from the observatory. Edmund Halley (1656-1742), who succeeded Flamsteed as the Astronomer Royal in 1720, re-equipped the observatory and continued observing the moon, the sun and the planets.[1] The observatory still preserves his 2.44 m quadrant and the transit instrument.

Astronomical Tables

Jai Singh is reported to have had in his possession European tables, prior to constructing his observatories. A number of famous astronomical tables were prepared in Europe before and during Jai Singh's time, such as the Alfonsine Tables (1252) and the Prutenic Tables prepared by Erasmus Reinhold using the values of Copernicus. In 1627, Kepler published his Rudolphine tables, which superseded all contemporary and previous tables. De La Hire of the Royal Observatory of Paris published his tables in 1687. The next edition of his tables came out in 1702. Cassini, who was also at the Royal Observatory of Paris, published his tables in 1740. In addition to these tables, numerous ephemerides had been produced before and during the period when Jai Singh built his observatories in India.[2] Jai Singh acquired some of these tables including those of de La Hire.[3]

The Astronomy of China

In order to gain a proper perspective of Jai Singh's efforts to revive astronomy in India, it is worthwhile to review astronomical progress in other countries of Asia during his time. In the late 17th and mid-18th centuries, observatories were built in China and Japan. In China, astronomy had a unique status and position.[4] There, astronomy was the property and monopoly of the government. The preparation of the Chinese calendar was the right of the Emperor just as the printing of money is the monopoly of governments today. In China, the calendarical calculations were kept secret from the general public.

In China, the government also controlled the study of mathematics and astronomy. A majority of the observers and mathematicians who carried out the

---

[1] The comet Halley is named after Edmund Halley. Recognizing the periodicity of the comet's appearance, he successfully predicted its next arrival in 1758.

[2] For a bibliography of ephemerides, see William D. Stahlman, and Owen Gingerich, *Solar and Planetary Longitudes for Years −2500 to +2000 A.D.*, p. xxiii-xxiv, Madison, 1963.

[3] See Ch. 13.

[4] For Chinese astronomy, see Joseph Needham, *Science and Civilization in China*, Vol. 3, pp. 171 ff., Cambridge Univ. Press, 1959.

calculations were in the state's bureaucratic service. For two thousand years, these officials were organized into a department having designations such as the Astronomical Bureau or the Astronomical Directorate.

The officials of the Astronomical Bureau were persistent and accurate observers of celestial phenomena and kept records of what they observed. They recorded novae, supernovas, comets, sun-spots, shooting stars, and eclipses. Records of Chinese observations date back to the fifth century B.C. Our information regarding the supernova of 1054 comes from these records.[1] Chinese observed the heavens for astrological purposes such as divining and foretelling the future.

Chinese astronomers borrowed astronomical elements from India and from Islamic countries. They developed observing techniques and instruments of their own. Some of these instruments were large masonry structures erected for greater accuracy. Jai Singh would build masonry structures centuries later.[2] The Imperial Observatory, when equipped in 1276-79, had the instruments listed in Table 1-2. The purpose of some of these instruments, however, remains uncertain.[3]

The Jesuits in China

The arrival of Jesuits in China changed the course of Chinese astronomy forever. The year 1601, in this regard, may be called the turning point, after which Chinese astronomy ceased to remain purely indigenous and began to assimilate Western elements. It is the year when the Jesuit scholar Matteo Ricci came in contact with the court of the Ming emperors in Peking.[4] Ricci was a scholar of great ability who translated European books on scientific subjects into Chinese. He was followed by other gifted scholars, who, with their knowledge of Western science, acquired positions of influence and power. Johann Adam Schall von Bell became the director of the Imperial Bureau of Astronomy in 1644. The directorship of this important bureau remained in Jesuit control thereafter for almost 150 years, until the end of the 18th century.

The successor of Adam Schall von Bell, Ferdinand Verbiest, a Belgian Jesuit, is credited with the renovation of the Peking observatory in 1673, which was originally built in 1442 or the 7th year of the Ming dynasty.[5] Although the renovation of the observatory was complete in about a year, its elaborate

---

[1] The remnants of the supernova of 1054, recorded by the Chinese, is the Crab Nebula in the constellation of Taurus with a spinning neutron star at its center.

[2] Needham, *Op. Cit.*, p. 296 ff.

[3] Needham, *Op. Cit.*, p. 369.

[4] The city is now called Beijing.

[5] Cui Shi Zhu, "Study of the History of Beijing Ancient Observatory," paper presented at the International Symposium on Indian and other Asiatic Astronomies, Hyderabad and Jaipur, Dec. 14-16, 1991.

equatorial armillary sphere was added much later, in 1744. A great deal of work was done at the observatory of which the publication of a catalog of 3,083 stars by Kögler and da Rocha in 1757, was the crowning feature.[1]

1.	*Ling lung i*	Ingenious Armillary Sphere
2.	*Chien i*	Simplified Instrument (equatorial torquetum)
3.	*Hun thien hsiang*	Celestial Globe
4.	*Yang i*	Upward-Looking Instrument, a hemispherical sundial[2]
5.	*Kao piao*	Lofty Gnomon
6.	*Li yün i*	Vertical Revolving Circle[3]
7.	*Chêng li i*	Verification Instrument[4]
8.	*Ching fu*	Shadow Definer[5]
9.	*Khuei chi*	Observing Table
10.	*Fih yüeh shih i*	Instrument to observe solar and lunar eclipses
11.	*Hsing kuei*	Star-Dial
12.	*Ting shih i*	Time-Determining Instrument
13.	*Chêng fang an*	Direction-Determining Table
14.	*Hou chi i*	Pole-Observing Instrument
15.	*Chiu piao hsüan*	Nine Suspended Indicators, probably a set of hanging plumb lines for checking gnomons
16.	*Chêng i*	Rectifying Instrument, purpose uncertain
17.	*Tso Chêng i*	Rectifying Instrument on Stand, purpose uncertain

Table 1-2 Chinese astronomical instruments.

Jai Singh probably came to know about the work at the Peking observatory at some later date in his career. He acquired a book on Chinese astronomy by Gaubil, a Jesuit astronomer at the observatory of Peking.[6,7]

---

[1] The instruments at the observatory were as follows: By Verbiest in 1673: 1. Simple ecliptic armillary sphere, 2. Simple equatorial armillary sphere, 3. Large celestial globe, 4. Horizon Circle for azimuth measurements, 5. Quadrant (movable about a vertical axis), 6. Sextant. By others: 7. Quadrant altazimuth, (by Stumpf, 1713-15), 8. Elaborate equatorial armillary sphere, (by Kögler et al), 1744, 9. Smaller celestial globe. See Needham, *Op. Cit.*, pp. 451-452.

[2] Needham erroneously compares *Yang i* with the Jaya Prakāśa of Jai Singh. Needham, *Op. Cit.*, p. 369.

[3] Perhaps, this instrument was an altitude measuring device, similar to the Cakra yantra described in Ch. 7.

[4] The purpose of this instrument is not clear.

[5] A pinhole image forming device to overcome the difficulty of penumbra. See Needham, *Op. Cit.*, p. 299.

[6] P. Gaubil, *Observations Chinoise*, Vol. 2. See Bahura II, p. 55, or No. 47, the Museum.

Either Johannes Terrentius or Johann Schereck brought the telescope to China in 1618. Refinements of the telescope such as the addition of the cross-hair and the micrometer reached China subsequently. In 1626, Adam Schall von Bell wrote a treatise in Chinese on the telescope. Records indicate that the Chinese Emperor received a telescope as a present in 1634. Although the Jesuit contributions to Chinese astronomy were noteworthy, there was a negative side to them. The Jesuits were responsible to a certain extent for preventing the Copernican revolution from entering the country.[1]

Astronomy in Japan

Up until the beginning of the 17th century, that is, up to the advent of the Jesuits in Japan, the astronomy of Japan was essentially Chinese, and there was little in it which had not been borrowed from its larger neighbor.[2] Then an ex-missionary, Chiara Giuseppe, settled in Japan and translated the first European book on astronomy into Japanese in 1650.[3] In 1720, under the patronage of the Shogun Yoshimune, the Japanese began to study Western sciences. Also about this time, a text was written based on European astronomy. Finally, in 1744, the Shogun founded a modern observatory, the instruments of which included a quadrant and an armillary sphere.[4] It is said that the director of the observatory, Kanane Genkei, was the first Japanese to learn about the Copernican system.[5]

---

[7] Claude Boudier, a Jesuit astronomer in India, used to communicate his astronomical data to Gaubil in China. See Antoine Gaubil, *Correspondence de Pekin*, 1722-1759, pp. 585, 655-656, Librainie Droz, Geneve, 1970. Jai Singh might have heard about Gaubil from Boudier whom he met in 1734. See Ch. 13.

[1] Virendra Nath Sharma, "The Impact of the Eighteenth Century Jesuit Astronomers on the Astronomy of India and China," *Indian J. History of Sci.*, Vol. 17, pp. 345-352, (1982). Also see Ch. 13.

[2] For the history of astronomy in Japan, see Shigeru Nakayama, *A History of Japanese Astronomy*, Cambridge, 1969.

[3] The book was originally written in Roman script and later transcribed into Japanese characters in 1677. Boleslaw Szczesniak, "The Penetration of the Copernican Theory into Feudal Japan," *J. Roy. Asiatic Soc.*, pp. 52-61, (1944).

[4] Legend has it that one of the Niyata Cakras the Miśra yantra at the Delhi observatory of Jai Singh represents the meridian half-circle of a Japanese observatory. However, there is no basis for this legend. See Ch. 5.

[5] Szczesniak, *Op. Cit.*, p. 55.

CHAPTER II

DECISION TO BUILD OBSERVATORIES

Petition to Muḥammad Shāh

Sawai Jai Singh noted that the astronomical calculations based on the existing planetary tables did not always agree with observations. In the preface to the *Zīj-i Muḥammad Shāhī* we read:[1,2]

> "He (Jai Singh) found that the calculation of the places of the stars as obtained from the tables in common use, such as the new tables of Sayyid Gurgānī and Khāqānī and the *Tas'hīlāt-i Mullā Chānd Akbar Shāhī*, and the Hindu books, and the European tables, in very many cases, give them widely different results than those determined by observation; [the problem was] especially [serious] with the appearance of the new moon—the computations of which did not agree with the observation.
>
> "Realizing that important affairs, both regarding religion and the administration of empire, depend upon these; and that in the time of the rising and setting of the planets, and the seasons of eclipses of the sun and moon, many considerable disagreements, of a similar nature, were being found; he represented it to his majesty of dignity and power, the sun of the firmament of felicity and dominion, the splendor of the forehead of imperial magnificence, the unrivalled pearl of the sea of sovereignty, the comparably brightest star of the heaven of empire, whose standard is the Sun whose retinue the Moon; whose lance is Mars and his pen like Mercury; with attendants like Venus; whose threshold is the sky, whose signet is Jupiter; whose sentinel Saturn; the Emperor descended from a long race of kings; an Alexander in dignity; the shadow of God the victorious king Muḥammad Shāh, may he ever triumph in battle.
>
> "He (Muḥammad Shāh) was pleased to reply, since you, who are learned in the mysteries of science and have a perfect knowledge of this matter, having assembled the astronomers and geometricians [of different schools of astronomy such as those] of the faith of Islam, the Brahmins and Pundits, astronomers from Europe, and having prepared

---

[1] ZMS, f. 0 ff.

[2] This translation is based on Hunter.

all the apparatus of an observatory, do you so labor for the ascertaining of the point in question that the disagreement between the calculated times of those phenomena and the times in which they are observed to happen may be rectified.

"Although this was a mighty task, which during a long period of time none of the powerful rajas had prosecuted; nor among the tribes of Islam since the time of the martyr prince, Mirzā Ulugh Beg, whose sins are forgiven, to the present, which comprehends a period of more than three hundred years, had any one of the kings possessed of power and dignity, turned his attention to this object; yet to accomplish the exalted command received, he undertook the task with a great determination and constructed here (at Delhi) several of the instruments of an observatory, according to the books of the Islamic School of astronomy such as the ones erected at Samarkand."

The Islamic School of Astronomy

Jai Singh's primary source of inspiration had been the Islamic school of astronomy. The school flourished in the Muslim countries all the way from Spain to Iran, and its astronomers hailed from all ethnic groups such as the Arabs, Persians, and Mongols living in countries controlled by the Muslim rulers. Most contributors to this school of science were followers of Islam. They also included Zoroastrians, Sabians (the star worshipers of Mesopotamia), Jews, and Christians. In its early stages, the Islamic school borrowed its astronomy from India and from the Greeks; later it made original contributions of its own.

Tradition has it that in the year 771 or 773, a man from India came to the court of the Caliph al-Manṣūr at Baghdad with some books on astronomy, with which he could calculate eclipses. The Caliph ordered the books translated into Arabic. The translation was called *Sindhind (Siddhānta)* by the Arabs.[1] Later, after Ptolemy's *Almagest* was rendered into Arabic sometime in the early 9th century, Islamic astronomy acquired more of a Greek and Alexandrian character.

The Islamic school produced many gifted scholars who composed original texts and commentaries on astronomical subjects. Māshā'Allāh (762- circa 815), a Jew, wrote a tract on the astrolabe, armillary sphere, and on astrology. Muḥammad ibn Mūsā al-Khwārizmī (circa 800-47) prepared astronomical tables for which the city of Ujjain in India was taken as the prime meridian. Abū ʿAbd Allāh Muḥammad ibn Jābir al-Battānī (d. 928), known in the West as

---

[1] The very first Sanskrit text translated into Arabic is said to be *Zīj al-Arkand* done in Sind shortly after 735. *Arkand* is apparently a corruption of the Sanskrit term *Ahargaṇa* meaning the number of solar days in a *yuga*. *The Encyclopedia of Islam*, New Edition, Vol. 3, p. 1136, London, 1971.

Albategnius, is considered to be the most outstanding astronomer of the Arab world. Al-Battānī was an accomplished observer. He produced a more accurate measure of the tropical year and of the rate of precession. In the mathematical aspect of astronomy, he introduced the modern concept of sines, replacing Ptolemy's chords. Abu'l-Ḥasan ʿAlī ibn ʿAbd al-Raḥmān ibn Aḥmad ibn Yūnus (d. 1000), observed at Cairo and produced planetary tables based on Ptolemy's theories. Abū Rayḥan Muḥammad ibn Aḥmad al-Bīrūnī (973-1050) determined the dates of solstices and equinoxes. He also wrote on the astronomy of India and on the construction and use of the astrolabe and sextant.

In the 11th and the 12th centuries, Spain had a high degree of art and culture. The Caliphate of Cordoba encouraged the study of sciences, including that of astronomy. Under the patronage of the Cordoban Caliphate, Ibrāhīm Abū Isḥāq, also known as Ibn al-Zarqālī (1029-87), took many observations and published astronomical tables known as the Toledo tables. The tables had a section on instruments including the astrolabe. He is said to be the inventor of an astrolabe which could be used at any latitude.[1] In the first half of the 12th century, Jābir ibn Aflah al-Ishbīlī wrote a general astronomical text which became widely used in Europe in its Latin translation. In the 13th century, following the tradition of the Cordoban Caliphate, a Christian king, Alfonso X, of Castille, also named el Sabio (1252-1284), assembled a number of astronomers and had them compile what are known as the Alfonsine tables. The Alfonsine tables were the last major contribution of the Iberian peninsula to astronomy. After publication of the tables, there was little original work done in Spain. It is said that after the Alfonsine tables, the study of astronomy disappeared from the Spanish peninsula.

Muḥammad ibn al-Ḥasan al-Ṭūsī (1201-74), was an astronomer associated with the observatory of Maragha in Iran. He built large instruments and later published the *Ilkhānik* tables. He also tried to improve upon the Ptolemaic theories of planetary motions but achieved little real success. ʿAlā al-Dīn Abu'l Ḥasan ʿAlī ibn Ibrāhīm ibn al-Shāṭir (1304-75) had more success than al-Ṭūsī by developing multiple epicycles rotating in opposite directions. For the planet Mercury, he had five epicycles. Finally, Ulugh Beg (1394-1449) built an observatory at Samarkand, and compiled the famous tables named after him. After Ulugh Beg there was no noteworthy astronomer of the Islamic school.

Observatories in the Islamic World

The interest in astronomy in the Islamic world was based on practical concerns. For example, Muslims had to face Mecca during prayer, and the mosques had to be built to accomodate this requirement. In addition, Islamic

---

[1] In the store house of the Jantar Mantar of Jaipur there is a large specimen of this type of astrolabe.

holidays were all geared to the lunar calendar. For instance, the celebration of *Īd* could only begin after the new moon had been sighted. The prayers were offered five times a day. Exact time-keeping was necessary so that the timings of the prayers could be kept with precision. Furthermore, the general belief in astrology and horoscopes by the population at large required accurate knowledge of the motions of the planets.

Realizing that these goals could only be fulfilled if there were better astronomical tables, rulers built observatories in different parts of the Islamic world. The most important of these were the observatory of Maragha built by Halāgū Khān (d. 1265) and the observatory of Samarkand in Central Asia constructed by Ulugh Beg.

The Observatory of Maragha

The Maragha observatory was located on a hill outside the city of Maragha, to the south of the city of Tabriz in modern Iran, and was founded around 1259 by Halāgū Khān, a Mongol ruler. The observatory was elaborately funded by the ruler, and it had a library with a large number of volumes.[1] However, the main function of the observatory was astrological. Halāgū, it is said, never took any step without consulting his chief astrologer, Naṣīr al-Dīn al-Ṭūsī. The exact number of instruments at the observatory complex is not known. Al-ᶜUrdī, an instrument designer for the observatory, describes the following instruments,[2] which might have been constructed for the observatory:

1. Mural quadrant of radius 430 cm, graduated in degrees and minutes
2. Armillary sphere with five rings
3. Solstitial armilla
4. Equinoctial armilla
5. Instrument with two holes
6. Azimuth ring with two quadrants
7. Parallactic ruler
8. Instrument for determining the azimuths and the sines of the complement of the angle of the elevation
9. Sine and versed sine instrument
10. The perfect instrument (*al-Ālā al-Kāmil*)

A large number of astronomers were associated with this observatory.[3] The crowning feature of the observatory was the publication of the *Ilkhānik* tables in

---

[1] Sayili, p. 193-94.

[2] Hugo J. Seemann, "Die Instrumente der Sternwarte zu Maragha nach den Mitteilungen von al Urdi," Sitzungsberichte der physikalischmedizunischenSozietat zu Erlangen, Vol. 60, pp. 15-126, (1928).

[3] Sayili names eleven astronomers at the observatory. See Sayili p. 205.

1271. Sayili believes that "the life of the observatory was at least 45 years and at the most about 55 or possibly 60 years." The observatory ceased to exist after 1316.[1]

The Observatory of Samarkand

The observatory of Samarkand was founded by Ulugh Beg, a grandson of Tamerlane, in 1424, and it remained in operation for about 30 years.[2] The remains of this observatory may still be seen at Samarkand in the former Soviet Socialist Republic of Uzbekistan. Although the exact nature and number of the instruments erected at the observatory is not known, it is reasonable to assume that some of them must have been similar to the ones erected at Maragha. The instruments might have included the following:[3]

1. Sextant or sixty-degree meridian dial (*Suds-i Fakhrī*)
2. Parallactic ruler revolving about a vertical axis (possibly *al-Ālā al-Kāmil*)
3. Mural quadrant (*libnā*)
4. Instrument with two sights, probably a parallactic ruler in the meridian, or triquetrum, (*Dhāt al-Shuᶜbatayn*)
5. Azimuthal quadrant
6. Armillary sphere
7. Sine and versed sine instrument

The most imposing structure of this observatory was a sextant, or sixty-degree masonry dial, in a dark chamber, which has been estimated as having a radius of 40 m.[4] Astronomers at the observatory completed the planetary tables, *Zīj-i Ulugh Begī*, or *Zīj-i Sulṭānī*, in 1437. Ulugh Beg himself was a man of learning, accomplished in both mathematics and astronomy. He himself took astronomical observations. Among the astronomers of the observatory, Giyāth al-Dīn al-Kāshī (d. around 1430), and later Qādīzādā-i Rūmī and ᶜAlī Khushjī (d. 1474) were the most prominent. Al-Kāshī wrote a treatise on astronomical instrumentation.[5]

The next observatory of consequence was completed at Istanbul in 1577 and destroyed three years later in 1580, at the reigning sultan's orders. It is doubtful that any meaningful tables were prepared from the data collected in the

---

[1] Sayili, p. 213.

[2] Sayili, p. 271.

[3] Sayili, pp. 283-287.

[4] Ernest W. Piini, "Ulugh Beg's Forgotten Observatory," *Sky and Telescope*, Vol. 71, No. 6, pp. 542-544, (June 1986).

[5] E. S. Kennedy, "Al-Kāshī's Treatise on Astronomical Observational Instruments," *J. Near Eastern Studies*, Vol. 20, pp. 98-107, (1961).

relatively short period of three years. During the heyday of Islamic astronomy, there were numerous *Zījes* or, astronomical tables, prepared throughout the Islamic world and in India.[1]

The Instruments According to the Islamic Books

Jai Singh in his *Zīj-i Muḥammad Shāhī* states that initially he constructed metallic instruments according to the books of the Islamic school of astronomy.[2] However, the *Zīj* does not reveal the books consulted for the purpose. It goes on to say though that the instruments constructed included the following:[3]

1.	*Dhāt al-Thuqbatayn*	(Dioptra)
2.	*Dhāt al-Shuᶜbatayn*	(Triquetrum)
3.	*Dhāt al-Ḥalaq*	(Armillary sphere)
4.	*Shāmlāh*	(Revolving parallactic ruler or a hemisphere with a revolving disk)
5.	*Suds-i Fakhrī*	(Sixty-degree meridian dial)

The instruments in the above list, except the *Shāmlāh*, are also included in the *Yantraprakāra*, a text on the instruments of Jai Singh written by one of his astronomers.[4] The above list could merely be an illustration and may not include all the instruments constructed in the beginning by Jai Singh according to the Islamic school of astronomy. The *Suds-i Fakhrī* in the above list is a masonry instrument, and it will be discussed in a later chapter along with Jai Singh's other masonry instruments.

Dhāt al-Thuqbatayn

Dhāt al-Thuqbatayn is the dioptra of the ancients. According to Ptolemy, Hipparchus (at Rhodes between 162-126 B.C.), used a version of it for measuring the apparent diameters of the sun and moon. Although Ptolemy does not discuss the instrument himself, it has been described in a commentary of his *Almagest* by Pappus (*fl.* 320).[5] The instrument consisted of a rectangular rod at least 4 cubits long and a vane having a pinhole fixed at one end.[6] Beginning

---

[1] Kennedy (1956), pp. 123-177, *Op. Cit.*, Ch. 1.

[2] There were two texts on Islamic instruments after al-Kāshī, one by al-ᶜĀmilī and the second by al-Khāzinī: 1. *Kitāb Taᶜlīm Ālāt Zīj*, of ᶜAbd al-Munᶜim al-ᶜĀmilī (16th century), *Op. Cit.*, Ch. 1; 2. A. Sayili, "Al-Khāzinī's Treatise on Astronomical Instruments," *Ankara Üniversitesi Dil ve Tarih-Cografya Fakültesi Dergisi*, XIV, pp. 18-19, (1956). It is not certain whether Jai Singh was directly or indirectly aware of the works of al-ᶜĀmilī or of al-Khāzinī.

[3] ZMS, f. 1.

[4] See *Yantraprakāra* (J) and *Yantraprakāra* (S).

[5] Derek J. Price in Singer et al (1957), p. 591, *Op. Cit.*, Ch. 1.

[6] The cubit, an ancient measure of length, was approximately 45-50 cm long.

at the end where the vane was located, and running along its entire length, there was a dove-tailed groove in which a vertical prism was loosely fitted. The observer, viewing through the pinhole, moved the prism back and forth until the disk of the moon was barely covered. The angular diameter of the moon was then calculated by dividing the width of the prism by the distance separating the prism and the pinhole.

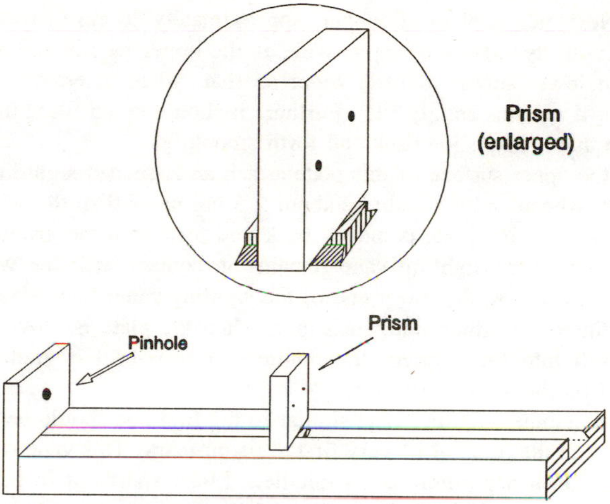

Fig. 2-1 The Dioptra (*Dhāt al-Thuqbatayn*) of Ptolemy.

Dhāt al-Thuqbatayn of Jai Singh

Regarding the *Dhāt al-Thuqbatayn* or dioptra which Jai Singh built, we get some insight from the *Yantraprakāra*.[1] The *Yantraprakāra* states:[2]

> "Now, for determining the diameter of the orb of the planets, an instrument is designed. Its name is *Jātuśśukavataina* (*Dhāt al-Thuqbatayn*) in Arabic-Persian."[3]

The text goes on to describe the construction of the instrument as follows:

---

[1] *Yantraprakāra* (J), p. 11, *Yantraprakāra* (S), pp. 20, 60-65.

[2] For a translation of the text see *Yantraprakāra* (S), pp. 60-61.

[3] On the margin of the manuscript, the author explains the meaning of the term "*Dhāt al-Thuqbatayn* (*Jātuh Śukavataina*)" as follows: He writes, "By the word *Jātuh* is meant the 'owner', and *śuka* means aperture." The "two apertures" are denoted by the expression *śukavataina*. Hence the meaning of the term is "the owner of two apertures."

"First, take a straight and smooth beam of wood with a square cross-section, 8-10 m long.[1] Make the width (*vistāra*) and the thickness (*piṇḍa*) of this beam such that they together measure ½ m (one-fourth of a *daṇḍa*).[2] Running through the middle of this beam, carve out a groove from one end to the other. The groove should be one-third as wide as the width of the beam. The depth of the groove should be kept about 2.5 cm, and it should be a little wider at the bottom than at the top. Next, take a plate of copper, approximately 20 cm (a *vitasti*) in length with its upper surface as wide as the upper part of the groove and the lower surface a little wider so that, when inserted into the groove, it remains snugly fit.[3] Further, it should be so fitted into the groove that it may slide back and forth smoothly.

"To the upper surface of this plate attach an L-shaped sighting vane (*bhujā*), whose width should be about 2.5 cm more than the width of the plate. As the plate is moved back and forth into the groove, the lower end of the sighting vane remains in contact with the wooden surface below. At the lower end of the sighting vane, there should be two pointers on either side, such that, when the plate is moved back and forth into the groove, the pointers slide over the graduations marked on the wooden surface.

"Next, another sighting vane similar to the first one should be made. Attach it to the beam at its very first division-mark. This vane remains fixed. Punch a pin-hole at its middle. Make apertures in both the sighting vanes such that they face each other and are in a line parallel to the groove.

"On either side of the groove, mark off 220 divisions of equal width on the surface of the beam. Starting from the bottom end, number these scale graduations as 1,2 etc.[4] Where the immovable sight is fixed, it is equal to this division.[5] Make an aperture in the movable sight situated in the groove so that its diameter is equal to one of the divisions made earlier on the scale."[6]

The description of the instrument ends abruptly at this point in the manuscript of *Yantraprakāra* and does not go into any detail regarding the support

---

[1] The length of the instrument is expressed as four or five *daṇḍas*. One *daṇḍa* equals approximately 6 ft or 2 m. For the length measures used in the *Yantraprakāra*, see P. K. Acharya, "Hindu System of Measurement," *University of Allahabad Studies*, Vol. 2, p. 69, (1926).

[2] The cross sectional area of the beam would be (25 cm)^2 or 625 cm^2. Perhaps, the perimeter of the beam is meant by the phrase "the width and thickness together," which is somewhat more reasonable. The cross sectional area in that case would be 1/4 as large or 156 cm^2.

[3] The plate is to be inserted length wise into the groove.

[4] The fixed vane is attached at the bottom end .

[5] The text is unclear here.

[6] *Yantraprakāra* (J), p. 11, and *Yantraprakāra* (S), pp. 60-61.

mechanism and the procedure for taking a reading. One can only surmise that the beam must have been supported by ropes and pulleys and moved around with the help of several men. The principal observer took the sighting from the end of the beam where the fixed vane was located. After the beam had been aligned with an object, for example, the moon, the observer directed his assistant to move the movable sight back and forth until the moon barely fitted into the aperture of the movable sight. The apparent diameter of the moon was then calculated from the distance between the sights and the diameter of the aperture.

A *Dhāt al-Thuqbatayn* made according to the *Yantraprakāra* with a 10 m long beam and 220 graduations etched on its scale would have had divisions about 4.5 cm wide. When the sighting vane is at the extreme end of the beam, it could measure the separation of two stars 5′ of arc apart. For measuring an angle of 30′ of arc, such as the diameter of the moon, the sighting vane would be roughly at the 115th division mark.

Analysis of the instrument shows that it was considerably over-designed. At the mid-range of the scale, where the diameter of the moon would have been measured, the vane should have moved about ±3.8 divisions or ±17 cm before the average eye could have detected any appreciable change in the lunar diameter. This is so because the resolving power of the eye is no more than 1′ of arc, and to record a change of ±1′ in the angle measured, the vane would have to be moved by ±17 cm in either direction.[1] The instrument could easily have been made 1/4 as large and yet have had the same degree of accuracy.

According to Jagannātha, the purpose of the *Dhāt al-Thuqbatayn* that Jai Singh built was to measure the altitude and the azimuth of an object.[2] The instrument could determine the altitude by measuring the height of the upper end of the beam over and above the lower end. However, to measure the angle of azimuth, a circular scale in the plane of the horizon would have been needed.

Dhāt al-Shuᶜbatayn

Dhāt al-Shuᶜbatayn is the *triquetrum*, or parallactic ruler of Ptolemy.[3] The instrument, designed to remain fixed in the plane of the meridian, measures the zenith distance of an object such as the moon.

---

[1] Let s be the width of the vane at a distance R from the eye. The angle θ subtended by it at the eye:

$$\theta = s/R, \text{ or } R\theta = s$$
$$R\,\Delta\theta + \theta\,\Delta R = \Delta s$$
$$\Delta\theta = \pm 1'$$

Thus δR turns out ±3.81 div. or 17 cm.

[2] SSRS, p. 1163; also SSMC, p. 39.

[3] Ptolemy, Book V, sec. 12, pp. 244-247.

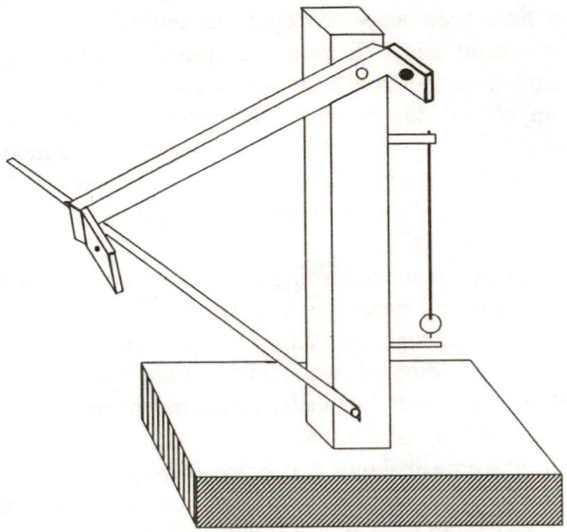

Fig.2-2 The Triquetrum (*Dhāt al-Shu^{c}batayn*) of Ptolemy.

The Iranian astronomer and mathematician, Giyāth al-Dīn al-Kāshī, who played a leading role at the Samarkand observatory of Ulugh Beg, gives detailed instructions for its construction.[1]

> "Take a ruler of no less than 2 1/2 cubits in length and fix it vertically into the ground. Take another ruler of equal length and mount one of its ends with a pin on the top of the first, like a hinged compass. Fix two sights on the second ruler. Take another ruler next, and mount it similarly with a pin near the base of the vertical ruler. The length of the third ruler should be such that when the free ends of the second and the third ruler are joined together, the three rulers form a right triangle—the third ruler being the hypotenuse of this triangle. The third ruler, in fact, represents the chord of an angle made by the first and the second rulers which are hinged together at the top."

Next, al-Kāshī instructs how to graduate the third ruler so that it measures degrees and minutes of an arc.

[1] Kennedy (1961), *Op. Cit.*

Dhāt al-Shuᶜbatayn of Jai Singh

In the *Yantraprakāra* we find the following description of the instrument.[1] Unfortunately, the *Yantraprakāra*'s author does not give any dimensions of the instrument constructed by Jai Singh.

"Now an instrument for measuring the Moon's parallax (*lambana*) is described. The name of this instrument is *Jātuśśuvatain* (*Dhāt al-Shuᶜbatayn*) in Arabic language.

"Prepare two bars,[2] [such that] their width and thickness are at right angles [rectangular cross section] and straight. Their length should not be less than four *daṇḍas* (approximately 8 m); if the length is chosen as suggested, the graduation marks will be quite distinct. The bars should not be too thick [or heavy] to bend under (their own) weight.

"Draw straight lines one each [lengthwise] in the middle of the two bars. To the ends of one of the bars, attach sighting vanes (*laghubhujā*) with an aperture each. They [the vanes] should be equal [in size] and parallel to one another. The aperture close to the [observer's] eye should be [kept] small, and the one towards the moon slightly large, so that its [the moon's] full orb is visible. For measuring zenith distance (*natāṁśa*), these (vanes) should be at the ends of the line [already drawn] in the middle of the bar.

"Hinge the two bars together with a pin so that they can be brought together or spread apart like a pair of compass (*karkaṭaka*). [Moreover,] for the bar with the vanes, the hinge should be toward that end where its large-aperture [vane] is located. [Further,] the lines on the two bars [should] intersect [at the hinge]. Now [starting] from the pin [where the two lines meet], mark divisions on the straight lines [already drawn] on the two bars.[3] These divisions should be equally spaced. On the line which has the two sights for measuring zenith distance, inscribe 60 divisions from one end to the other. [Further,] divide these divisions into as many subdivisions of minutes (*kalās*) as possible.

"Spread the two bars apart in the plane of the meridian in such a way that one of them [the one without the sighting vanes] is vertical. Fix it [the bar without the vanes] into the ground firmly, such that the first division of its straight line is above the floor. The hinge [on this bar then] will be on the other end above the floor. At the upper end of this fixed [vertical] bar attach a small arm. At the lower end also, attach

---

[1] *Yantraprakāra* (J), pp. 9-10; and *Yantraprakāra* (S), pp. 18-19, 58-60.

[2] Perhaps of some metal such as brass.

[3] The object of these divisions is not clear.

a [second] arm equal in size and parallel to the top one. The [vertical] bar should be so fixed [into the ground] that a plumb line dropped from the top arm [to the bottom] remains parallel [to the bar].

"Next, prepare a third bar, somewhat thicker and longer than the first two. This [third bar] should be pivoted to the fixed bar at the bottom [end] of the line [drawn on the bar]. Moreover, it should be pivoted such that it can be opened or closed like a pair of dividers. It [the third bar] should be long enough to become the hypotenuse of a right triangle when the first two bars are spread to form a right angle between them. With this (third) bar the distance between the ends of the first two bars is measured, when they are spread apart [to form an angle].

"When the moon is on the meridian, its position should be sighted with the bar endowed with two sighting vanes. Move the third bar so that it is in contact with the [unhinged] end of the bar [with vanes]. The linear distance between [the end of bar with the vanes] and the bottom end [of the vertical bar] is the chord (*pūrṇajyā*) of the zenith distance. The magnitude of this chord should be known from the line with sixty divisions. Its arc [the arc formed by the chord] is the zenith distance. [The third bar is also graduated in the same way as the second for this purpose.]

"Observations should be taken when the moon is located in the Capricorn or Cancer and transits the meridian. Then the meridian circle, the secondary to the ecliptic or latitude circle (*kadambavṛtta*), the secondary to the celestial equator or declination circle (*dhruva vṛtta*), and the circle passing through the center of the moon and zenith (*dṛk-maṇḍala*), all are in one plane. Thus the latitude (*akṣāṁśa*); the Moon's declination, (the Moon's) celestial latitude, zenith distance—all these will be arcs of the same circle."

The procedure for measuring the latitude of the moon, given in *yantraprakāra*, is similar to the one described by Ptolemy in his *Almagest*.[1] Ptolemy also prescribes measurements when the above conditions are met or nearly met. However, these conditions are seldom met because a simultaneous occurrence of solstitial points and the moon on the meridian is a rare event.

Dhāt al-Ḥalaq

Another instrument of Jai Singh was the *Dhāt al-Ḥalaq*, or armillary sphere. The armillary sphere is a very old instrument. Hipparchus and Ptolemy are said

---

[1] Ptolemy, pp. 244-247.

to have used it. In his *Almagest*, Ptolemy describes a six-ring armillary.[1,2] Scholars coming after Ptolemy have discussed armillaries with rings anywhere from five to nine in number. For example, al-Kāshī describes a seven-ring armillary, whereas al-ᶜUrḍī calls for only five.[3] Greek writers such as Theon, Pappus, and Proclus describe an instrument very similar to a nine-ring armillary sphere, known as the meteoroscope. Ptolemy, in his *Geography*, also describes this instrument.[4] Although the total number of rings in various armillaries might have been different, they all had rings for the ecliptic, solstitial colure, and the meridian, and a sighting arrangement which could either be a ring with sights or an alidade.

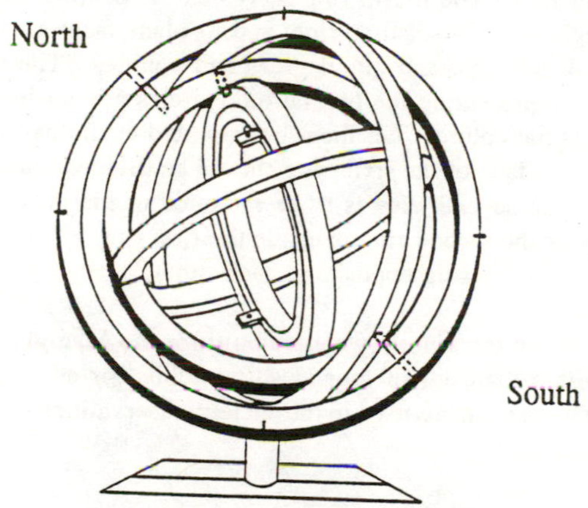

Fig. 2-3 *Dhāt al-Ḥalaq* or Armillary Sphere.

---

[1] Ptolemy is ambiguous about the number of rings. The number of rings, according to him, may also be interpreted as seven in all. Ptolemy, Book V, sec. 1, pp. 217-219. In his description, Ptolemy calls the armillary sphere an astrolabe. In fact, it is not an astrolabe as we know the instrument today.

[2] Also see, D. R. Dicks, "Ancient Astronomical Instruments," *J. British Astronomical Association*, pp. 77-85, (January 1954).

[3] Seemann, *Op. Cit.*, p. 35. The outer ring of al-ᶜUrḍī's armillary sphere had its inner diameter approximately 1.75 m (three cubits). However, al-Kāshī says that it was four cubits. See Kennedy, *Op. Cit.*, p. 104.

[4] Dicks, *Op. Cit.*

Armillary Sphere of al-Kāshī

Al-Kāshī in his treatise on astronomical instruments describes the instrument as follows:[1]

"An Armillary Sphere or 'the instrument with rings' is composed of seven rings of square cross-section. Set up the first ring in the meridian plane on a support, and let the second ring be positioned inside the first such that its convex surface remains in contact with the inner surface of the first. On both sides of the first ring let there be two or three guide-plates fastened so that the second ring may slide within the first ring freely. The third ring serves as a solstitial colure and it should be mounted at the two poles which are evenly spaced on the second ring. The fourth ring serves as the ecliptic. It is fixed at a right angle with the solstitial ring, and its plane faces are divided into twelve zodiacal signs and into degrees and minutes. The fifth and the sixth rings representing the two latitude circles are set inside the ring of the solstitial colure. Let them be mounted at the two poles of the ecliptic. One face of the sixth ring should be divided into degrees and minutes. The seventh ring is to be mounted inside the sixth ring the same way as the second ring is inside the first ring. Two sights are to be attached to this ring opposite to each other."

Al-Kāshī says that the armillary sphere built for the Maragha observatory had a first ring with a diameter of four Hāshimī cubits, or about 2.3 m. He also remarks that the instruments used in the ancient observations were not more than a cubit across.[2]

Armillary Sphere in Hindu Astronomy

The armillary sphere, or *gola yantra*, is not new to Hindu astronomy. The *Sūryasiddhānta*, written sometime around 400 A.D., describes a rudimentary armillary sphere. Lalla, writing in the 8th century, describes a fairly elaborate armillary primarily for telling time.[3] He says:

"Now in order to determine the *lagna* (the longitude of the rising point of the ecliptic) and the *kāla* (the time elapsed since sunrise in the forenoon, or remaining until sunset in the afternoon), one should get a *Bhagola* (sphere of asterisms) constructed with the help of nine circles,

---

[1] See Kennedy (1961), *Op. Cit.*

[2] A Hāshimī cubit was about 23 inches or 58 cm long. See Kennedy (1961), p. 104, *Op. Cit.*

[3] *Śiṣyadhīvṛddhāda Tantra* of Lalla, Part II, tr., Bina Chatterjee, pp. 280-281, Indian National Science Academy, New Delhi, 1981.

the prime vertical, etc. The ecliptic should also be exhibited in it, in the manner stated before."

The Armillary Sphere in Yantraprakāra

The *Yantraprakāra* calls the armillary sphere a *Gola yantra* and describes two versions of it—a five-ring armillary and a six-ring armillary.[1] Unfortunately, the *Yantraprakāra* does not give the dimensions of either instrument. However, from the *Zīj-i Muḥammad Shāhī* we learn that the diameter of the instrument constructed by Jai Singh was about three meters (three *gaz*), slightly larger than the instrument of Maragha mentioned by al-Kāshī. The six-ring instrument is described in the *Yantraprakāra* as follows:

"Now an armillary sphere known as *Jātulhalak* (*Dhāt al-Ḥalaq*) is being written. Take two rings of metal or wood of equal size and of rectangular cross-section. Join them diametrically opposite such that they form four right angles. Consider one of them as the ecliptic and the other the solstitial colure. At a distance of 90 degrees from the points of intersection, on the solstitial ring, let there be the (two) pole(s) of the ecliptic. At the poles fix two pins projecting on either side (of the ring). Next, at a distance equal to the maximum declination from the poles of the ecliptic and on the solstitial ring, fix two pins projecting a little outward. [These pins represent the poles of the celestial sphere.]

"At the two pins of the poles of the ecliptic (as pivot points), insert two rings, one on the inside and the other on the outside of the solstitial ring, skillfully made, such that while turning these, the (two) rings remain in contact with the ecliptic. These two represent latitude circles—one on the inside and the other on the outside (of the ecliptic).

"Similarly, pivot a ring, a little bigger than the others, to the pins at the poles (of the celestial sphere) so that all the other rings are within it. Let this ring be known as the meridian. Similarly, within the interior latitude circle insert a smaller ring that turns, remaining in contact with the circle. [That is, the outer surface of the inner ring remains in contact with the inside surface of the inner latitude circle] This ring which rotates within the latitude circle is for observing the celestial latitude. On this ring, let there be two sighting vanes fixed rigidly, and diagonally opposite to each other, (or) spaced six signs of the zodiac apart and having an aperture each. In this manner, an instrument of six rings is completed.

"Now, graduate the ecliptic ring and also the latitude ring within it

---

[1] *Yantraprakāra* (S), pp. 55-58.

with as many degrees and minutes as possible. During a rotation of (any) ring, if the slack caused by the pins at the poles is relatively small, the instrument is well made. Next, make two meridian rings [or one in addition to the one already made]. Within the outer ring, the inner one should be adjoined without any pins, [so that it is free to rotate just as the ring within the inner latitude circle]. Graduate the outer meridian ring with degrees and minutes.

"By rotating the meridian ring, the [zenith distance] of the celestial pole (of the instrument) may be made equal to the [colatitude] of the place. This way the instrument is built with seven rings. However, by eliminating the outer one of the two meridian rings and, secondly, eliminating the latitude ring on the outside of the solstitial ring, the instrument may be made smaller (or with fewer rings). That is, the instrument thus becomes a five-ring unit.

"The meridian circle of the instrument should be fixed along the meridian in the sky. By raising the north celestial pole (marked on the instrument) equal to the latitude of the place, (or) lowering the south pole [by the same amount], the rotation of the instrument (then) becomes identical with that of the sky.

"If both the sun and the moon are visible, let the ecliptic ring be rotated so that its shadow falls onto itself. Next rotate the outer latitude ring so that its shadow also falls onto itself. The point of intersection of the two rings (the ecliptic and the latitude) is the *spaṣṭa sūrya*, (or indicates the location of the sun on the ecliptic.) Next, at the same time, rotate the inner latitude (sighting) ring so that the moon is seen through the two sighting apertures on the vanes. In this manner, the point of intersection of the inner latitude ring and (the section of) the ecliptic facing the moon indicates the *spaṣṭa candra* (moon's longitude). From the ecliptic, up to the place of the sight, above or below the ecliptic, the angular distance indicates the north and south latitude (of the moon) respectively.

"Now the determination of the latitude and longitude of planets and stars at night by the armillary sphere: (First), calculate the sign and the angular distance of a selected star from one of the equinoctial points (on the ecliptic) and mark this on the ecliptic ring. Align the outer latitude circle along with it (the point just marked) and clamp it there.[1] Next sight the selected planet or star with the latitude circle. (That is) move the latitude circle such that its lower and the upper edges (i.e., its plane) are in alignment with the (center of the orb of the) planet or star. [This aligns the ecliptic ring of the instrument with the ecliptic in the sky.] At the same time another person should observe the (desired)

---

[1] The text misleadingly states the "inner latitude circle."

planet or star with the inner latitude circle which has two diagonally opposite sighting vanes. When this is accomplished, the sign or the longitude (of the desired object) is given by the intersection of the latitude circle and the ecliptic ring. If the sighting point is beyond the (first six signs), six signs should be subtracted in order to determine the sign and the longitude (of the object).[1] The celestial latitude is given by (the angular distance of) the point of intersection of the latitude circle and the ecliptic to the sighting aperture. The latitude is considered 'north' if it (the sighting vane) is to the north (of the ecliptic) and 'south' if it is to the south."

Fig. 2-4 An Armillary Sphere of Jai Singh-Madho Singh period. The azimuth ring of the instrument is 53 cm in diameter.

The Havamahal Museum of Jaipur has an armillary sphere of eight rings including an azimuth ring of about 53 cm in diameter. The instrument has a rod at the center along the axis of one of the rings. As judged from the script of the inscription on the instrument, the instrument seems to be of the post Jai Singh-

[1] Here the longitude is measured from an equinoctial point or from the intersection of the ecliptic and the equator and not necessarily from the vernal equinox.

Madho Singh period. Moreover, it has been made neither according to the description of al-Kāshī nor according to the *Yantraprakāra* as described above. Apparently, it was fabricated according to Hindu books, as it has a small sphere at the center. It also does not have any sights. As a result, the value of the instrument as an observational tool is rather limited.

Shāmlāh

Joseph Frank discusses an Arabic instrument *al-Ālā al-Schāmila*, which is phonetically close to Shāmlāh. The instrument consists of a hemisphere and a revolving disk for the ecliptic.[1] According to Frank, "one may consider *al-Schāmila* as a combination of the quadrant on the reverse side of an astrolabe, with the celestial sphere."[2] The *Shāmlāh* could also have been the *al-Ālā al-Kāmil*, meaning the "perfect instrument," of al-ᶜUrḍī, which, according to al-ᶜUrḍī's own description, was a revolving parallactic ruler measuring the azimuth and altitude of a star.[3] For measuring the azimuth of a star, the instrument must have had a circular scale in the horizontal plane. Ghulām Ḥusain in his *Jamiᶜ-i Bahādur Khānī* draws a picture of a *Shāmlāh* which is similar in principle to the *al-Āla al-Kāmil* of al-ᶜUrḍī.[4]

Difficulties with Wood and Metal Instruments

The metal instruments made according to the books of the Islamic school of astronomy did not measure up to Jai Singh's expectations. In the *Zīj-i Muḥammad Shāhī*, we read that the "brass instruments did not come up to the ideas which he had formed of accuracy." The sizes of these instruments, such as the armillary sphere, were relatively small. As a result, their scales could not be divided into minutes.[5] On the other hand, they were too heavy. Moreover, the shaking and wearing down of their axes led to the displacement of the centers and shifting of the planes of the instruments, causing error in the measurements.

Jagannātha confirms that the early instruments were indeed too heavy and unwieldy.[6] He says that in the *Dhāt al-Ḥalaq*, or armillary sphere, its ecliptic

---

[1] Joseph Frank, "Über zwei astronomische arabische Instrumente," *Zeitschrift für Instrumentenkunde*, pp. 193-200, July 1921.

[2] *Ibid*.

[3] Seemann, *Op. Cit.* pp. 96-99. Sayili translates *volkommene Instrument* as "the perfect instrument," or *al-Ālā al-Kāmil*. Sayili, p. 200.

[4] Ghulām Ḥusain, *Jamᶜ-i Bahādur Khānī*, p. 519, Calcutta, 1835. Cited by Sarma in *Yantraprakāra of Sawai Jai Singh*, *Op. Cit.*, p. 50.

[5] The *Zīj-i Muḥammad Shāhī* gives the diameter of the armillary as about 3 meters. A degree division on a circle of 3 m diameter would be 2.6 cm. If the degree divisions are further divided into minutes, each minute-division would be about 0.44 mm wide.

[6] SSRS, p. 1162; and SSMC, p. 38.

ring bent at its poles or the pivot points. Because of this bending, errors on the order of 30′ of arc were noted. A large astrolabe was no better either, because it also gave inaccurate readings for similar reasons. If the surviving specimens at the Jaipur Jantar Mantar are indicative of Jai Singh's attempts to build metal instruments, it is obvious that his early instruments were too heavy indeed, with little effort made to eliminate excess weight. They were simply enlarged versions of the small medieval instruments of the East that may still be seen in museums around the globe. The technology for building large metal instruments had not yet developed in India.

For the *Dhāt al-Thuqbatayn*, or dioptra, Jagannātha writes:

> "Now there is an instrument for determining the azimuth and altitude, called *Dhāt al-Thuqbatayn* by the Muslims. Its sighting rod is extremely huge, and requires many persons to manipulate it; therefore, sighting is done with great difficulty. Even this does not take place at the right moment. Because of its huge size, the rod tends to bend downwards. Hence sighting is never correct. There also occurs difference in the zenith distance."[1]

The instrument, *Dhāt al-Thuqbatayn*, as described in the *Yantraprakāra*, must have been heavy and quite unwieldy, as a 10 m long beam of wood with a 625 cm^2 cross section will weigh almost 300 kg. On the other hand, if the cross section had been 1/4 as large, as suggested earlier, the weight would have been 75 kg.[2] The metal instruments of Jai Singh were also quite heavy and troublesome.

Accordingly, Jai Singh discarded his wood and metal instruments and concluded that the problem he confronted was not unique and that astronomers in the past has also faced this problem. In the *Zīj-i Muhammad Shāhī*, Jai Singh writes:

> "Determinations of the ancients such as Hipparchus and Ptolemy have proved inaccurate, . . . he constructed in *Dāru'l Khilāfat Shāhjahānābād* (Delhi), which is the seat of empire and prosperity, instruments of his own invention, such as Jaya Prakāśa, Rāma yantra, and Samrāṭ yantra, the semidiameter of which is of eighteen cubits, and one minute on it is a barley-corn and a half; of stone lime, of perfect stability, with attention to the rules of geometry, and adjustment to the meridian, and to the latitude of the place, and with care in the measuring and fixing of them; so that the inaccuracies, from the shaking of the circles, and the wearing of their axes, and displacement

---

[1] SSRS, p. 1163 and SSMC, p. 39.

[2] Ref. 22 and 23.

of their centers and the inequality of the minutes, might be corrected."[1]

Confirming the above, Jagannātha writes that recognizing the defects of metal instruments in general, Jai Singh built Jaya Prakāśa in place of the *Dhāt al-Ḥalaq* (armillary sphere), and invented an immovable instrument, the Rāma yantra, for determining the azimuth and altitude in place of the *Dhāt al-Thuqbatayn* (dioptra).[2]

The Intended Accuracy of Measurements

From the size of the instruments, particularly that of the *Dhāt al-Thuqbatayn*, it appears that Jai Singh was aiming, initially, to achieve an accuracy of less than a minute of arc in the measurement of angles. The Jaya Prakāśa and the Rāma yantra, which he designed to replace his metal instruments, have accuracies of $\pm 3'$ of arc, at best.[3] However, with his large Samrāṭs and Ṣaṣṭhāṁśas, this accuracy was brought down to $\pm 1'$ of arc—the very limit of the naked eye observations.[4]

---

[1] ZMS, f. 1.
[2] SSRS, p. 1163, also SSMC, p. 39.
[3] See Ch. 3.
[4] See Ch. 3.

CHAPTER III

MASONRY INSTRUMENTS I

The metal instruments, constructed according to the texts of the Islamic school of astronomy, did not measure up to Jai Singh's expectations. Jai Singh, therefore, discarded them in favor of the instruments of stone and masonry that he himself designed. He tried to achieve the desired precision through their large size, and steadiness from the relative inflexibility of their structures. A French Jesuit, du Bois, writes that Jai Singh prepared wax models of these instruments with his own hands.[1]

Jai Singh's instruments at his observatories may be classified into three main categories based on the instruments' precision. In Table 3-1, we present an inventory of his masonry instruments according to their precision.

Masonry Instruments

Before constructing their masonry instruments, Jai Singh and his associates selected a suitable ground first and leveled it with water standing in masonry channels built just for this purpose.

Instrument	Number	Location
1. Dhruvadarśaka Paṭṭikā (North star indicator)	1	Jaipur
2. Nāḍīvalaya (Equinoctial dial)	5	Jaipur (2), Varanasi, Ujjain, Mathura
3. Palabhā (Horizontal sundial)	2	Jaipur, Ujjain
4. Agrā (Amplitude inst.)	5	Delhi, Ujjain, Mathura Mathura (2), Jaipur
5. Śaṅku (Horizontal dial)	1	Mathura
6. Unknown Instrument	1	Varanasi

Table 3-1a Sawai Jai Singh's Low Precision Masonry Instruments

---

[1] Joseph du Bois, Introduction to de La Hire's *Tabulae Astronomicae*, ms., The Museum, Jaipur.

Instrument	Number	Location
1. Jaya Prakāśa (Hemispherical inst.)	2	Delhi, Jaipur
2. Rāma yantra (Cylindrical inst.)	2	Delhi, Jaipur
3. Rāśi valaya (Ecliptic dial)	12	Jaipur
4. Śara yantra[1] (Celestial latitude dial)	1	Jaipur
5. Digaṁśa (Azimuth circle)	3	Jaipur, Ujjain, Varanasi
6. Kapāla (Hemispherical dial)	2	Jaipur

Table 3-1b Sawai Jai Singh's Medium Precision Masonry Instruments

Instrument	Number	Location
1. Samrāṭ (Equinoctial sundial)	6	Delhi, Jaipur (2), Ujjain, Varanasi (2)
2. Ṣaṣṭhāṁśa (60 deg meridian chamber)	5	Delhi, Jaipur (4)
3. Dakṣinottara Bhitti (Meridian dial)	6	Delhi, Jaipur, Ujjain Varanasi (2), Mathura

Table 3-1c Sawai Jai Singh's High Precision Masonry Instruments

Instruments added to the observatories afterwards are given in Table 3-2.

Instrument	Number	Location
1. Miśra yantra (Composite instrument)	1	Delhi
2. Śaṅku yantra (Vertical staff)	1	Ujjain
3. Horizontal Scale (known as the seat of Jai Singh)	1	Jaipur

Table 3-2 Instruments added after Sawai Jai Singh

---

[1] Little is known about the Śara yantra or latitude measuring instrument of Jai Singh which is briefly mentioned in the *Yantraprakāra*. In Ch. 7, we attempt to identify it with the remains of an unfinished instrument of the Jaipur observatory. If our conjecture is correct, the instrument had medium precision only.

Such masonry channels may still be seen at Jaipur and Ujjain. Next, north-south and east-west directions were marked on the ground, the procedure for which had been known for a long time and described in standard texts such as the *Sūryasiddhānta*.[1]

Just as in any other observatory of medieval times, Jai Singh's observatories primarily housed time-measuring and angle measuring devices. Jai Singh graduated his time measuring devices according to the sexagesimal scheme in which a 24-hour day is divided into 60 parts or *ghaṭikās*, and the *ghaṭikās* are in turn divided into 60 subparts known as *palas*. For measuring angles, he divided a circle into four quadrants of 15 main divisions each. He partitioned the main divisions further into six subdivisions of one *aṁśa* or one degree each. He divided the degree-divisions, as the scale permitted, into minutes or *kalās*. Sixty *kalās* equalled one *aṁśa* or degree. Any deviation from this division-scheme seen presently at his observatories is purely a modern adaptation. It has nothing to do with Jai Singh. The scale markings on Jai Singh's time-measuring instruments had zeros on the meridian line, and 15-*ghaṭikā* marks along the east-west line. These instruments thus measured time from midday or midnight. The angle-measuring instruments had their zeros on the east-west line.

SAMRĀṬ YANTRA

The Samrāṭ yantra or the "Supreme Instrument" is Jai Singh's most important creation. The instrument is basically an equinoctial sundial, which had been in use in one form or the other for hundreds of years in different parts of the world. In India, Brahmagupta (b. 598) describes *Kartarī* yantra, an equinoctial sundial, which operates on the same principle as the Samrāṭ.[2] Lalla (*fl.* 750 A.D.) describes a modified version of the *Kartarī* yantra.[3] An Egyptian astronomer of the late 13th century, al-Maqsī, has also written about equinoctial sundials erected in the plane of the equator, similar to a Samrāṭ.[4] Jai Singh,

---

[1] On a surface of stone leveled with water or on a leveled floor of lime plaster, describe a circle of several digits. Place a vertical gnomon of 12 digits at its center and mark off the two points where the shadow of the gnomon, before and after the noon hour, meets the circumference of the circle: these two points are called the west and the east points. Draw a line between them. This is the east-west line. Next, draw two intersecting arcs with the east and the west points as centers. Draw another line through the points of intersection of the arcs. This line represents the north-south direction. In the same manner, determine the intermediate directions through the arcs formed between the points of the already determined directions. *Sūryasiddhānta*, Ch. 3, 1-4.

[2] Sreeramula Rajeswara Sarma, "Astronomical Instruments in Brahmagupta's Brāhmasphuṭasiddhānta," *Indian Historical Review*, Vol. 13, Nos. 1-2, pp. 63-74.

[3] See *Śiṣyadhīvṛddhida Tantra* of Lalla, pp. 284-285, *Op. Cit.*, Ch. 2.

[4] David King, "On the Astronomical Tables of the Islamic Middle Ages," *Studia Copernicana*, Vol. 13, pp. 37-56, (1975).

however, turned the simple equinoctial sundial into a tool of great precision for measuring time and the coordinates of a celestial object.

The Samrāṭ in Literature

The *Yantraprakāra*, which describes a number of Jai Singh's instruments, omits the Samrāṭ. However, Jagannātha, the principal astronomer of Jai Singh, provides the following instructions for its construction.[1,2]

> "On a ground leveled with (standing) water and after marking the cardinal points, draw a north-south line. On this line erect a triangular gnomon wall with its top surface pointing toward the celestial pole. On the eastern and the western sides of this right-triangular wall, with their centers at the middle of the top edges, construct roughly finished quadrants of lime mortar and brick etc.—like a crescent moon. The upper edge of this construction should be like two quadrants. Having constructed this semicircle (of the two quadrants put together) 4 *aṅgulas*, (about 10 cm) thick and a *hasta* (about 50 cm) wide, make its edges strong and smooth with lime plaster or stone.[3] On these quadrants inscribe 15 *ghaṭikās* each and as many small divisions of *palas* as possible. In the middle of the gnomon wall pointing toward the pole, construct steps (all the way to the top). Now graduate the two upper edges of the gnomon wall with altogether 60 divisions in accordance with the tangent of the angles of declination. These divisions should be to the north and to the south of the center. This instrument is also called a Nāḍīvalaya.
>
> "Next, for observing, prepare a narrow plate of the same curvature as the quadrants' edge. To this, after filing out an appropriate section of it, fasten a 2 *aṅgula* (about 5 cm) long iron strip with a slit in the middle. This is then the sighting device."

As Jagannātha points out, the other name for the instrument had been Nāḍīvalaya. During Jai Singh's own lifetime, and long thereafter, the instrument was known both as Nāḍīvalaya and Samrāṭ yantra. A map of the Jaipur observatory drawn sometime after Jai Singh's death also labels the two instruments there as Nāḍīvalayas.[4,5] As late as 1876, in his *Jeypore Guide*,

---

[1] SSMC, p. 8, or SSRS, p. 1039.

[2] The instrument is said to have been described by Lakṣmīpati (1740 A.D.) also, but we have not been able to locate the manuscript. S. N. Sen, *A Bibliography of Sanskrit Works on Astronomy and Mathematics*, part I, p. 124, National Institute of Sciences of India, New Delhi, 1966. The author is indebted to S. R. Sarma to draw his attention to the reference.

[3] The author has changed Jagannātha's measures into metric units.

[4] Map No. 23, The Museum.

Hendley describes the two Samrāṭs as Nāḍīvalayas.[1] Today the name Nāḍīvalaya is used for the equinoctial double dials only. It is Jagannātha who is probably responsible for popularizing the name "Samrāṭ yantra" for the equinoctial dial Jai Singh so successfully modified for his work.

Figure 3-1 illustrates the principle of a Samrāṭ yantra. The instrument consists of a meridian wall ABC, in the shape of a right triangle, with its hypotenuse or the gnomon CA pointing toward the north celestial pole and its base BC horizontal along a north-south line. The angle ACB between the hypotenuse and the base equals the latitude λ of the place. Projecting upward from a point S near the base of the triangle, are two quadrants SQ_1 and SQ_2 of radius DS. These quadrants are in a plane parallel to the equatorial plane. The center of the two "quadrant arcs" lies at point D on the hypotenuse. The length and radius of the quadrants are such that, if put together, they would form a semicircle in the plane of the equator.

The quadrants are graduated into equal-length divisions of time measuring units, such as *ghaṭikās* and *palas*, according to the Hindu system, or hours, minutes and seconds, according to the Western system. The upper two ends Q_1 and Q_2 of the quadrants indicate either the 15-*ghaṭikā* marks for the Hindu system, or the 6 A.M. and the 6 P.M. marks according the Western system. The bottom-most point of both quadrants, on the other hand, indicates the zero *ghaṭikā* or 12 noon. The hypotenuse or the gnomon edge AC is graduated to read the angle of declination. The declination scale is a tangential scale in which the division lengths gradually increase according to the tangent of the declination angle as illustrated in Fig. 3-2.[2] The zero marking of this scale is at point D. Further, the gnomon scale AC is divided into two sections, such that the section DA reads the angle of declination to the north of the celestial equator, and the section DC reads the declination to the south, as illustrated in Fig. 3-2.

Time Measurement

The primary object of a Samrāṭ is to indicate the apparent solar time or local time of a place. On a clear day, as the sun journeys from east to west, the shadow of the Samrāṭ gnomon sweeps the quadrant scales below from one end to the other. At a given moment, the time is indicated by the shadow's edge on a quadrant scale.

In order to find time with the instrument at night, one has to know the time of the meridian transit of a prominent star. This may be obtained from appropriate

---

[5] The instrument, currently known as the "Nāḍīvalaya," is labeled "Dakṣinottara Nāḍīvalaya Yantra," or "Viṣuvate-saṅkrāntijñārtham Nāḍīvalaya Yantra." The Museum, map No. 23.

[1] Thomas Holbein Hendley, *The Jeypore Guide*, pp. 32-34., Jaipur, 1876.

[2] The distance from the zero marking for an angle of declination equals r tan δ, where r is the radius of the quadrant and δ the declination.

tables or calculated from the knowledge that a star returns to the meridian after 23 h, 56 min, 4.09 sec, the length of a sidereal day. The time at night is measured then by observing the hour angle of the star or its angular distance from the meridian. The readings then may easily be converted into the mean solar time. For measuring time at night, a tube or slit is used as a sighting device as Jagannātha suggests. With the quadrant edge as the vantage point, the observer looks at a prominent star through the device and moves the device back and forth along the quadrant edge SQ (Fig. 3-1), until the star appears to graze the gnomon edge AC. The vantage point V on the quadrant edge then indicates the hour angle or meridian distance SV of the star, which after proper conversion gives the apparent solar time.[1]

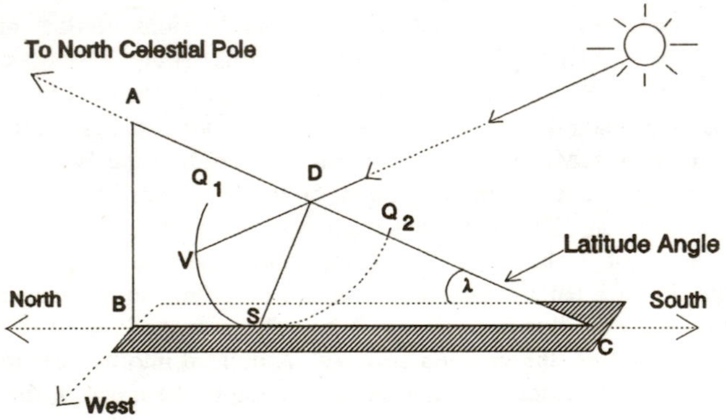

Fig. 3-1 Samrāṭ Yantra: Principle and Operation.

Because a Samrāṭ, like any other sundial, measures the local time or apparent solar time and not the "Standard Time" of a country, a correction has to be applied to its readings in order to obtain the standard time. The correction for Indian Standard Time is as follows.

Indian Standard time = Local Time $\pm$ Equation of Time $\pm$ Longitude difference

[1] For the conversion, the observer has to consult an ephemeris or tables for the right ascension of the star, and that of the sun that day. Next, by adding or subtracting the difference between the two right ascension values to the hour angle of the star, the apparent solar time is determined.

The longitude difference is obtained by adding or subtracting the longitude of the place in time units to 5 h, 30 min East, the longitude chosen for Indian Standard Time. For example, the longitude difference for Jaipur, converted into time units, is +26 min 43 sec and for Varanasi −2 min 3.4 sec.[1] For the Equation of Time, elaborate tables are available in books such as the Indian Astronomical Ephemeris.[2] A typical table is given in Appendix I. The *Zīj-i Muḥammad Shāhī* of Jai Singh contains a table for the equation of time which his astronomers probably consulted for their work.[3]

Right Ascension

The right ascension or RA of an object is determined by simultaneously measuring the hour angle of the object and the hour angle of a reference star. From the measurements, the difference between the right ascensions of the two is calculated. By adding or subtracting this difference to the right ascension of the reference star, the RA of the object is determined.

Declination

To measure the declination of the sun with a Samrāṭ, the observer moves a rod over the gnomon surface AC up or down (Fig. 3-1) until the rod's shadow falls on a quadrant scale below. The location of the rod on the gnomon scale then gives the declination of the sun. Declination measurements of a star or a planet requires the collaboration of two observers. One observer stays near the quadrants below and, sighting the star through the sighting device, guides the assistant, who moves a rod up or down along the gnomon scale. The assistant does this until the vantage point V on a quadrant edge below, the gnomon edge above where the rod is placed, and the star—all three—are in one line. The location of the rod on the gnomon scale then indicates the declination of the star. In this exercise, in addition to two observers, a torch bearer may also be necessary to shine light on the rod on the gnomon edge for the principal observer near the quadrants below to see it clearly.[4]

---

[1] The longitude of Jaipur observatory is 77;49,18.7 E, and that of Varanasi 83;0,46.1 E. See Kaye, pp. 51 and 61.

[2] The Indian Astronomical Ephemeris, published yearly, Govt. of India Press, Delhi.

[3] ZMS, f. 146.

[4] Garrett suggests using a tube or a string stretched between the gnomon and the quadrant of the Great Samrāṭ. See Garrett, p. 42. However, a string for an instrument as large as the Great Samrāṭ can never be satisfactory. The length of the string would have to be at least 15 m or more. A string of such a length will invariably sag. Besides, even a mild wind will cause the string to deviate from a straight line, causing additional error. Finally there is a problem of illuminating the string so that an observer can see it properly in dark. Jagannātha's suggestion of using a small plate with a small hole is the most appropriate procedure. SSRS, p. 1039, or SSMC, p. 8.

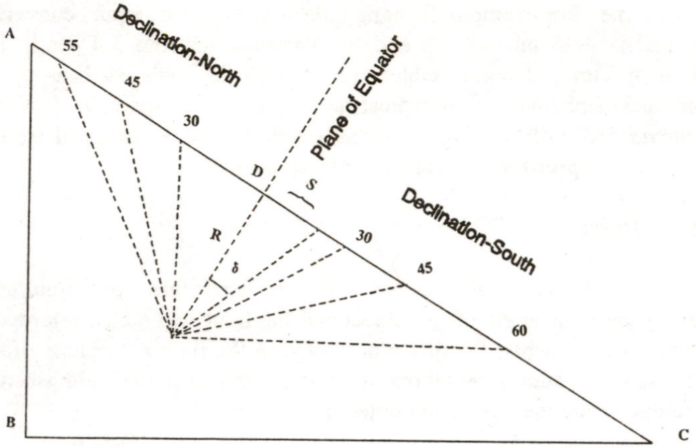

Fig. 3-2 The Tangential Scale over a Samrāṭ Gnomon. The scale indicates the angle of declination, δ.

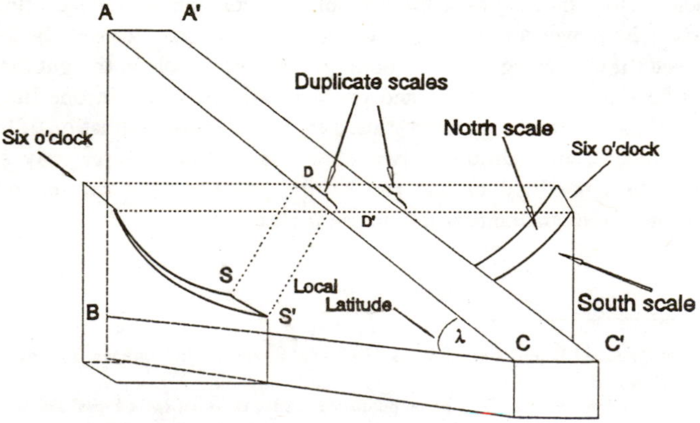

Fig. 3-3 Samrāṭ Yantra, Perspective. The gnomon and quadrants of a Samrāṭ are walls of finite width.

In actual practice, the Samrāṭ's quadrants are not thin arcs as illustrated in Fig. 3-1, but curved surfaces of finite width over thick walls engraved along their edges with two parallel time-measuring scales. With this arrangement, time can be measured by two observers simultaneously and then compared for accuracy. The observer uses the northern edge of the quadrants for measuring the declination of a star located to the south of the equator, and the southern edge if it is to the north. Further, because of the finite width of the quadrant surface, duplicate scales of length equal to the quadrants' width become necessary on the gnomon surface above. These scales, as shown in Fig. 3-3, are engraved at the zero markings on the gnomon.

Place	Latitude λ	Height lower	Height upper	Base BC	Hypotenuse AC	Thickness AA′	Least Count
Delhi[1]	28;37	2.34 m	20.73 m	34.59 m	39.2 m	3.20 m	1′
Jaipur[2] (small)	27;30[3]	0.00 0.00	6.9 5.73	13.22 6.53	14.73 8.59	1.29 1.29	2′
Jaipur (great)	26;53[4]	3.5	22.6	44.6	50.1	2.9	1′
Varanasi (small)	25;19[5]	1.65	2.53	3.06[6]	3.39	0.28	10′
Varanasi (large)	25;14[7]	1.64	6.80	10.92	12.9	1.39	6′
Ujjain	23;10[8]	0.6	6.7	14.66	16	1.51	2′

Table 3-3 The Samrāṭ Yantra gnomons

[1] The dimensions of the Delhi Samrāṭ are based mostly on Kaye's book. Refer Kaye pp. 36, and 41-42.

[2] The instrument is unique as it has a double hypotenuse. The "gnomon" of the instrument as a result is a quadrangle and not a right triangle as elsewhere.

[3] Kaye p. 36.

[4] Kaye, p. 36.

[5] *Ibid.*

[6] *Ibid.*

[7] *Ibid.*

[8] *Ibid.*

Jai Singh built his Samrāṭs in several sizes with varying degrees of accuracy in mind. His smallest unit, with a least count of one minute, is located at Varanasi, and the largest one, almost ten times as big in size and having a least count of 2 sec, is at Jaipur. In Table 3-3, the dimensions of the Samrāṭs at different locations are compared. For the symbols used in the table, refer to Fig. 3-3.

DESIGN, CONSTRUCTION, AND PRECISION

The precision or accuracy of a Samrāṭ depends on a number of factors inherent in its design and on other factors that depend on the care taken in construction.

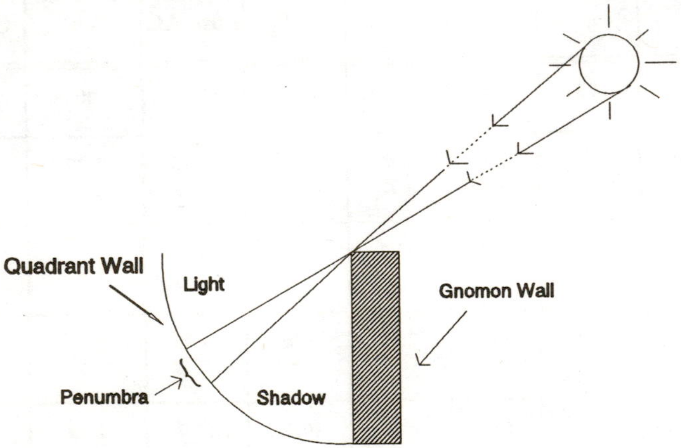

Fig. 3-4 The penumbra of the sun is one-half degree wide. The Figure shows a highly exaggerated version of the effect.

Precision in the Measurement of Time

The Penumbra

The first and foremost factor affecting the precision of time measurements, and which is inherent in the very design of the instrument, is the width of the penumbra. For large Samrāṭs, such as those at Delhi and Jaipur, the penumbra could be several centimeters wide. The sun, because of the finite width of its disc, casts diffused shadows of objects. As a result, pinpointing the exact location of the shadow-edge becomes a difficult as well as a subjective matter. The penumbra at the Great Samrāṭ of Jaipur, at mid-morning on a clear day, can be as wide as 3 cm or more, making it difficult to read time with accuracies better than ±15 seconds. However, when the sun is strong, the difficulty may be eliminated and the procedure of measuring time made objective. This is done by superimposing on the penumbra the shadow of a thin object such as a needle or a string.

By holding a one to two cm long taut string parallel to the shadow edge, about one cm or so above the instrument's surface, and reading the scale where the string's shadow merges with the shadow of the gnomon edge, we could repeat our readings with an accuracy of ±3 sec or better.[1] The author believes that the astronomers of Jaí Singh must have used some device similar to that of the string for overcoming the problem of the penumbra.[2]

Place	Width SS′	Radius DS	Least Count
Delhi[3]	2.32 m	15.1 m	2 sec
Jaipur (small)	0.98 0.98	2.78 2.78	20
Jaipur (great)	2.84	15.15	2
Varanasi (small)	0.52	0.97	1 min
Varanasi (large)	1.76	2.79	15 sec
Ujjain	0.92	2.78	20 sec

Table 3-4 The Samrāṭ Quadrants

Penumbra and the Quadrants

Even after eliminating the penumbra-caused uncertainty with the "string-method" described above, a second difficulty remains. Because of the penumbra, the two quadrants of the instrument indicate time that differs from

---

[1] Blanpied, basing his criticism on the problem of the penumbra encountered when reading time, concludes that the large Samrāṭs of Jai Singh are over-designed and their subdivisions exceed the intrinsic precision of the instrument. However, as we have indicated, this problem can be easily eliminated with a string. William A. Blanpied, "Raja Sawai Jai Singh II: An 18th Century Medieval Astronomer," *Am. J. Physics*, Vol. 43, No. 12, 1025-1035, (1975).

[2] At the observatory of Jaipur, the author saw an experienced guide achieving almost similar results with the tip of his index finger.

[3] The dimensions of the Delhi Samrāṭ are based mostly on Kaye's book. Refer Kaye pp. 36, and 41-42.

the true local time. The west quadrant of the instrument of Jaipur, for instance, read in the morning hours, indicates time ahead of the true local time, whereas the east quadrant indicates time behind the true local time. In other words, a Samrāṭ clock runs fast in the morning and slow in the afternoon.

Penumbra and Noon-Hour

The penumbra effect becomes quite evident and at the same time perplexing when one observes the noon hour at the quadrants. Theoretically, at the noon hour, the shadow of the gnomon, disappearing from the west quadrant, must reappear immediately on the east quadrant. However, in actual practice this does not occur. At the Great Samrāṭ of Jaipur, the shadow, after disappearing from the west quadrant, takes about 80 sec to reappear on the east.[1]

The lapse of 80 seconds or so the shadow takes to travel from one side to the other and the discrepancy of time-readings at the two quadrant-scales may be understood as follows: The width of the penumbra in mid-January produced by the entire disc of the sun should be as wide as the disc of the sun, which is 1/2 degree across. (See Fig. 3-4.). A one-half degree wide penumbra would be about 14.3 cm across at the Great Samrāṭ of Jaipur,[2] and, ideally, the midpoint of this penumbra, corresponding to the midpoint of the sun, should indicate the true local time. However, the visible part of the penumbra in midmorning, say, around 10 o'clock, is only about 3 cm wide. It is not produced by the entire disc of the sun but by a part of it or, more precisely, by its western 1/5 part only.[3] Any point selected in this 3-cm-wide region, as a result, does not correspond to the midpoint of the sun and, therefore, does not indicate the true local time. The point that we selected in the penumbra, while reading time with a string in the morning hours, corresponds to a point somewhere within the one-fifth of the disc measured from the sun's western edge. As a consequence, time readings of the western quadrant are always ahead of true local time. In the afternoon the situation becomes quite the reverse. In the afternoon, the visible penumbra is cast by the eastern 1/5 of the disc, and then any point within it, selected as the edge of the shadow to indicate the time is behind the true local time.

---

[1] The large Samrāṭ of Varanasi appears to be an exception to the "delay of the shadow" effect and needs to be investigated further. At that Samrāṭ, at noon hour, the shadow on the east quadrant can be seen almost 2 min before it disappears from the west quadrant.

[2] As the disc of the sun in mid January is 0;32,35 of arc, its shadow should be $\{(0;32,35)/30''\} \times 2.2$ mm or 14.3 cm wide. The average length of a small division measuring 30″ of arc on the quadrant scale is 2.2 mm.

[3] A 3-cm penumbra corresponds to approximately one-fifth of the angular diameter of the sun which measures 14.3 cm on the quadrant scale.

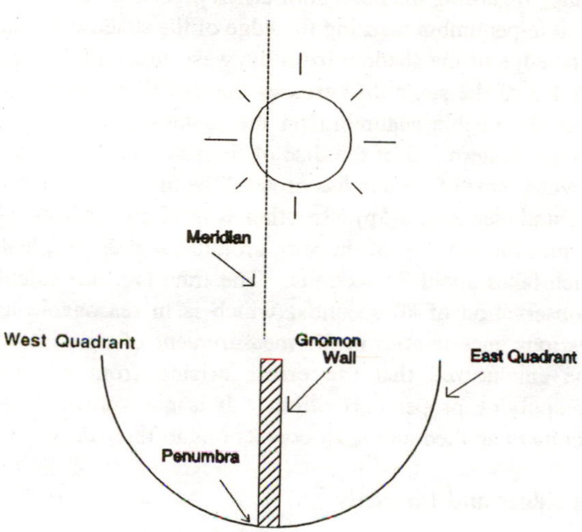

Fig. 3-5a Noon-hour at the West Quadrant of a Samrāṭ. The penumbra indicating the shadow-edge corresponds to the western one-fifth disc of the sun.

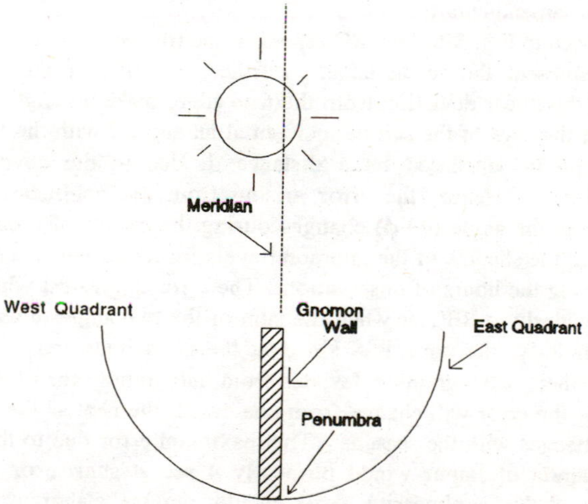

Fig. 3-5b Noon-hour at the East Quadrant of a Samrāṭ. The penumbra indicating the shadow-edge corresponds to the eastern one-fifth of the sun's disc.

The discrepancy regarding the noon hour arises precisely due to this penumbral effect. The visible-penumbra marking the edge of the shadow is cast by 1/5 disc of the sun. The edge of the shadow from the west quadrant disappears as soon as a little over 1/5 of the sun's disc crosses the meridian (see Fig. 3-5a). The shadow-edge or the visible-penumbra on the eastern quadrant is cast, on the other hand, by the eastern 1/5 of the disc of the sun. This edge appears on the east quadrant when most of the sun has crossed the meridian and only its eastern 1/5 remains behind (see Fig. 3-5b). In other words, no shadow appears on the east quadrant until the 3/5 disc of the sun, around its midpoint, has crossed the meridian, which takes about 72 seconds. The time lag thus calculated comes close to our observation of 80 seconds, which is in reasonable agreement in view of the various uncertainties in the measurement of time intervals.

It should be emphasized that the errors arising from penumbra can be eliminated by applying proper corrections. It is not known, however, if Jai Singh's astronomers applied any such corrections to their data.

The Gnomon Edges and Linearity

At every Samrāṭ we examined, the gnomon edges AC or A′C′ deviated from a straight line by several millimeters (see Fig. 3-3). In most cases, these deviations from linearity were due to shifted stones or to poor maintenance and might not have been present when the instruments were newly constructed. We noted that at the Great Samrāṭ of Jaipur this deviation was ± 3 mm laterally and also ± 3 mm perpendicularly.

With reference to Fig. 3-6, let ABC represent the true edge of the gnomon and let A′B′B represent the actual edge. Further, let BB′, or the distance S, representing the linear deviation from the true edge, make an angle ϕ with the vertical. Let the rays of the sun be incident at an angle θ with the vertical. If the solar rays are displaced by a distance d, due to the deviation, then $d = S \sin(\theta+\phi)$. Hence, the error arising from the nonlinearity changes continuously as the angle $(\theta+\phi)$ changes during the course of a day. Or the error varies as the shadow of the gnomon travels from one end of a quadrant to the other during the hours of observation. The error is greatest when the rays fall perpendicularly to BB′, or when the sum of the two angles θ and ϕ equals 90 deg. Similarly, the error is $S \sin \phi$ at the noon hour and $S \cos \phi$ at 6 o'clock. Further, if the gnomon deviates from one point to the other along its entire length, the error will change from one day to the next as the declination of the sun changes with the seasons. The maximum error due to this effect at the Great Samrāṭ of Jaipur would be nearly 4 sec at sunrise or sunset. A rectification of the nonlinearity error would require elaborate calibration curves—one curve for each day of the sun's journey from one solstice to the other.

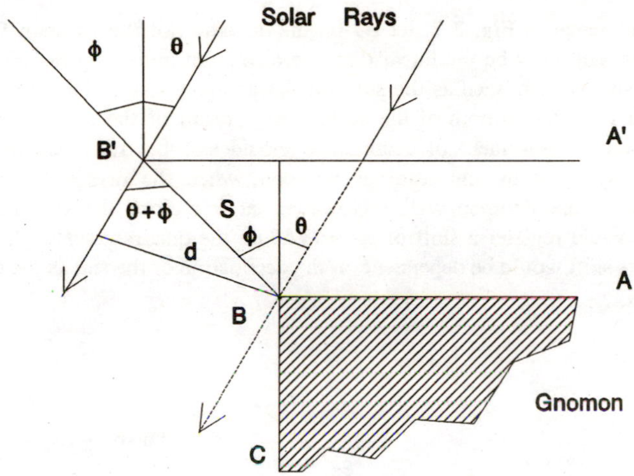

Fig. 3-6 Deviation of Gnomon Edge from Linearity. In the Figure, B is the true edge and B′ the deviated edge.

The 6 O'Clock Marks and the Gnomon Plane

Another defect that we noted concerns the line joining the 6 o'clock marks on either side of the gnomon wall (see Fig. 3-3). We noted that in a number of instruments, contrary to the requirement of the theory of the instrument, the line was non-tangential to the scales on the gnomon surface. At the Delhi Samrāṭ, the line was observed about 3-4 cm above the surface of the scales. There could be two reasons for this defect. Either the quadrant markings of these instruments have been inaccurately done, or the surfaces with scales do not have proper elevation. At the Delhi Samrāṭ the defect causes errors of the order of 27 to 36 sec in time readings around 6 o'clock.[1] We did not notice this defect at the Great Samrāṭ of Jaipur, however. There, the plane of the gnomon does not lie below the line joining the two 6 o'clock marks.

---

[1] At 6 o'clock the shadow will be shifted by 3 to 4 cm. For a 4 cm shift, the error in time measurement would be (4 cm/2.2 mm) x 2 sec, or 36 sec. The average length of a small division at the Delhi Samrāṭ is 2.2 mm.

The Gnomon and the Polar Axis

With reference to Fig. 3-7, let $\Delta\theta$ be the deviation of the gnomon from the polar axis, and let y be the linear distance of a point indicating the declination of a celestial object, such as the sun, on the gnomon scale. The distance y is measured from the bottom of the scale. As a result of the deviation $\Delta\theta$, the error in the measurement of declination should also be $\Delta\theta$. One would not observe any error in time-readings at noon, when the rays of the sun are tangential to the gnomon wall. However, at 6 o'clock the shadow of the gnomon would register a shift of $\Delta s = y\Delta\theta$ on the quadrant scale. The error due to this shift would be dependent on the declination of the sun as the equation indicates.

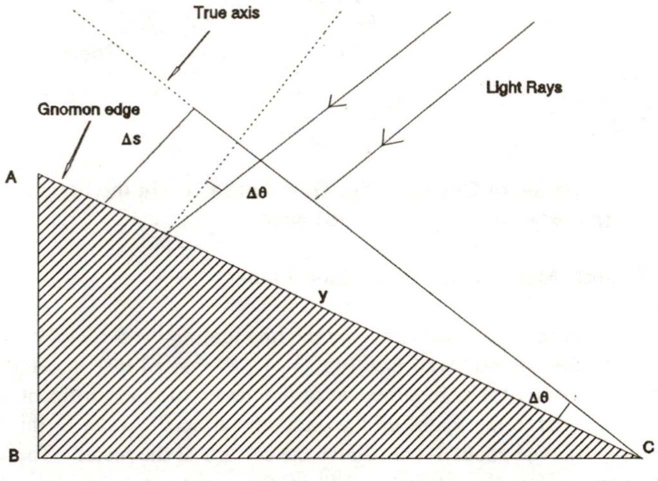

Fig. 3-7 Deviation of a Gnomon Edge from the Polar-Axis.

Lieutenant A. ff. Garrett, who supervised the restoration of the Jaipur observatory, writes that the gnomon of the Great Samrāṭ there deviates less than 1′ of arc from the polar axis.[1] At the Great Samrāṭ of Jaipur, a 1 arc-minute deviation will contribute to an error of about 7 sec in time-readings taken around 6 o'clock at the time of equinox.[2] The declination readings will have an error of 1 arc-minute, and they will be independent of the time of measurement.

---

[1] Garrett, p. 42.

[2] Taking $y = (50.15 \text{ m})/2$ or approx. 25 m, and $\Delta\theta$ equal to 1 arc-minute (2.9×10^{-4} rad), the Δs turns out to be 6.6 sec.

The Radius of Curvature of the Quadrants

We also investigated the error caused by variations in the radius of curvature of a quadrant from one point to another. We noted that such deviation at the Great Samrāṭ of Jaipur, which has a radius of 15.15 m, is less than 1 cm. We found that the error caused in time measurement by such small variations would be negligible.[1]

Precision in the Measurement of Declination

Graduations on the Gnomon

The precision of a Samrāṭ's declination measurements depends on the graduations on its gnomon scale. As shown in Fig. 3-2, the declination angle δ is measured as the length s along the gnomon edge. Since $s = R \tan \delta$, the uncertainty $\Delta s = R \sec^2 \delta \, \Delta\delta$. At equinox when the rays of the sun are perpendicular to the gnomon, and δ is zero, the uncertainty Δs becomes $R \, \Delta\delta$. For measurements with the unaided eye, the uncertainty $\Delta\delta$ equals 1′ of arc or 3.9×10^{-4} rad. Taking R = 15.15 m, the uncertainty in length Δs equals 0.44 cm. The uncertainty increases with $\sec^2 \delta$, and thus at 53 deg, the uncertainty is 1.21 cm.[2]

At the Great Samrāṭ of Jaipur, the division-length for one arc-minute at the zero point of gnomon scale is about 4 mm, whereas the division-length at 53 deg is 1.2 cm. Hence, division-lengths of the Great Samrāṭ's gnomon are compatible with the uncertainty of naked eye observations.

The Quadrant Plane and the Equatorial Plane

According to the theory of the instrument, the plane in which the quadrants are built should be parallel to the equatorial plane. If this is not the case, the horizontal line joining the two 6 o'clock marks on the quadrants on either side of the gnomon may not pass through the zero markings of the declination scale on the gnomon (see Fig. 3-3). The horizontal line would then be either to the north or to the south of the zero markings on the gnomon scale. This fact allows us to calculate the deviation of the quadrant plane from the true equatorial plane. An estimation of this deviation is described below.

---

[1] Let θ be the angle along the quadrant scale where the radius varies by ΔR. Now $s = R\theta$. Therefore, $\Delta s = R\Delta\theta + \theta\Delta R$. For the graduations of equal length $\Delta s = 0$. Thus $R\Delta\theta = -\theta\Delta R$. If the variation ΔR does not last more than a degree or two, and if ΔR is no more than 2 cm at the most, then for the Great Samrāṭ of Jaipur (radius 15.15 m), the variation, $\Delta\theta = (\theta\Delta R)/R = 9.6''$ of arc, which is less than 1 sec of time.

[2] Blanpied has also carried out similar analysis. See Blanpied (1975), *Op. Cit.*

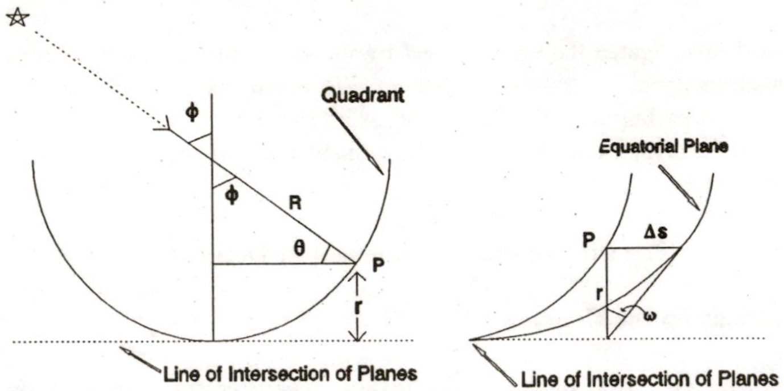

Fig. A Fig. B

Fig. 3-8 Deviation of the Quadrant Plane from the Equatorial Plane. See the text for details.

As shown in Fig. 3-8, let the light from a point source such as a star fall on point P of the quadrant scale. Let ϕ be the hour angle of the star. Let ω be the angular separation between the equatorial plane and the plane of the quadrant. Let Δs be the linear separation between the two planes at P. If r is the perpendicular distance of point P from the horizontal line passing through the lower end of the quadrant, then as a first order approximation, $\Delta s = r\omega$. The linear separation Δs will show up as an error in the measurement of declination. For an object on the equator, such as the sun at equinox, the value of r is equal to $(R - R\cos\phi)$ where R is the radius of the quadrant. Therefore, Δs would be largest for objects at the horizon for which $\phi = 90$ deg. Under these conditions, the error Δs, becomes equal to $R\omega$. This shows up as a shift of the zero markings of the declination either to the north or to the south of the line clock marks. By measuring this shift, the angle ω (equal to $\Delta s/R$) can be determined.

At the Great Samrāṭ of Jaipur, the zero of the declination scale of the gnomon corresponding to the west quadrant was noted to be shifted by about 3 cm toward the south. This suggests that the quadrant plane deviates from the true equatorial plane by 3 cm/15.15 m = 8 arc-minute. The corresponding numbers for the east quadrant are 5 cm to the south and 13′ of arc. We have noted defects of this kind and magnitude in almost all of the Samrāṭs we have examined.

Declination of the Sun

The measurement of the declination of the sun presents another difficulty. The declination of the sun is measured by holding a rod over the gnomon scale in such a way that its shadow falls on a quadrant edge below. Because of the great distance of the gnomon scale from the quadrants of large Samrāṭs, the shadow is often so weak that it is very easy, particularly for an inexperienced observer, to miss it completely or to misjudge its midpoint. The problem becomes serious when the sun is near an equinox, and its declination is measured by that section of the gnomon scale where divisions are close together. At the Great Samrāṭ of Jaipur these divisions are only 4 to 5 mm apart.

The difficulty may be overcome if the observer, with his vantage point at the quadrant edge, looks directly at the sun through a smoked glass or some such device to filter out the harmful rays of the sun. Looking through the filter, the observer aligns a pencil-thin rod, which is held by an assistant at the edge of the gnomon, with the center of the sun, or better still, with the lower edge of the sun.[1] Using this procedure one may obtain readings of declination to the nearest small division which is $1'$ of arc for the Great Samrāṭ of Jaipur (see Ch. 6).

Effect of the Deviations on Precision

Because of these defects and deviations, the present instruments do not come anywhere near their intended accuracy. The great Samrāṭ of Jaipur, which is supposed to be the most sensitive of the line, in its present form measures time with accuracies of between ± 10 and ± 90 seconds only, depending on the sections of the quadrants' scales and the gnomon's scales used. Similarly, the errors in measurement of declination for this instrument may be anywhere from $\pm 1'$ to $\pm 15'$ of arc, depending on the location of the object in the sky, as well as on the time of observation. For declination measurements, maximum uncertainties would be encountered for objects close to the horizon.

It is difficult to ascertain if the defects elaborated above were present in the original construction or are the results of the numerous repairs the instrument has undergone from time to time. Garrett reports seeing two scales superimposed upon each other on the gnomon of the Great Samrāṭ of Jaipur, indicating that some errors were indeed made originally and rectified later on.[2]

---

[1] For a filter, the author used an exposed black and white photographic film.

[2] Garrett, *Op. Cit.*, p. 42.

ṢAṢṬHĀṀŚA

A Ṣaṣṭhāṁśa yantra is a 60 deg arc built in the plane of meridian within a dark chamber. The instrument is used for measuring the declination, zenith distance, and the diameter of the sun. Figure 3-9 explains the principle of its operation. The figure represents a dark chamber with a 60-degree arc facing south. The arc is divided into degrees and minutes. High above the arc, at its center on the south wall, is a pinhole to let the sunlight in. As the sun drifts across the meridian at noon, its pinhole image falling on the Ṣaṣṭhāṁśa scale below enables the observer to measure the zenith distance, declination, and the diameter of the sun. The image formed by the pinhole on the scale below is usually quite sharp, such that at times even sunspots may be seen on it.

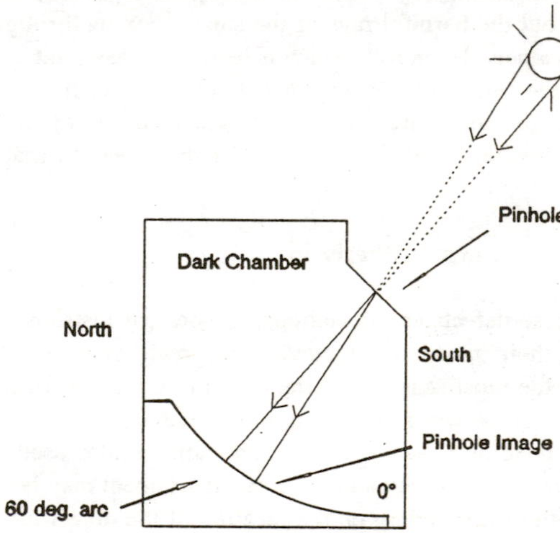

Fig.3-9 The Principle of Ṣaṣṭhāṁśa Yantra. The instrument displays a pinhole image of the sun over a sixty-degree meridian scale. The image of the sun is highly exaggerated in the diagram.

History

The Ṣaṣṭhāṁśa has its origin in the Islamic tradition of astronomy. According to Kennedy, the instrument was invented by Abū Maḥmūd al-Khujandī in the late 10th century. Al-Khujandī named it *Suds-i Fakhrī* or Fakhrī sextant, after his patron, Fakhr al-Daulā. With the instrument, al-Khujandī carried out observations at Rayy, near modern Teheran, in 994.[1]

The instrument has also been described by writers, such as Abū al-Ḥasan ʿAlī al-Marrākushī (c. 1250).[2] The instrument was erected at the celebrated observatory of Ulugh Beg founded in 1424 at Samarkand, where its ruins may still be seen.[3] Jamshīd Giyāth al-Dīn al-Kāshī (d. 1429-30) elaborates upon it in a short tract he wrote on instruments before the Samarkand observatory was built.[4] Al-Kāshī writes:

> "The *Fakhrī Sextant*, is a sextant of a circle, set up in the plane of the meridian circle, it being graduated in seconds, and it is such that a wall is to be erected, of stone and plaster, such that the length of its base is eighty cubits and its thickness four cubits, and the length of its height in the direction of north forty cubits, and to the south one cubit. Make (it) such that from the south side, from the base of the wall, to the northern side, from the top of the wall, it is a sextant of the concavity of a ring, so that if a perpendicular from its center is erected to the horizon plane it will pass through one side of the sextant, and that concavity is to be faced with worked stone, and in the middle of that, lengthwise, let a hollow be cut out, the width of which will be four digits, and its depth one digit. Outside that, plates of copper or brass are placed so that its visible surface is as accurately circular (as possible). Graduate it (the sextant) in degrees, minutes, and seconds. This (instrument) can sometimes be so constructed that the meridian line is determinable with extreme accuracy."

Although al-Kāshī does not specifically mention a pinhole arrangement at the center of the arc, it is clearly prescribed by Jagannātha in his *Samrāṭ Siddhānta*.[5] A pinhole gives a better image than a hole of large size.

Ṣaṣṭhāṁśa in Sanskrit Literature

The Ṣaṣṭhāṁśa is described both in the *Samrāṭ Siddhānta* and the *Yantraprakāra*. In the *Samrāṭ Siddhānta*, Jagannātha explains:

> "On a ground leveled with water [standing in channels], and having marked the cardinal directions on it, erect a rectangular wall in the meridian just as done before [for the Dakṣinottara Bhitti yantra]. To

---

[1] Sayili, pp. 118, 120.

[2] Kennedy (1961), p. 105, *Op. Cit.*, Ch. 2.

[3] Piini, *Op. Cit.*, Ch. 2.

[4] Kennedy (1961), *Op. Cit.*

[5] SSMC, pp. 7-8, or SSRS, pp. 1038-1039.

> the south of it, leaving some space, draw a vertical line. Similarly, at the top end of the wall, draw a north-south line. With the intersection of these (two lines) as center, draw an arc a little larger than one-fourth of a circle. Touching this arc, construct four *aṅgulas* (10 cm) wide arc of lime mortar and bricks. Having made it [the arc's surface] smooth, inscribe on it *aṁśas* and *kalās*. Again, similar to the wall erect another one parallel to it at a distance of 2 *hastas* (about a meter). There also draw arcs. Erect a door to the north and close the area between the walls. Similarly close the space to the south between the walls. In the same manner, cover up the roof with slabs etc. At the upper end of the two lines drawn at the south corner of the two walls, cut a diagonal space and cover it with a plate. On the north-south line (meridian) passing through two diagonal cuts [on the southern enclosing wall], close to the vertical line (between the vertical lines), drill a pinhole. When the rays of the midday sun fall on the arcs, take a reading. Having constructed this instrument and with its doors closed, and observing the solar image formed by its rays passing through the pinhole every day, the two extreme values of the sun's declination should be determined. The rest should be understood just as for the Dakṣinottara Bhitti yantra. The larger this instrument is made in size, the greater is the precision achieved. This is the instrument which the *yavanas* (Muslims) call *Suds-i Fakhrī*."

The *Yantraprakāra* of Jai Singh also describes a Ṣaṣṭhāṁśa. However, that description is not as elegant as the one in the *Samrāṭ Siddhānta*.[1] Jagannātha describes two arcs to be constructed side by side. However, he does not explain their purpose, because one arc could have served the purpose equally well. It is noteworthy that Jagannātha's description in this respect agrees with the instrument excavated at Samarkand. At Samarkand also there were two arcs separated by less than a meter.[2] However, at Jaipur their separation is only a few centimeters.

Jai Singh built his Ṣaṣṭhāṁśas only at Delhi and Jaipur. At Delhi the instrument was built in a chamber supporting the east quadrant of the large Samrāṭ. However, now it is in ruins. Even the chamber that once housed it is open and not closed to the outside as it had once been. An arched opening has been built into it facing west. Because of a water seepage problem at the observatory, the instrument was not restored along with the others in 1909-10.[3] A close examination of the chamber, where the instrument had once been located, indicates that the radius of the arc must have been approximately 6 m.

---

[1] *Yantraprakāra* (J), p. 17; *Yantraprakāra* (S), pp. 24-25, 71-73.

[2] The arc had a 40 m radius. Piini, *Op. Cit.* Ch. 2.

[3] Bhavan, pp. 72-73.

At Jaipur, there are four units of the Ṣaṣṭhāṁśa. They have been built in the lofty chambers flanking the Great Samrāṭ and are in an excellent state of preservation. The units are identical and have a radius of 8.65 m.

Precision

Deviation from Meridian Plane

The four units of Ṣaṣṭhāṁśa at Jaipur have two scales each parallel to one another and separated by a distance of 7.5 cm—one measuring declination and the other the zenith distance. A single pinhole serves them both. In the unit we examined closely, the pinhole above the instrument and the scale below do not lie in the plane of the meridian of either two scales. The pinhole lies 15 mm to the east of the vertical plane of the declination scale. However, the error due to this shift should be negligible in the measurements.[1]

The Pinhole

Aperture Broadening

The pinhole of the Jaipur instrument has a diameter of 3.8 mm.[2] Because of this width of the pinhole, the true image of the sun should be surrounded by a halo of light one-half as wide.[3] Theoretically, the sharpness of the image should taper off to zero at a distance of one-half of 3.8 mm, or 1.9 mm, from the edge of the true image. However, in actual practice, it is virtually impossible to pinpoint the edge of the true image or to pinpoint the end of the aperture broadening. Therefore, the uncertainty in the readings of the angle of declination at the Jaipur Ṣaṣṭhāṁśa is 1.9 mm or nearly 3/4 of an arc-minute.[4]

---

[1] The declination angle δ is related to the angle of azimuth by the following relation: $\sin \delta = \sin \lambda \cos Z + \cos \lambda \sin Z \cos A$, where Z = zenith distance, λ = latitude, and A = azimuth. At equinox, declination $\delta = 0$, and for the latitude of Jaipur, which is 27 deg, the zenith distance is equal to 27 deg. For a 1/2 deg deviation from the true meridian plane, i.e., for δA equal to 0.5 deg, the variation in δ, the declination, is found to be about 0.1 arc-minute with a computer. On the other hand, at winter solstice, the same would be 0.2 arc-minute.

[2] The plate with a circular aperture was put in 1902. Garrett says that he had found a different plate with a square aperture at the site which he replaced with the present one. See Garrett, p. 37. It is not certain when the plate with a square aperture was put in, because the *Yantraprakāra* does not recommend a square aperture as such. See *Yantraprakāra* (J), p. 17, also *Yantraprakāra* (S), pp. 24-25, 71-73.

[3] The image is, in fact, a superposition of a multiple images formed by each point of the pinhole. The edges of this image, because they receive light from fewer points of the pinhole, are not as bright as the middle. We will refer to this broadening as aperture broadening.

[4] The 1′ divisions of the Ṣaṣṭhāṁśa of Jaipur are 2.51 mm wide. See Ch. 6.

Diffraction Broadening

Along with the aperture broadening, the image of the sun is also affected by the diffraction phenomenon at the pinhole edges. The diffraction effect begins to predominate if the size of the pinhole is reduced beyond a certain limit in order to minimize the aperture broadening. Because of diffraction, the first minimum of the diffracted image would lie at a certain angle, say, θ from the edge. The angle θ is given by the relation, $\theta = 1.22\ \lambda/d$, where λ is the mean wavelength of light and d is the diameter of the hole.[1] For a 3.8 mm wide hole, and for the mean visible length λ of 560 nm, the angle θ equals 0.6′ of arc. The image of the sun, in other words, is broadened by a total of 1.2 arc-minute. For optimum results, the aperture broadening and the diffraction broadening should be equal. It is interesting to note that the two broadenings are nearly equal at the Jaipur Ṣaṣṭhāṁśa.

Although the penumbra and the diffraction broadenings will lead to erroneous results regarding the diameter of the sun, they should cause relatively small errors in declination measurements, provided one chooses two equally bright points to mark the ends of the image on either side and then takes their midpoint as the midpoint of the sun.

Precision of the Jaipur and Delhi Ṣaṣṭhāṁśas

The Ṣaṣṭhāṁśa of Jaipur is Jai Singh's one of the most accurate instruments that still maintain the precision inherent in its design.[2] The Ṣaṣṭhāṁśa can measure the zenith distance or declination with an accuracy of $\pm 1'$ of arc.

There are reasons to believe that in its original state the Delhi Ṣaṣṭhāṁśa was an extremely accurate instrument. It approached or surpassed the limit of naked-eye observations, which is generally accepted to be 1 minute of arc.[3] We will discuss the Delhi Ṣaṣṭhāṁśa in detail in Ch. 5.

---

[1] We have adapted the relation from Raleigh's criterion for the resolution of diffraction patterns.

[2] We disagree with Prahlad Singh, who states that planets such as Venus can be observed with this instrument. Venus can never be observed with a Ṣaṣṭhāṁśa, because in order to be observed the planet should have to be at the meridian. But that happens only in broad daylight. See Prahlad Singh, *Stone Observatories in India*, p. 110, Varanasi.

[3] An instrument such as the Ṣaṣṭhāṁśa can have a precision better than 1 arc-minute. This is because the readings are taken off the pinhole image of the sun, and thus the resolving power of the eye does not come into picture.

DAKṢINOTTARA BHITTI YANTRA

A Dakṣinottara Bhittị yantra consists of a graduated quadrant or a semicircle inscribed on a north-south wall. At the center of the arc is a horizontal rod. The instrument is used for measuring the meridian altitude or the zenith distance of an object such as the sun, the moon, or a planet.

Dakṣinottara Bhitti yantra is a modified version of the meridian dial of the ancients, which was a portable instrument.[1] Ptolemy in his *Almagest* describes two versions of the portable meridian dial: one, a meridional armillary, and the other a meridian quadrant on a solid block called the *Plinth* in his *Almagest*.[2,3] In *Zīj-i Khāqānī fi Takmīl-i Ilkhānī*, al-Kāshī mentions a mural quadrant or *Libnā*.[4,5] A copy of al-Kāshī's *Zīj-i Khāqānī* may be seen at Jai Singh's library in Jaipur.[6] The meridian dial was popular in Europe also. Tycho Brahe, for example, constructed a mural quadrant of radius 194 cm at Hven, a Danish island, in 1582.[7]

In Hindu astronomy the meridian quadrant, described as *Turīya* yantra, is also a portable device. The *Yantraprakāra* discusses a portable meridian dial, *Yāmyottara yantra*, and a fixed mural quadrant, or *Yāmyottara Bhitti yantra*, both of which are described in Naṣīr al-Dīn al-Ṭūsī's Arabic version of the *Almagest*.[8] Jagannātha describes the construction of the instrument as follows:[9]

> "Having constructed a square north-south wall, draw a quadrant with its radius equal to the length of the wall and its center at the upper end of the wall towards its south end. Graduate [the arc] into degrees and minutes as permissible. Fixing an iron style at the center, take readings of the shadow cast by the midday sun. From the shadow down to the point [on the arc] intersected by a plumb line, measure (then) the zenith angle. In this manner observing everyday, determine the maximum and the minimum declination [of the sun].

---

[1] The other names in Sanskrit for the yantra are, *Dakṣinodag-bhitti*, *Yāmyodag-bhitti* and *Yāmyottara-bhitti*.

[2] Ptolemy, book 1, ch. 12.

[3] For a discussion of the instrument, see Dicks, *Op. Cit.*, Ch. 2.

[4] Sayili, p. 286.

[5] In his treatise on instruments, al-Kāshī does not mention a mural quadrant or *Libnā* but discusses a "double ring," which Kennedy interprets as the solstitial or meridional armillary. Kennedy (1961), p. 105, *Op. Cit.*, Ch. 2. A solstitial armillary has also been discussed by al-ᶜUrḍī. Seemann, p. 53, *Op. Cit.*, Ch. 2.

[6] *Zīj-i Khāqānī* of Giyāth al-Dīn al-Kāshī, 184 pp., copied ca. 1600, acquired 1728. Bahura II, p. 72. Also David A. King, "A Handlist of the Arabic and Persian Astronomical Manuscripts in the Maharaja Man Singh II Library in Jaipur," J. Arabic Hist. Sci., Vol. 4, pp. 81-86, (1980).

[7] Raeder et al, pp. 29-31, *Op. Cit.*, Ch. 1.

[8] *Yantraprakāra* (S), p. 52-53.

[9] SSMC, pp. 6-7, or SSRS, pp. 1036-1038.

"One half of the difference between the maximum zenith distance [while the sun is] on the south and the minimum zenith distance [when the sun is to the north], is the obliquity of the ecliptic. This [value] is [correct] if the latitude of the place [of observation] is greater than the obliquity. At a location where (the sun's) northern zenith distance is minimum [while it is to the north of the zenith], extend the arc slightly toward south [of the plumb line]. The maximum declination then is one-half the sum of the maximum and the minimum zenith angles. At a location where the obliquity is larger than the latitude, the latitude equals the obliquity minus the minimum zenith distance. If the obliquity is smaller than the latitude, the latitude then equals the obliquity plus the minimum zenith distance.

"With this instrument the latitude of Delhi was determined to be 28;39 and the obliquity 23;28.[1] In the *Yavana* country (Greece), Hipparchus and other *Yavana* masters obtained the obliquity of 23;51,19 with their observations. Again in Greece, at 36 deg latitude, Ptolemy found the obliquity to be 23;51,15.[2] Further, at Samarkand, at 39;37 deg latitude, Ulugh Beg observed the obliquity to be 23;30,17. We (also) with this instrument found obliquity of 23;28 in 1651 S.E. (1729 A.D.) at Delhi."

Dakṣinottara Bhitti Yantras at Jai Singh's Observatories

With a Dakṣinottara Bhitti instrument, the meridian altitude of the sun is discerned from the shadow of the rod cast by the sun on the instrument scale at noon on a clear day. Some practice is necessary, however, for working with the large units of this instrument, such as the ones at Jaipur and Ujjain, because the shadows of their rods are often indistinct from their surroundings. For measuring the altitude of a star or a planet at meridian, the observer, sighting the star, operates a slit or a viewing device around the circular scale such that the object in the sky, the rod at the center of the arc, and the slit in his hand—all three—fall in one line. The location of the slit on the scale then indicates the meridian altitude of the object in the sky.

Although, in principle, a Dakṣinottara Bhitti can be used for measuring the altitude of any object at the precise moment of its meridian transit, in practice it is most suitable for measuring the altitude of the sun. For the sun, the observer has only to locate the shadow of the rod at noon the moment the rod becomes fully lighted with the rays of the sun. For other objects that do not cast a shadow and for which the observer places his viewing device at the scale, some difficulty may be encountered. Because the scales of the instruments have

---

[1] The SSMC has a misprint on p. 6, where the obliquity is reported as 23;39.

[2] Ptolemy observed at Alexandria, Egypt, at latitude 31;13 and not 36;00.

been constructed flush with the surface of the walls they are inscribed upon and do not project out, measurement at the precise moment of the object's meridian-transit is difficult. However, similar to the Ṣaṣṭhāṁśa, the error because of this shortcoming is negligible, so long as the object is within one-half degree of the meridian.[1]

From the meridian-altitude measurements, a number of astronomical parameters, such as the local latitude and the obliquity of the ecliptic, may be determined. Jai Singh apparently thought quite highly of the instrument as he built Dakṣinottara Bhitti yantras at all of his observatories. The observatories had several versions of the instrument, such as a double arc intersecting at bottom, a semicircle concave upward, and a semicircle concave downward. In Table 3-5, the instruments at his observatories are compared.

The instruments listed in the table below are not all extant today. The instruments at Mathura were reported by Hunter in 1799 but have disappeared totally with all other instruments of that observatory. Hunter also described a large Dakṣinottara Bhitti at Delhi, which also has long since disappeared.

Location	No. of Arcs	Radius	Least Count
Jaipur	3	6.07 m	2′ of arc
Ujjain	2	6.13	2′
Varanasi I (large)	2	3.21	6′
Varanasi II (small)	2	2.36	10′
Delhi I	1	1.6	1 deg
Delhi II[2]	1	4.123	1′ (probably)
Mathura[3] (large)	1	1.18	unknown
Mathura[4] (small)	1	0.61	unknown

Table 3-5 The Dakṣinottara Bhitti Yantras at Jai Singh's Observatories

[1] As the altitude of objects at the meridian changes rather slowly, the error caused by the deviation of the instrument from the meridian plane will be well within the precision of the instrument. See Ref. 48.

[2] Calculated from the data given by Hunter. Hunter, p. 191.

[3] Hunter, p. 200.

[4] *Ibid.*

According to Jagannātha, this was the instrument with which Jai Singh determined the obliquity of the ecliptic in 1729.[1] The present instrument of Jaipur is of relatively recent origin. It was constructed in 1876 to replace the old one that had to give way to a road passing next to the observatory.

[1] SSMC, pp. 6-7, or SSRS, p. 1037.

CHAPTER IV

MASONRY INSTRUMENTS II

Kapāla yantras and Jaya Prakāśa

The Kapāla yantras of Jai Singh are two concave hemispherical instruments, built side by side on a masonry platform at the Jaipur observatory. We will designate them Kapāla A and Kapāla B. The yantras have a diameter of 3.46 m each and are so named because of their remote resemblance to the brain cover of a human skull. In the chronological development of Jai Singh's instruments, Kapāla A came first before his Jaya Prakāśa which played an important role in Jai Singh's astronomical program,[1] according to Jagannātha. In a map of the observatory, drawn sometime before 1728, a pair of instruments resembling the Kapālas may be clearly identified. The map does not show the Jaya Prakāśa. The Kapalas are built in the same general location where they are today, suggesting that the Kapālas predate the Jaya Prakāśa of Jaipur. Jaya Prakāśa is evolved from the Kapāla A. In fact, Jaya Prakāśa may be looked upon as the Kapāla A built into two complementary halves.

Jaya Prakāśa and Kapāla A

Because there is a great deal of similarity between the Jaya Prakāśa and the Kapāla A, it is worthwhile to discuss them together. The Jaya Prakāśa and the Kapāla A are both multipurpose instruments consisting of hemispherical surfaces of concave shape and inscribed with a number of arcs. These arcs indicate the local time, and they measure various astronomical parameters, such as the coordinates of a celestial body. One difference between the two instruments is that the Kapāla A indicates the ascendants, while the Jaya Prakāśa reveals the sign on the meridian. Another is that the Jaya Prakāśa is built in two complementary halves, giving it the capacity for night observations.

Coordinate Systems

The engravings for coordinate measurement are identical on the surfaces of both Jaya Prakāśa and Kapāla A. Their surfaces, in this regard, represent inverted images of two spherical coordinate systems, namely, the azimuth-altitude, or horizon system, and the equatorial. In the azimuth-altitude

---

[1] The phrase "Jaya Prakāśa" translates literally as the "Light of Jai or of Jai Singh."

system, the rim of each hemispherical surface represents the local horizon and the bottommost point, the zenith. Cardinal points are marked on the rim, and cross wires are stretched between them. A great circle drawn between the north and the south points and passing through the zenith on the instrument's surface represents the meridian. From the zenith point a number of equal azimuth lines are drawn up to the rim or horizon. Next, a number of equally-spaced circles with their centers on the vertical axis passing through the zenith are inscribed on the surface. These circles are parallel to the rim and intersect the equal azimuth lines at right angles.

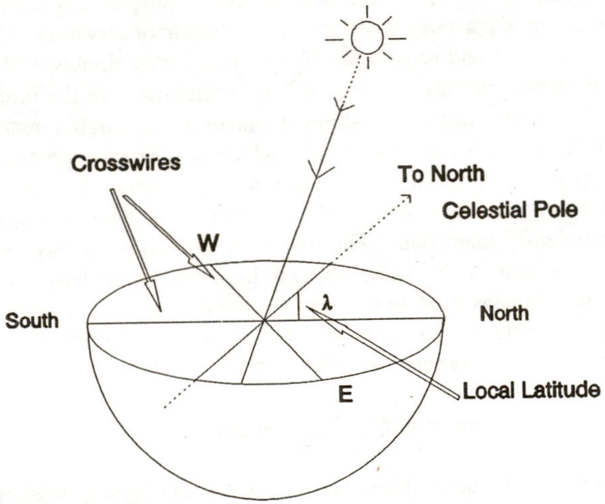

Fig. 4-1 The Principle of Jaya Prakāśa and Kapāla Yantras.

For the equatorial system, a second set of coordinates is inscribed. For these coordinates, a point on the meridian, at an appropriate distance below the south-point on the rim, represents the north celestial pole. At a distance of 90 deg further down, a great circle intersecting the meridian at right angles represents the equator. On both sides of the equator, a number of diurnal circles are drawn. From the pole, hour circles radiate out in all directions up to the very rim of the instrument.

On a clear day, the shadow of the cross-wire falling on the concave surface below indicates the coordinates of the sun. These coordinates may be read either in the horizon or in the equator system as desired. The time is read by the shadow's angular distance from the meridian measured along a diurnal circle. In this regard the two instruments work as hemispherical sundials.

The concept of using a hemisphere as a sundial for indicating local time is not new. Berosus, a Greco-Babylonian astronomer, is said to have made a hemispherical sundial in about 300 B.C.[1] During the early Middle Ages, masonry hemispheric dials, *skaphes* or hemicycles, were built on church buildings in Europe.[2] A sundial, *Yang i*, constructed at the Chinese imperial observatory at Nanking in the late thirteenth century is also said to have been of hemispherical shape. Needham has compared *Yang i* with Jaya Prakāśa.[3] However, Jai Singh's Jaya Prakāśa is more elaborate, complex, and versatile than the Bowl of Berosus, the *skaphe* of the Middle Ages, or the *Yang i* of the Chinese. It is an over-simplification to compare Jaya Prakāśa with any of these simple devices.[4,5]

Ascendants and Signs on the Meridian

For determining ascendants, the instrument Kapāla A has a set of 12 curves inscribed on its surface. The curves represent the 12 ascendants. On a clear day, the shadow of the cross-wire falling on one of the curves indicates the ascendant or sign emerging at the horizon at that very moment. The Jaya Prakāśa, on the other hand, indicates the sign approaching the meridian. A number of curves are engraved on the instrument's surface for this purpose. As the shadow of the cross-wire passes over one of these curves the sign designated by the curve is on the meridian at that very moment.

The Complementary Halves of Jaya Prakāśa

The Jaya Prakāśa consists of two complementary hemispheres. In the two halves the area between alternate hour circles is removed and steps are provided in its place for the observer to move around freely for his readings. The space between identical hour circles of the two hemispheres is not removed, however. The sections left behind in the hemispheres complement each other. They do so in such a way that, if put together, they would form a complete hemispherical surface similar to that of the Kapāla A. By constructing the instrument in two complementary halves and by providing space between the sections, Jai Singh gave his Kapāla yantra the capacity for night-time observations. For night observations the observer sights the object in the sky from the space between the

---

[1] The sundial of Berosus was a hemisphere of wood or stone on which a pointer was fixed, such that the pointer's other end coincided with the center of the hemisphere. The surface of the instrument was inscribed to indicate 12 hours between sunrise and sunset. According to al-Battānī (Albategnius, c. A.D. 858-929) the dial was still in use in Muslim countries during his days. *Encyclopedia Britannica*, "Clock, Watches, and Sundials," Vol. 4, p. 743, (1980).

[2] Derek J. Price, in *History of Technology*, ed., Singer et al (1957), pp. 594-596, *Op. Cit.*, Ch. 1.

[3] Needham, p. 369, *Op. Cit.*, Ch. 1.

[4] Kaye erroneously calls Jaya Prakāśa as "practically the hemisphere of Berosus, somewhat elaborated." Kaye, p. 86.

[5] For a detailed discussion of this subject see *Yantraprakāra* (S), pp. 48-51.

sections. The observer obtains the object in the sky and the cross-wire in one line. The coordinates of the vantage points are then the coordinates of the object in the sky. For details of nighttime measurements, the reader is referred to Chapter 7 of this book, where the instruments of the Jaipur observatory are described. Jai Singh built his Jaya Prakāśas only at Delhi and Jaipur. These instruments survive in varying degree of preservation. The instrument of Delhi has a diameter of 8.33 m and that at Jaipur, 5.4 m.

Jaya Prakāśa in Samrāṭ Siddhānta

In his *Samrāṭ Siddhānta*, Jagannātha tells us that when portable instruments of metal were found unsatisfactory, Jai Singh, adopting a new approach, invented the Jaya Prakāśa.[1] Jagannātha gives a detailed description on how to construct the instrument:[2]

> "First, on a ground levelled with water [standing in channels], draw a circle of desired radius. On this circle, draw north-south and east-west lines. Excavate a concave pit within the circle. Cut out a semicircle of wood or metal of the same radius as the circle, and turn it within the excavated pit. If the semicircle turns all around while remaining in contact with the surface, the hemispherical pit is correctly done. As an alternative, the concave hemisphere may also be made out of metal or wood.[3] Next, on the hemispherical concave surface inscribe the meridian and prime vertical lines. The intersection of these two circles is the zenith point, and the rim of the instrument is the horizon. Graduate the horizon into degrees and minutes.
>
> "On the meridian, below the southern end of the horizon, at a distance equal to the latitude of the place, mark a point. This point is the south pole.[4] Similarly, to the north, on the meridian above the north point, at a distance equal to the latitude-angle, is the north pole.[5] With these two poles as centers, draw two circles with their radii equal to the maximum declination of the sun. These are the ecliptic-pole circles. [On these circles, the poles of ecliptic revolve.] Graduate the circles into degrees and minutes. The degrees and minutes are marked on those sections of the circles only which are below the south pole and

---

[1] SSMC, p. 39, and SSRS, p. 1162.

[2] SSMC, pp. 11-13, and SSRS, pp. 1042-1046.

[3] The library of Jai Singh did have once a Jaya Prakāśa of Metal or wood. R.S.A., Pothikhana records.

[4] In reality the point represents the north pole and not the south pole as the text says.

[5] The so called "north pole" and the structure surrounding it is a temporary construction for the purpose of inscribing various circles on the instrument surface. Once the instrument has been constructed, the structure is dismantled.

above the north pole, in proper order. Mark on the ecliptic-pole circles, the twelve positions of the pole of the ecliptic corresponding to the rising signs or the ascendants. The markings should be westward from the meridian on the ecliptic-pole circle on the north end and to the eastward on the circle toward the south end. With its center on an ecliptic-pole circle, draw the ecliptic passing through the east and the west points on the horizon. Again taking the rising points of the signs already marked on ecliptic-pole circles as centers, draw twelve semicircles. These circles represent twelve different positions of the ecliptic—one [position] for each ascendant respectively.

"Now draw another circle with its radius equal to that of the instrument with one of celestial poles as the center. This circle is the celestial equator. Similarly, draw six diurnal circles.[1]

"On the equator, both on its eastern side and also on its western side, beginning at the point where it intersects the meridian, mark 15 *ghaṭikās* each. Divide the *ghaṭikās* into as many *palas* as possible. Draw hour circles passing through each and every [of these] *ghaṭī* marking.[2] [These circles will pass through the two celestial poles.] The declination should be measured on either side of the equator along the meridian. Taking the zenith point as center, draw altitude circles at intervals of a degree. Similarly, beginning at the horizon, draw equal azimuth circles one degree apart.

"In daytime hours, an ascendant or *lagna* is indicated by the shadow of the cross-wire falling on an ecliptic curve. On whichever ecliptic arc the shadow of the cross-wire falls, the sign indicated by that arc is then the one emerging on the horizon. Further, determine the time elapsed by the cross-wire's shadow. Also read the zenith distance and the azimuth from the shadow on the respective altitude and azimuth circles. If the shadow falls between two adjacent circles, determine the time by interpolation.

"Remove the section of the instrument ten degrees beyond the Tropic of Capricorn to the south, keeping the rest intact.[3] Again, remove a *ghaṭī* wide section to the east of the meridian. Construct a passage in this section for a man to walk around. Next, leaving out a four-*ghaṭī*-wide section intact, remove a *ghaṭī*-wide section and construct in its

---

[1] The centers of these circles lie on the vertical line passing through the zenith. The radii of these circles equal "R cos δ" where R is the radius of the hemisphere and δ the declination. By declination angles, Jagannātha, probably, means the declination of the first point of the signs of the zodiac.

[2] The text is unclear here.

[3] Strictly speaking, it is the section below the Tropic of Cancer that has to be removed. The engravings on the instrument surface, as pointed out earlier, are the reverse of those on the celestial sphere.

place a passage waist-deep. Similarly, construct two passages to the west of the meridian. Thus there will be five passages in the instrument altogether. The edge of the surface of the instrument should be about 4 *aṅgulas* (10 cm) thick on either side of a passage. The instrument should not be cut completely; its horizon should be left intact. In the passages construct stairs for descent. Having made the instrument in this manner, the observation should be next conducted."

Jagannātha next explains the preparation of a sighting device for nighttime observation. His description, except for the removal of certain sections from the instrument, agrees well with the Kapāla A constructed at Jaipur. The Kapāla A of Jaipur indicates ascendants, whereas the Jaya Prakāśa there shows the sign on the meridian.

Accuracies and Limitations of Jaya Prakāśa and Kapāla A

The accuracies of the Jaya Prakāśa and the Kapāla A depend on the location of the object being observed. In the horizon system of coordinates, the two instruments are most sensitive for the objects near the horizon for which the division length is widest. The instruments are least sensitive when the object is at the zenith for which the division-lengths narrow down to a point. Similarly, for the equator system, the readings are most sensitive if the object is at or near the equator, and they are least accurate if the object is near the north pole. A Jaya Prakāśa of the type at Jaipur can measure coordinates with an accuracy of $\pm 3'$ of arc if the object is ideally located. Kapāla A of Jaipur also has accuracies of the order of $\pm 3'$ of arc.

The time readings at the two instruments are more precise when the sun is near or approaching an equinox and least sensitive when it is near a solstice. At an equinox the instruments of the type built at Jaipur may measure time with an accuracy of $\pm$ ½ minute.

By the very nature of their construction, the Kapāla A and the Jaya Prakāśa both have certain limitations. Kapāla A can be used only in daytime. Jaya Prakāśa may be used day or night, but nighttime observations taken at an arbitrary moment can be quite cumbersome.[1]

Kapāla B

Although the two Kapālas, designated at the Jaipur observatory as the Kapāla A and the Kapāla B, are of the same dimensions, they differ considerably with each other in engraving and, hence, in function. The function of the Kapāla B is to transform graphically the horizon system of coordinates into the equator

---

[1] See Ch. 7.

system and vice versa. It is the only instrument at the Jaipur observatory which is not meant for observing, and will be discussed in detail in Chapter 7.

Kapāla in the Yantraprakāra

The *Yantraprakāra* describes a Kapāla yantra which is somewhat similar in function to Kapāla B, although it is not identical to it. The function of the instrument as described in the Yantraprakāra is to transform the equatorial system of coordinates into the ecliptic system.[1] The text calls the instrument *Sarvadeśīya Kapāla Yantra*, or a Universal Hemispherical Dial, and emphasizes that its function is not to conduct observations. The description is as follows:[2]

> "On a lêvel ground dig out a hemispherical-shaped pit and mark cardinal points on its rim. Call the bottom-most point of the hemisphere the zenith. Draw both the meridian and the prime vertical passing through the zenith and divide them into degrees. With the zenith as center, draw altitude circles passing through the degree markings just inscribed. Next, divide the rim into degrees and draw circles of equal azimuth which pass through the zenith point.
>
> "Next, on the meridian, toward the south from the zenith, at a distance equal to the obliquity, mark the celestial pole. With the pole as center, draw diurnal circles on the surface. The circle drawn at a distance of 90 deg from the pole is the celestial equator. Now emanating from the pole in all directions, draw hour circles."

For transforming the equatorial system of coordinates into the ecliptic system, the rim of the instrument represents the ecliptic and it is divided into zodiacal signs. The so-called zenith point on the surface of the instrument in reality is the pole of the ecliptic.

It is interesting to note that about a century and half earlier, Tycho Brahe had an idea similar to that of Kapāla described in the *Yantraprakāra*. Around 1570, Tycho constructed a device for graphically transforming coordinates. His device was a sphere of wood covered with brass, about 1.5 m in diameter. Later, he used the sphere to depict the stars of which he had measured the coordinates.[3]

Kapāla Yantras in Hindu Literature

An instrument by the name of Kapāla has been described in *Sūryasiddhānta*, the canon of Hindu astronomy. However, the instrument it describes is no more

---

[1] The function of the Kapāla B, on the other hand, is to transform the horizon system of coordinates into equatorial coordinates.

[2] *Yantraprakāra* (J), p. 48., or *Yantraprakāra* (S), pp. 27, 78-79.

[3] Thoren (1973), pp. 25-45, *Op. Cit.*, Ch. 1.

than a clepsydra for measuring time and has nothing in common with the Kapāla yantras of Jai Singh, which are his own inventions. A Kapāla has been mentioned by Brahmagupta in his *Brāhmasphuṭasiddhānta* and also by Lalla in his *Śiṣyadhīvṛddhida Tantra.*[1] However, Brahmagupta's and Lalla's Kapālas are horizontally placed dials for telling time in daylight.[2] Nanda Rāma in his *Yantrasāra* also describes a Kapāla yantra somewhat similar to the one of Jai Singh's. However, his work was published after Jai Singh and thus was not available to Jai Singh.[3]

Nāḍīvalaya

A Nāḍīvalaya consists of two circular plates fixed permanently on a masonry stand of convenient height above ground level. The plates are oriented parallel to the equatorial plane, and iron styles of appropriate length pointing toward the poles are fixed at their centers. The Nāḍīvalaya is not a new device to the astronomers of India. Instruments similar to the Nāḍīvalaya in concept are found in Hindu astronomical literature written long before Jai Singh. The *Bhagaṇa* yantra of Lalla of the 8th century has elements of it.[4] Bhāskarācārya II (b. 1114), in his *Siddhānta Śiromaṇi* describes a version of it.[5] The instrument Nāḍīvalaya is, in fact, an equinoctial sundial built in two halves, indicating the apparent solar time of the place.

Yantraprakāra, the treatise on Jai Singh's instruments, describes an early version of Nāḍīvalaya, which is more like an equatorial armillary with a quadrant. In the text we find the following information regarding its construction and operation.[6]

> "On a level ground, over a diagonal [right-angled] stand, construct a circular wall parallel to the plane of the equator and mark east-west and north-south lines on its surface. Divide the circumference of the wall into appropriate time-measuring units. Next, pivot a metallic strip radially at the axis of the wall such that it revolves freely around over the surface. Take the length of the strip equal to the radius of the wall and, on its free end, mount a graduated quadrant perpendicular to the plane of the wall. The quadrant should be of the same radius as the

---

[1] *Śiṣyadhīvṛddhida Tantra* of Lalla, p. 285, *Op. Cit.*, Ch. 2. For the Kapāla in *Brāhmasphuṭasiddhānta* and in *Siddhāntaśiromaṇi*, see footnote on the same page of the citation.

[2] For a discussion on Brahmagupta's instruments see Sreeramula Rajeswara Sarma, "Astronomical Instruments in Brahmagupta's *Brāhmasphuṭasiddhānta*," *Op. Cit.*, Ch. 3.

[3] *Yantrasāra* of Nandarāma, V.S. 1845 (1788), The Museum, No. 14.

[4] *Śiṣyadhīvṛddhida Tantra* of Lalla, p. 286, *Op. Cit.* Ch. 2.

[5] *Siddhānta Śiromaṇi* of Bhāskarācārya, Ch. 11, verse 5-7, tr., Lancelot Wilkinson, Calcutta, 1860-1862; reprint, Osnabrück, 1981.

[6] *Yantraprakāra* (J), pp. 18-19, *Yantraprakāra* (S), pp. 26, 73-76.

wall and concave inward. Attach a sighting device with a small tube between the quadrant and its center.

"In order to measure time during the day, rotate the strip along with the quadrant mounted over it about the axis of the wall until the shadow of the sighting device falls radially on the circular wall. The angular distance of the radial strip measured along the rim of the wall from the north-south line, then, gives the time from mid-day. For measuring the declination of the sun, the sighting tube is manipulated until the rays of the sun pass through it unhindered. The angle of declination is read off the quadrant scale.

"At night the time is measured with a star or *nakṣatra*. The radial strip along with the quadrant mounted over it is rotated until the selected star or *nakṣatra* lies in the plane of the quadrant. Next, the angular distance of the plane of the quadrant from the north-south line or meridian is read off the rim of the wall. With the angular distance known, the sidereal time or the time elapsed since the sunset or the time until the next sunrise may be computed using appropriate charts or by calculation."

The *Yantraprakāra* does not explain how to find the declination at night. As a matter of fact, the design of the instrument does not lend itself easily to such measurements. That may be why it was used mainly for reading time during the day. In his *Samrāṭ Siddhānta*, Jagannātha gives the following instructions for constructing the instrument:[1]

"At a desired location, construct a column about one-half meter high and on it fix two parallel wooden boards circular in shape along an east-west line, separated by an appropriate distance and parallel to the plane of the equator.[2] The axis of the tilted surfaces should definitely point toward the poles. On the edge of the tilted surfaces, with the axis as center, draw circles for easy sighting.

"Next, after making the round surfaces smooth, graduate the circles into 60 *ghaṭikās*. Divide the circle into four quadrants, by east-west and up-and-down lines. At the center tie a thread by something such as a style which points toward the pole. With a strip (or slit) held at the edge of the surface as a vantage point, the planets and the stars should be sighted along the thread [held taut between the style and the rim]. The angle between the vantage point on the rim and the upper end of the scale indicates the hour angle of the object. With this

---

[1] SSMC, p. 2, or SSRS, pp. 1032-1033.

[2] By the "east-west" line, the orientation of the horizontal diameter of the boards along the east-west direction is implied.

instrument one observes the planets and the stars to the north of the celestial equator and not to the south of it. Because of this the instrument has received little recognition."

Jagannātha's criticism that the Nāḍīvalaya cannot be used for objects below the equator is unwarranted. The instrument can be used with equal facility for objects in the southern or northern halves of the celestial sphere, provided the instrument has a south facing scale. Possibly, the instrument was made according to the specifications of the *Yantraprakāra* and had a north facing scale only.

The Nāḍīvalaya is an effective tool for demonstrating the passage of the sun across the celestial equator.[1] On the vernal equinox and the autumnal equinox the rays of the sun fall parallel to the two opposing faces of the plates and illuminate them both. However, at any other time, only one or the other face remains in the sun. After the sun has crossed the equator around March 21, its rays illuminate the northern face for six months. After September 21, it is the southern face that receives the rays of the sun for the next six months. Jai Singh built Nāḍīvalayas at each of his observatory sites except Delhi. We summarize these instruments in Table 4-1.

Fig. 4.2 The Nāḍīvalaya Yantra at Ujjain.

[1] The Jaipur instrument is indeed identified as "the equatorial ring to indicate the passage of the sun across the equator" in the map No. 23 of the Museum.

Location	North Scale			South Scale	
	Plate Separation	Diameter	Least Count (time units)	Diameter	Least Count
Jaipur	5.37 m	3.62 m	1 min	0.84 m 1.09 3.63	4 min 5 1
Varanasi	none	1.08 1.38	1 1	0.54 0.66	4 2.5
Ujjain	2.16	0.52	3	1.12	6
Mathura[1]	none	2.79	48 sec	none	—

Table 4-1 The Nāḍīvalaya yantras

A Nāḍīvalaya is a low precision instrument. At Jaipur, where its largest specimen exists, the time measurements are no better than accuracies of ± 1 minute and the angles that of ± 15′ of arc.

Digaṁśa yantra

A Digaṁśa yantra, consisting of two cylindrical walls surrounding a central pillar, measures the angle of azimuth of a celestial body. Its central pillar as well as its walls are engraved in degrees and minutes at their top surfaces. Cross wires are stretched between the cardinal points marked over the outer wall. The observer uses one or more strings with one end tied to a knob on the pillar and the other end to stone pebbles suspended over the walls. With these strings, the observer defines a vertical plane containing the cross-wire and the object in the sky. The angular distance of the vertical plane from the north point, read on the scales, indicates the azimuth of the body.

The well known texts of Hindu astronomy do not mention any Digaṁśa yantra similar to that of Jai Singh's. The Samarkand observatory of Ulugh Beg also did not have any instrument like it. However, Ibn Sīnā (980-1037) devised a masonry instrument with circular walls for measuring azimuth.[2] It is uncertain whether Jai Singh knew about Ibn Sīnā's instrument. In the list of instruments Jai Singh built according to the Islamic books, no instrument similar to the

---

[1] Hunter, p. 200.

[2] Sevin Tekeli, "The Observational Instruments of Istanbul Observatory," pp. 33-44, in *International Symposium on the Observatories in Islam, 19-23 September, 1977*, ed., M. Dizer, Istanbul, 1980.

Digaṁsa is mentioned. It appears that Jai Singh developed the instrument independently. Both the *Yantraprakāra* and the *Samrāṭ Siddhānta* of Jagannātha have an identical description of Digaṁśa. Jagannātha describes the instrument as follows:[1]

> "On a ground, levelled with [standing] water, and marked with the cardinal directions, inscribe a circle of desired radius. This circle is called horizon. Similarly, inscribe altogether three different [concentric] horizons. Next, construct a column of desired height over the first circle. Over the second circle, construct a cylindrical wall of the same height as the column. On the third circle, construct a ring-shaped surface over a cylindrical wall which is twice as high as the inner. On all three horizons mark the north-south and the east-west lines and inscribe degrees and minutes of arc. Over the third circle, stretch cross wires [between its cardinal points] as described earlier. At the center of the first circle, fasten one end of a string. Fasten the other end of the string to a stone and suspend it over the surface of the third wall. This string is called *dṛk maṇḍala*.[2]
>
> "Now the observing procedure: Sight the desired planet or star from the outer edge of the second wall as the vantage point, such that the cross-wire and the *dṛk maṇḍala* line up with it. Move the *dṛk maṇḍala* string around over the third wall (as necessary) to achieve this. Once this is done the angular distance between the east-west line and the string on the third wall indicates the azimuth of the object."

With the Digaṁśa of the type described by Jagannātha, the azimuth of the sun is measured by overlapping the shadow of the cross-wire with that of the *dṛk maṇḍala* string. Because of the penumbra of the sun, the shadow of the cross-wire is often indistinct and the procedure may require some practice.

According to the description of Jagannātha, the observer, while sighting the object, defines a vertical plane with the help of the cross-wire and a string, which passes through both the object and the center of the instrument. The point where the plane intersects the circular scale of the instrument indicates the *digaṁśa* or the azimuth of the object. The plane may be defined equally well in several other ways. If the instrument has an upright rod fixed at its center, the plane may be defined with a string tied to its upper end and then stretched over the outer wall as explained in Chapter 9.[3] It may also be defined with a pair of strings tied to the top and at the bottom ends of the rod. The other two

---

[1] SSMC, p. 4, or SSRS, pp. 1032-1033.

[2] The *dṛk maṇḍala* defines a vertical plane containing the object in the sky.

[3] See the Digaṁśa yantra at Ujjain, Ch. 9.

ends of the strings are tied to stones passing over the two walls as described in Chapter 8.[1]

Although Jagannātha recommends the outer wall to be twice as high as the inner, in practice the height is not critical at all. As a matter of fact, the outer wall can be of any height greater than that of the inner wall. Further, the circular scales on the central pillar and on the inner wall are superfluous. With a Digaṁśa, following the procedure of Jagannātha, one cannot measure the azimuth of objects too close to the horizon, because the sighting is done from the outer edge of the inner wall, which is lower than the wall surrounding it. For example, with the Digaṁśa of Varanasi, despite its being at an elevated location, one cannot measure the azimuth of an object until it is at least ten degrees above the horizon.[2]

Jai Singh built his Digaṁśa yantras at Jaipur, Varanasi, and Ujjain only. The instruments at these three locations have roughly the same dimensions, and the details of their construction are given in Table 4-2.

Fig. 4-3 Digaṁśa Yantra at Jaipur.

---

[1] See Digaṁśa at Varanasi, Ch. 8.

[2] See Ch. 8, Varanasi Observatory.

	Jaipur	Ujjain	Varanasi
Central Pillar			
Diameter	1.04 m	1.24 m	1.08 m
Height	0.96	1.32*	1.26†
Mid-Circular Wall			
Height	0.96	2.57	1.26 m
Dia. (Inner)	4.47	6.28 ± 0.06	5.47
Dia. (Outer)	5.33	7.42	6.4
Width	0.43	0.57	45.5 cm
Outer Circular Wall			
Height	1.95	2.57	2.52 m
Dia. (Inner)	7.77	9.86 ± 0.6	7.48
Dia. (Outer)	8.29	11.32 ± 0.04	8.6
Thickness	0.61	0.73	56 cm
Least Count	6′	6′	3′

* The central pillar has a rod at the center 1.23 m high and approximately 6 cm in diameter.
† The central pillar has a rod at center 1.28 m high and about 4.3 cm in diameter.

Table 4-2 The Digaṁśa yantras of Jai Singh's observatories

Although the small divisions of the Digaṁśa instruments at Jaipur and Ujjain measure 6′ of arc, with proper interpolation one can easily read angles with an accuracy of ±3′.

Rāma Yantra

The Rāma yantra, probably named after Rāma Singh, the grandfather of Jai Singh, is a cylindrical structure in two complementary halves that measure the azimuth and altitude of a celestial object, for example the sun. Hunter calls the instrument *Ustuanah*,[1] which Kaye identifies with *Ustuwanī*, or cylindrical (orthographic) projection of a globe, as described by al-Bīrūnī in his *Chronology of Nations*.[2] The function of the Rāma yantra is much different from that of a simple orthographic projection, however, and it would be a mistake to identify the Rāma yantra with the *Ustuwanī* of al-Bīrūnī.[3] In the Islamic school of astronomy, as well as in Hindu books on astronomy, there is no instrument similar to the Rāma yantra. The instrument is indeed an invention of Jai Singh.

---

[1] Hunter, p. 191-92.
[2] Kaye, p. 37.
[3] Al-Bīrūnī, *Chronology of Nations*, pp. 357-358, tr. and edited, C. E. Sachau, London, 1879, reprint, Frankfurt, 1969.

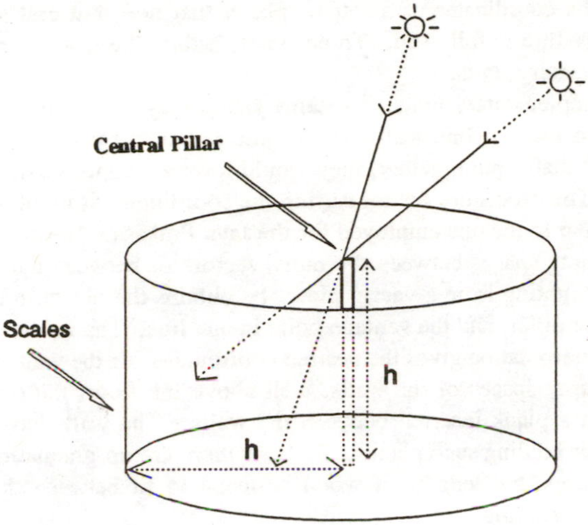

Fig. 4-4 The Principle of a Rāma Yantra.

The cylindrical structure of Rāma yantra is open at the top, and its height equals its radius. Figure 4-4 illustrates its principle and operation. To understand the principle, let us assume that the instrument is built as a single unit as illustrated in Fig. 4-4. The cylinder, as in the figure, is open at the top and has a vertical pole or pillar of the same height as the surrounding walls at the center. Both the interior walls and the floor of the structure are engraved with scales measuring the angles of azimuth and altitude. For measuring the azimuth, circular scales with their centers at the axis of the cylinder are drawn on the floor of the structure and on the inner surface of the cylindrical walls. The scales are divided into degrees and minutes. For measuring the altitude, a set of equally spaced radial lines is drawn on the floor. These lines emanate from the central pillar and terminate at the base of the inner walls. Further, vertical lines are inscribed on the cylindrical wall, which begin at the wall's base and terminate at the top end. These lines may be viewed as the vertical extension of the radial lines drawn on the floor of the instrument. The radial and the vertical lines are inscribed with scales for measuring the altitude of a celestial object and are graduated such that the zero mark is at the top, and the 90-deg mark is at the base of the central pillar. If the preference is to measure the zenith distance, then the markings would have to be reversed, i.e., the top end would now denote a 90-deg mark and the base of the pillar, zero.

In daytime the coordinates of the sun are determined by observing the shadow of the pillar's top end on the scales, as shown in the figure. The coordinates of the moon, when it is bright enough to cast a shadow, may also be read in a similar manner. However, if the moon is not bright enough, or if one wishes to measure the coordinates of a star or planet that does not cast a shadow, a different procedure is followed. To accomplish this, the instrument is built in two complementary units.

The two complementary units of a Rāma yantra may be viewed as if obtained by dividing an intact cylindrical structure into radial and vertical sectors. The units are such that if put together, they would form a complete cylinder with an open roof. The procedure for measuring the coordinates at night with a Rāma yantra is similar to the one employed for the Jaya Prakāśa. The observer works within the empty spaces between the radial sectors or between the walls of the instrument. Sighting from a vacant place, he obtains the object in the sky, the top edge of the pillar, and the vantage point in one line. The vantage point after appropriate interpolation gives the desired coordinates. If the vantage point lies within the empty spaces of the walls, well above the floor, the observer may have to sit on a plank inserted between the walls. The walls have slots built specifically for holding such planks. Because there are no graduations between the empty spaces, arc lengths of wood or metal to fit between the walls are necessary for a reading.

The Rāma yantra's accuracy varies because its division lengths are not uniform. It is most sensitive around the base line of the walls where the altitude reads 45 degrees. Near the 45-deg marks, both the azimuth and the altitude divisions are of maximum width. Theoretically, around the 45-deg marks, one should be able to obtain accuracies of $\pm 1'$ of arc with the instruments of the size built at Delhi or Jaipur. In daytime, however, because of the finite width of the penumbra of the shadow of the pillar, accuracies of this order are not possible. Therefore, a precision of $\pm 6'$ is the most that one may expect with these two instruments. Furthermore, for zenith angle readings smaller than 45 degrees, or for altitude readings greater than the same value, the accuracy is much lower and, in fact, deteriorates to ± 1 degree or worse as the shadow of the pillar reaches the base of the pillar. Furthermore, near the base of the pillar the readings are practically meaningless.

Elaborating upon the reasons for building the Rāma yantra, Jagannātha gives a brief description of the instrument in his *Samrāṭ Siddhānta*.[1] He writes:

> "Realizing . . . [problems with *Dhāt al-Thuqbatayn*], we invented the sturdy Rāma yantra for the purpose of knowing the azimuth and the altitude.[2] With this instrument the observing of these two is properly done. Two Rāma yantras should be constructed. In them opening(s), as many degrees wide as necessary, should be established for

---

[1] SSMC, p. 39, or SSRS, p. 1163.

[2] For Dhāt at-Thuqbatayn, see Ch. 2.

observation. Construct an [observational] surface [wall] of lime mortar and stone equal to the width of the opening. Similarly, a second Rāma yantra should also be constructed, but where there is an observational surface in the first Rāma yantra, construct an opening in the second. Whenever observing is necessary, it should be done [with the instrument]. If there is an obstruction to the sighting of a planet or star in one Rāma yantra, the observation should be made from the opening of the second instrument. By following this procedure no inaccuracies result."

Jai Singh built Rāma yantras only at Delhi and Jaipur. The Jaipur instrument is still intact today, whereas the one at Delhi is deteriorating. In Table 4-3, we compare the dimensions of the instruments at the two places.

Location	**Delhi**	**Jaipur**
Wall Height	7.5 m	3.45 m
Height of Pillar	7.48	3.45 (approx.)
Effective Radius	7.5	3.45
Diameter of Pillar	1.61	8 cm (approx.)
No. of Sectors	30	12
Least Count	6′	6′

Table 4-3 The Rāma Yantras at Delhi and Jaipur

Difficulties with the Instrument

While working with the instrument, a casual observer may encounter a number of difficulties, such as the penumbra of the sun which makes it difficult to repeat observations from one day to the next with the same degree of certainty. The penumbra difficulty may be eliminated, however, by superimposing the shadow of a cross-hair over the shadow edge of the pillar and reading the scales at a point where the shadow of the cross-hair barely disappears into the penumbra. It is likely that Jai Singh's astronomers used some such device, though not exactly a cross hair, to overcome the difficulty.

Rāśivalayas

Rāśivalayas were also invented by Jai Singh. The Rāśivalayas are a set of 12 instruments based on the principle of the Samrāṭ, which measure the latitude and longitude of a celestial object. A particular Rāśivalaya instrument becomes operative when the first point of the sign of the zodiac it represents approaches the meridian. At that moment, its "gnomon" points toward the pole of the ecliptic and its quadrants become parallel to the plane of the ecliptic. Because there are 12 signs of the zodiac, there are 12 Rāśivalayas, representing each sign. These instruments were built at Jaipur only and will be discussed in detail in Chapter 7.

Minor Instruments

Agrā, Śaṅku, Palabhā, Dhruvadarśaka Paṭṭikā, and Śara yantra may be classified as Jai Singh's minor instruments. The Agrā and Śaṅku are simple upright styles fixed at the center of a circular scale. An Agrā's purpose is to measure the azimuth of an object as it rises above the eastern horizon. A Śaṅku measures the length of the sun's shadow. The Palabhā is a horizontal sundial. Dhruvadarśaka Paṭṭikā is nothing but a North Star indicator. These instruments will be discussed in detail later as part of the general discussion of the observatories in which they are located.

Jai Singh's Śara yantra, listed in the Yantraprakāra, no longer exists. The instrument might have been a simplified torquetum built over an equinoctial masonry base of circular shape. The incomplete Krāntivṛtta yantra at Jaipur could be the remains of this instrument.

Observatories

Jai Singh built his observatories at Delhi, Jaipur, Mathura, Varanasi, and Ujjain. The observatories were completed between 1724 and 1735. Currently, his observatories, except for that of Mathura, exist in varying degrees of preservation. The observatory of Mathura disappeared about 1850, a few years before the unsuccessful uprising of 1857 against the British.[1]

---

[1] See Ch. 9.

CHAPTER V

DELHI OBSERVATORY

Within the city limits of New Delhi, on Parliament Street, among high-rise office buildings is the Delhi observatory of Jai Singh. The observatory is popularly known as the Jantar Mantar of Delhi.

The parameters of the observatory are as follows:[1]

Latitude:	28;37,35 N
Longitude:	77;13,5 E
Local time:	Indian Standard Time −21 min, 8 sec; or Universal Time +5 hr, 8 min, 52 sec
Height above sea level:	220 m.

Date of construction

The date of construction of the Delhi observatory is uncertain. Garrett suggests that the observatory was constructed in 1710.[2] A plaque on the Samrāṭ yantra of the observatory, erected in 1910, also specifies this date. The year 1710, however, seems unlikely, because around this time Jai Singh, being in disfavor with the imperial court, had stayed away from the capital. In fact, along with other Rajput princes, he was in open insurrection against the imperial authority, and his attention was occupied with the problem of reacquiring his ancestral lands, which had been annexed by the imperial court as a punishment for supporting the losing side during the war of succession to the Delhi throne.[3]

According to the *Zīj-i Muḥammad Shāhī* the observatory was built during the reign of Muḥammad Shāh, who ascended the throne in 1719. The newly crowned ruler, in a conciliatory mood, invited Jai Singh to Delhi for a meeting which took place in November 1720, and some important decisions were made in that meeting.[4] It is reasonable to assume that the question of the observatory was also brought up during the meeting. If so, construction could have begun soon thereafter in 1721.

The 1721 date is further supported by another statement in the *Zīj*. We are told in the *Zīj* that seven years had elapsed in observatory-related work, when news of similar work being conducted in Europe was received, and Jai Singh

---

[1] Kaye, p. 41.

[2] Garrett, p. 14.

[3] See Ch. 1.

[4] Bhatnagar, pp. 38-92.

dispatched a fact-finding mission to Portugal shortly thereafter. Because the delegation left Amber for Europe in 1727, this fact further confirms that construction of the observatory was initiated as early as 1721. Its masonry instruments, however, might not have been completed before 1723-24, and that is why a number of visitors have narrated that the observatory was built in 1724. The initial experimentation at the observatory had been primarily with metal instruments, which were discarded later in favor of those of masonry.[1]

VISITORS' ACCOUNTS OF THE OBSERVATORY

Delhi, because it was the imperial capital and the most important city of India, was frequently visited by travelers, some of whom have left valuable accounts of the observatory.

Tieffenthaler

Tieffenthaler, a native of Botzen in the Austrian Tyrol, arrived in India in 1743 as a Jesuit missionary sometime after Jai Singh's death and traveled extensively in the country.[2] He visited the observatory of Delhi in 1747, about four years after the death of Jai Singh.[3] He gives a brief account of the observatory. Tieffenthaler writes:

> "To the south-west, at the edge of the town, is located the observatory built by Jai Singh, a good king of Jaipur. It barely differs at all from the one you can see at Jaipur, for here you find the same sort of parallactic or equatorial (*axis mundi*) machine, a gnomon [Samrāṭ yantra], and three large astrolabes. But what this observatory has is unique. There are two round buildings constructed in the form of circles [cylindrical shape] and equipped with windows [Rāma yantra], in the midst of which is set a cylinder itself divided (into sidereal times or) into the hours of the prime mover (sun). Since this observatory is located on a plain, the trees and the roof lines of the neighboring buildings make it impossible to enjoy an unrestricted view. Besides, all the instruments except for the astrolabes are so patched and firmed up with lime that it is impossible to make any accurate observations there."

---

[1] See Ch. 3.

[2] Noti alleges that Tieffenthaler came to participate in the astronomical pursuits of the king. However there is no independent confirmation of this. S. Noti, "Joseph Tieffenthaler, S.J., A Forgotten Geographer of India," pp. 142-152, East and West, Vol. V, 1906.

[3] Tieffenthaler, pp. 125-126.

W. Franklin

Franklin in his article, "An Account of the Present State of Delhi," written for *Asiatic Researches*, gives a brief account of the observatory.[1] Franklin had gone to Delhi in 1793 with a party of surveyors dispatched by the East India Company Government of Bengal. His narration is as follows:

> "It (the observatory) was built in the third year of the reign of Mohummud Shah, by the Rajah Jeysing, who was assisted by many persons celebrated for their science in astronomy from Persia, India, and Europe; but died before the work was completed: and it has since been plundered, and almost destroyed by the Jeits [Jats], under Juwaher Sing."

William Hunter

A somewhat detailed study of Jai Singh's Delhi observatory was carried out by William Hunter in the 1790's. He published his report in *Asiatic Researches* of Calcutta in 1799.[2] Hunter writes:

> "The observatory at Dehly (Delhi) is situated without the walls of the city, at the distance of one mile and a quarter (2 km); it lies S. 22 deg. W. [22 deg south-west] from the Jummah Musjid, at the distance of a mile and three quarters (2.8 km), its latitude 28 deg. 37 min. 37 sec. N., longitude 77 deg. 2 min. 27 dec. E. of Greenwich; it consists of several detached buildings."

Hunter goes on to describe the dimensions of the instruments, and also gives their least counts.

Samrāṭ Yantra

About the Samrāṭ yantra Hunter says:

> "Its form was pretty entire, but the edges of the gnomon, and those of the circle on which the degrees were marked, were broken in several places. . . . The instrument was built of stone, but the edges of the gnomon and of the arches, where the graduation was, were of white marble, a few small portions of which only remain."

---

[1] W. Franklin, "An Account of the Present State of Delhi," pp. 417-431, *Asiatic Researches*, Vol. 4, 1807.

[2] Hunter, pp. 190-194.

Miśra Yantra

Hunter describes the Miśra yantra next.

> "At a little distance from this instrument [Samrāṭ yantra] towards the N.W. is another equatorial dial, more entire, but smaller, and of a different construction. In the middle stands a gnomon, which, as usual in these buildings, contains stairs up to the top. On each side of this gnomon are two concentric semicircles, having for their diameters the two edges of the gnomon: they have a certain inclination to the horizon. . . . The length of a degree on outer circle is 3.74 inches (9.5 cm). The distance between the outer and inner circle is two feet, nine inches (84 cm). Each degree is divided into ten parts, and each of these is subdivided into six parts or minutes. . . . On each side of this part is another gnomon, equal in size to the former; and to the eastward and westward of them are the arches on which the hours are marked."

Hunter could not learn from his guides the function of the central part, or Niyata Cakras as they are known today. In his account, he is not very specific about the second arc on the north-west corner of the structure. In addition, he fails to report the inclination of the Karkarāśi Valaya on the north wall of the instrument building. Apparently he did not learn the function of the Valaya instrument. He writes:

> "The north wall of this building connects the three gnomons at their highest end; and on this wall is described a graduated semicircle, for taking the altitudes of bodies, that lie due east, or due west, from the eye of the observer."

Dakṣinottara Bhitti

About the Dakṣinottara Bhitti, Hunter writes:

> "To the westward of this building, and close to it, is a wall, in the plane of the meridian, on which is described a double quadrant, having for centers the two upper corners of the wall, for observing the altitudes of bodies passing the meridian, either to the north or south of

> the zenith. One degree on these quadrants measured 2.833 inches (7.2 cm),[1] and these are divided into minutes."

Although Hunter does not give the radius of the quadrants, from the degree width measured by him it can be deduced that the radius must have been 13 ft, 6.32 inches or 4.123 meters.

Rāma Yantra

Hunter does not call the instrument by its well known name, Rāma yantra, but by the name of *Ustuanah*. He says:

> "To the southward of the great dial are two buildings, named *Ustuanah*. They exactly resemble one another, and are designed for the same purpose, which is to observe the altitude and azimuth of the heavenly bodies."

He goes on to describe the instrument in detail and gives dimensions. Hunter was doubtful about the complementary nature of the two yantras.

> "I do not see how observations can be made when the shadow [of the sun or the moon] falls on the spaces between the stone radii or sectors; and from reflecting on this, instead of being duplicates, may be supplementary one to the other; . . . This point remains to be ascertained."

Jaya Prakāśa

Hunter gives a very brief description of Jaya Prakāśa which he calls *Shamlah* (*Shāmlāh*).[2] He says:

> "Between these two buildings (*Ustuanah*) and the great equatorial dial, is an instrument called *Shamlah*. It is a concave hemispherical surface, formed of mason work, to represent the inferior hemisphere of the heavens. It is divided by six ribs of solid work, and as many hollow spaces,[3] the edges of which represent meridians at the distance

---

[1] By reporting the width of the degree divisions to the third decimal place, Hunter is indicating the average value of a degree division and not stating that the degree divisions were accurate to within thousandths of an inch. Hunter, p. 191.

[2] Jaya Prakāśa is much different than the *Shāmlāh* described in the Islamic texts. See Ch. 2.

[3] Hunter is in error here. There were only five open sectors and not six as he says.

of fifteen degrees from one another. The diameter of the hemisphere is twenty-seven feet, five inches (8.36 m)."[1]

Hunter describes only one unit of the Jaya Prakāśa, which has led some to believe erroneously that the second unit of the instrument is a later addition. Also, Hunter missed the Ṣaṣṭhāṁśa yantra which was definitely there and most likely in poor condition.

William Thorn

Major William Thorn visited Delhi a few years after Hunter, sometime in September of 1803 during Lord Lake's campaign to conquer the independent principalities of north India.[2] Thorn writes:

"In our way back to the camp, we stopped to view the celebrated observatory called the *Gentur Muntur*, erected in the third year of the reign of Mohammed Shah, or 1724, by the famous astronomer, Jeysing, or Jayasinha, Rajah of Ambhere, and founder of principality of Jeypoor. This monument of oriental munificence and science, is situated without the walls of the city, near two miles (3.22 km) from the Jumma Musjid; but work was never completed, on account of the death of the projector, and the subsequent confusions of the empire. The observatory was, however, sufficiently advanced to mark the astronomical skill and accuracy of the prince by whom it was designed, though it has suffered severely from the ravages of the Jauts [Jats], who, not content with carrying off all the valuable materials which were portable, committed many wanton excesses upon the finest parts of the edifice. The great equatorial dial is still nearly perfect, but the gnomon and the periphery of the circle on which the degrees are marked have been injured in several parts. The length of this gnomon is one hundred and eighteen feet seven inches (36.14 m); the base one hundred and four feet one inch (31.7 m); and the perpendicular fifty-six feet nine inches (17.3 m). A flight of stone steps leads up to the top of the gnomon, the edges of which, as well as the arches, were white marble."

---

[1] Hunter is silent about the second Jaya Prakāśa. Kaye's suggestion that perhaps the other Jaya Prakāśa had disappeared by then is a plausible explanation. Kaye, p. 43.

[2] William Thorn, *Memoir of the War in India, conducted by General Lord Lake*, pp. 171-173, London, 1818.

Next, Thorn gives a brief description of the Miśra yantra. He says:

> "Besides this stupendous instrument [Samrāṭ yantra], . . . there are two others of a similar construction and materials, but on a smaller scale. The three gnomons are connected by a wall, on which is described a graduated semicircle for measuring the altitudes of the objects lying due east or west from hence."

About the Rāma yantra Thorn writes:

> "In a southerly direction from the great equatorial dial are two buildings exactly alike, and adapted for the same purpose, which was that these duplicate structures were designed to prevent errors, by obtaining different observations at the same time, and comparing the results. . . ."

Thorn apparently misunderstood the complementary nature of the structures. About the Jaya Prakāśa he writes:

> "Between these buildings [Rāma yantra] and the great equatorial dial is a concave of stone-work, representing the celestial hemisphere, twenty-seven feet five inches (8.4 m) in diameter. it is divided by six lines of masonry, at the distance fifteen degrees from each other, and intended as delineations of so many meridians."[1]

About the *Zīj-i Muḥammad Shāhī*, the astronomical tables prepared by Jai Singh, he says that it was completed in 1728.

Fanny Parks

Fanny Parks also confirms that the observatory instruments were crumbling away when she visited the site in the early 19th century. She states that the observatory consisted of several detached buildings.

> "(Its Samrāṭ yantra) had its form pretty entire, but the edges of the gnomon, and those of the circle on which the degrees were marked, were broken in several places. . . . It was built of stone, but the edges of the gnomon, and of the arches where gradation was, were white marble; a few small portions of which only remain."[2]

---

[1] A great deal of similarity between Thorn's and Hunter's accounts suggests that Thorn borrowed some of his material from Hunter's paper published only a few years earlier in 1799.

[2] Fanny Parks, *Wanderings of a Pilgrim in Search of the Picturesque*, Vol. II, p. 209-212, reprint, Oxford U. Press, 1975.

Parks next describes other instruments in some detail. Her description, however, is remarkably similar to Hunter's and probably was copied from him.

Thomas and William Daniell

In 1816, Thomas and William Daniell published two engravings of the observatory in their book *Oriental Scenery.*[1] Their drawings, however, do not factually represent the instruments at that time, as they show the quadrants of the Great Samrāṭ smooth, complete, and in good condition. In 1816, the Great Samrāṭ was in ruins. Apparently, the authors drew their sketches to illustrate how the quadrants must have appeared originally. Their engravings are useful, nonetheless, for comparing the instruments' environs with their present settings.[2]

Leopold Von Orlich

Von Orlich writing sometime before 1845, also repeats that "the observatory was built in 1724, under Mahomet Shah, by his minister and favorite, Jeysingh, Rajah of Jeypoor, who was celebrated for his love of astronomy. . . ." He continues:

> "This observatory is about two miles (3.2 km) west of the city; it lies in the midst of many ruins; but it was never completed, and has been unhappily, so wantonly dilapidated by the Juts (Jats) that the shattered ruins alone are to be seen. However, enough remains to show the plan of this fine building: the colossal sun-dials and quadrants, which rest upon large arches, are formed of red sandstone and bricks, and the ascent to them is by handsome winding marble staircases. . . ."[3]

Garcin de Tassy

Garcin de Tassy's account is based on an article of Sayyid Aḥmad Khān, a resident of Delhi who later founded Aligarh Muslim University. De Tassy published his account in the *Journal Asiatique* in 1860.[4]

---

[1] Thomas and William Daniell, *Oriental Scenery*, Fifth series, plates 19 and 20, and p. 31, London, 1816. Daniell's drawings were made in 1790, according to Fanshawe. H.C. Fanshawe, *Delhi Past and Present*, pp. 247, London, 1902.

[2] H.C. Fanshawe, *Op. Cit.*, pp. 247-248.

[3] Leopold Von Orlich, *Travels in India*, Vol. 2, pp. 19-20, London, 1845.

[4] Joseph H. Garcin De Tassy, "Description des monuments de Delhi en 1852, d'apres le texte Hindoustani de Saiyid Ahmad Khan," *Journal Asiatique*, Cinquieme serie, Tome 16, pp. 536-543, (Dec. 1860).

Fig. 5-1 Delhi Observatory According to an Engraving Published by Thomas and William Daniell in 1818.

Fig. 5-2 Aerial View of the Delhi Observatory.

He writes:

> "The observatory . . . was built by Sawai Jai Singh, prince of Jaipur, in accordance with the orders of the emperor Muḥammad Shāh, in the seventh year of his reign, corresponding to the year 1137 of the Hegira (1724 A.D.).[1] . . . Most of the devices in this observatory were made of limestone and rocks in order to prevent variations in the observations. Now this observatory has fallen into total ruin, all the devices are broken and every trace of divisions lines has disappeared. There no longer remains any instrument that one can make use of. However, of all these devices built of lime and rock, there remain three, quite broken and damaged."

De Tassy then goes on to describe the three instruments. Unfortunately, he confuses the Rāma yantra with the Miśra yantra, and the Jaya Prakāśa with the Rāma yantra, respectively. His description of the instruments is sketchy and inaccurate in places. Similarly, his account of the involvement of the Europeans in Jai Singh's astronomical efforts is also inaccurate and misleading.[2] It is noteworthy, however, that he reports two complementary concave hemispheres of radius 7.92 m (26 ft) south of the large Samrāṭ. These hemispheres represented the Jaya Prakāśa as we know them today and had been there since the beginning.[3]

De Tassy also gives excerpts from the *Zīj-i Muḥammad Shāhī*, observing that the beginning of the Muḥammad Shāhī era took place on the 1st of *Rabī al-Sānī* 1131 of the Hegira, a Monday, which, according to him, corresponds to the year A.D. 1718.[4] He confirms that the Samrāṭ yantra was repaired at the request of the Archaeological Society of Delhi in 1852. According to de Tassy, however, the restoration of the rest of the observatory was not completed.

---

[1] De Tassy is inaccurate here. The epoch date according to the Gregorian calendar was February 20, 1719, and the seventh Islamic year of the reign of Muḥammad Shāh, 1725.

[2] For instance, he says that the English began to establish most of the rules of modern astronomy at this observatory.

[3] Hunter reported only one bowl of the Jaya Prakāśa which has led to the theory that originally there had been only one hemisphere of the Jaya Prakāśa at Delhi, and that the second hemisphere is a later addition. It should be emphasized, however, that between the visit of Hunter to the observatory some time in the late 1790's and that of de Tassy just before 1860, there had been no additions to the observatory.

[4] The date should be February 21, 1719, according to the Gregorian calendar, and not 1720.

Carr Stephen

Stephen wrote a book in 1876 on the archaeology and monuments of Delhi,[1] in which he included a description of the observatory. His description, however, has been taken directly from Thorn, and there is little new information in it.

H. C. Fanshawe

Fanshawe published an account of Delhi in 1902.[2] Although his account lacks any new information, the description is important, nonetheless, because of a photograph and a map of Delhi included with it. The photograph shows the observatory prior to its restoration of 1910, and the map shows the area surrounding the observatory, where, according to Fanshawe, existed a palace of the "Raja of Jeypur," of Madho Singh, perhaps.[3] The map also shows a nearby area marked as "Jai Singh Pura" which had been the property of Jai Singh. The entire area surrounding the observatory was once known as the Jai Singh Purā. According to *Savāī Jaya Siṁha Carita*,[4] and also according to Bhavan, the observatory was built within the locality Jai Singh Purā.[5]

Restorations

According to visitors to the observatory, as well as according to Jagannātha Samrāṭ, the principal assistant of the Raja, the Delhi observatory had, at one time or another, the following seven instruments:

1. Samrāṭ yantra
2. Jaya Prakāśa
3. Rāma yantra
4. Ṣaṣṭhāṁśa yantra
5. Dakṣinotara Bhitti
6. Miśra yantra
7. Agrā or horizontal sundial[6,7]

---

[1] Carr Stephen, *The Archaeology and Monumental Remains of Delhi*, pp. 269-270, reprint, New Delhi.

[2] Fanshawe, *Op. Cit.*

[3] Fanshawe writes, "A palace and stables of the Raja of Jeypur once existed in Madhoganj, a village just east of the observatory, and this village is still held in *jāgīra* by the Jeypur State." Fanshawe, *Op. Cit.* The name Madhoganj suggests that the building was erected by Madho Singh, a son of Jai Singh who later ruled from the throne of Jaipur.

[4] *Savāī Jaya Siṁha Carita* of Ātmā Rāma, pp. 77, ed., Gopal Narain Bahura, Jaipur, 1979. Ātmā Rāma was a court poet of Jai Singh.

[5] Bhavan, p. 66.

[6] This instrument was perhaps built over a pillar which may be seen atop the Samrāṭ in the drawing of Von Orlich. See Von Orlich, *Op. Cit.*

[7] A Palabhā yantra is a horizontal sundial.

It should be noted that Jagannātha does not mention the Miśra yantra in his *Samrāṭ Siddhānta*, nor does the *Yantraprakāra* include it in the list of instruments by Jai Singh. Tieffenthaler, who visited Delhi just four years after Jai Singh's death, is also silent about it. It is probable, therefore, that the instrument was added to the observatory afterwards, most likely by Madho Singh, the second son of Jai Singh, who ascended the Jaipur throne in 1750. Madho Singh also had some interest in astronomy. Madho Singh must have built the Miśra yantra between 1750 and 1754 when he was in the good graces of the emperor Aḥmad Shāh. Strong political motives cannot be ruled out in Madho Singh's construction of the yantra at Delhi, the seat of the imperial power. The emperor, however, was deposed in 1754, after which the political situation in Delhi deteriorated rapidly. Madho Singh himself got entangled in numerous conflicts of his own, with the result that the Delhi observatory was completely neglected.[1]

Less than two decades after Jai Singh's death in 1743, the Delhi observatory ceased to function as a place of astronomical activity. Anarchy swept the country, and safety was sometimes to be found only within city walls. The central administration from Delhi became weak, and a Jat attack on Delhi created chaos. In 1764, a Jat army led by Javāhara Singh bombarded Delhi.[2] As pointed out by Franklin, the observatory was looted and reduced to ruins during this time.

As a result of the anarchy, the operation of the observatory ceased entirely, and the instruments themselves were vandalized for their material, as Thorn observes. Decades later, when peace returned, attention began to be paid once again to the historical treasures of the country. In 1852, the Archaeological Society of Delhi, with funds from the Jaipur Raja, repaired the Great Samrāṭ.[3] The repair of the Samrāṭ in 1852 is confirmed by Sayyid Aḥmad Khān.[4] The funds received by the Archaeological Society of Delhi, however, were insufficient to carry out a complete restoration of the rest of the observatory.

---

[1] For a history of the Madho Singh period see Tikkiwal, pp. 116-136, *Op. Cit.*, Ch. 1. Also see Vincent A. Smith, *The Oxford History of India*, 3rd ed., pp. 437, 443, Oxford, 1958.

[2] Tikkiwal, p. 135, *Op. Cit.*, Ch. 1.

[3] The *Proceedings of the Delhi Archaeological Society*, for January 6, 1853, state, "It having been stated that the large gnomon of the Junter Munter had been repaired at a cost of Co.'s Rs.442-1-10, leaving a balance of Co.'s Rs.157-14-2 of the sum presented to the Society by the Rajah of Jeypore, for the repairs of that Observatory, and this being much too small a sum to enable the Society to complete the repairs, or even to build around a compound wall, which is absolutely necessary, for the security of the remains from further dilapidation, it was unanimously resolved that the Agent to the Lieutenant-Governor, Delhi, be requested to make known to the Rajah of Jeypore, through the proper authorities, the inability of the Society to complete the contemplated work without further funds." See Kaye, p. 48.

[4] De Tassy, p. 539, *Op. Cit.*

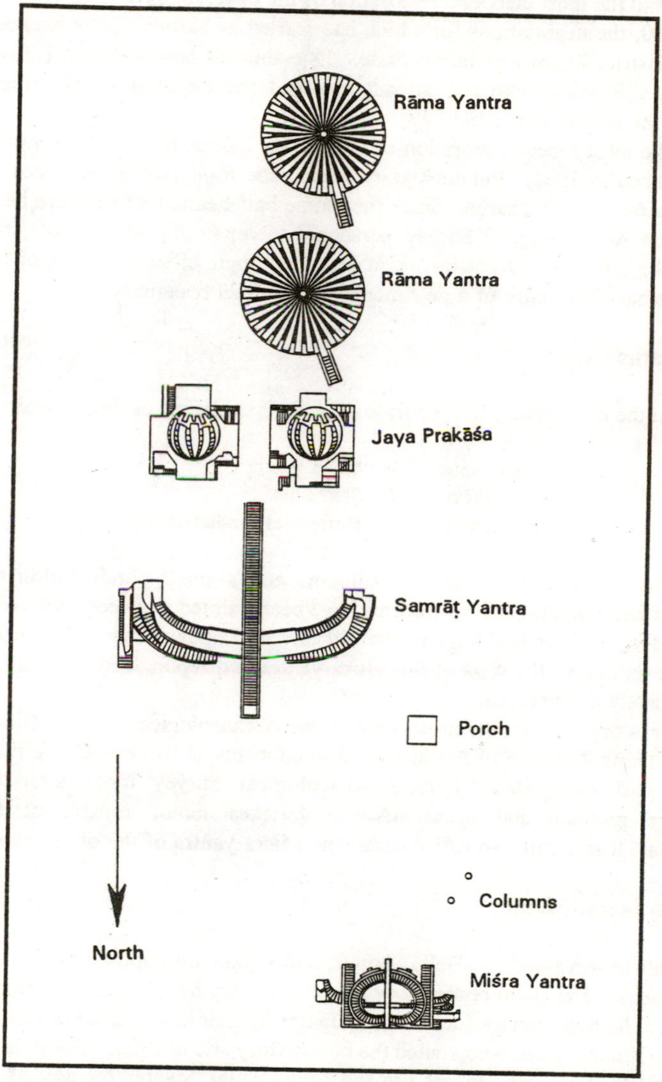

Fig. 5-3 Architectural Plan of the Delhi Observatory.[1]

---

[1] based on Kaye, plate XIII.

The next and the most elaborate restoration of the observatory was carried out in 1909-1910, the negotiations for which had started as early as 1889 between the Delhi District Board and Jaipur State. Its estimated cost was Rs.11,364.[1] Gokul Chand Bhavan, who had already restored the observatory of Jaipur, supervised the restoration.

Finally, the most recent restoration of the Niyata Cakras of the Miśra yantra was undertaken in 1951. Pundit Kedāra Nātha, the *Rāja Jyotiṣī* (Astronomer Royal) of Jaipur, was in charge. Since then some half-hearted efforts have been made by the Archaeological Survey personnel to repair the instruments, but these may be classified as patchwork at best, and their effect on some of the instruments has been more of a detrimental nature than beneficial.

Present Status

At present the observatory has the following six instruments or their remains:

1. Samrāṭ yantra
2. Jaya Prakāśa
3. Rāma yantra
4. Miśra yantra
5. Ṣaṣṭhāṁśa
6. Horizontal sundial or Agrā

In addition, there are two vertical columns and a small porch within the observatory compound. The instruments have been painted in a deep pink color similar to the color of buildings in the old city of Jaipur. The Dakṣinottara Bhitti once located to the west of the Miśra yantra and reported by Hunter has now completely disappeared.

The observatory is under the supervision of the Archaeological Survey of India which maintains it as one of the historical monuments in its charge. A small caretaker staff designated by the Archaeological Survey looks after the observatory grounds and occasionally undertakes minor repairs of the instruments. It maintains an office under the Miśra yantra of the observatory.

The Name Jantar Mantar

Jai Singh's observatories of Delhi, Jaipur, and Ujjain are popularly known as Jantar Mantars. The Delhi observatory has been known by this name for almost 200 years. The first reference to the name Jantar Mantar is found in the account of Major William Thorn, who visited the observatory site in 1803. The archives of Jaipur State, however, do not use the name Jantar Mantar for any of the observatories. In the archives, the observatory of Jaipur, for example, is merely

---

[1] The authorization for the repairs was signed on September 28, 1908. R.S.A., Jaipur State records, serial Nos. 147, 555, 579, in file Nos. 1853/83 and 1861/6.

referred to as the *Jantra*,[1] which colloquially corrupts to *Jantar*.[2] The genesis of the word *Mantar* combined with the word *Jantar* lies in the colloquial practice of adding a second word rhyming with the first for emphasis.[3] The second word added to the first usually has no meaning, but here it has a similar meaning.[4]

The Samrāṭ Yantra

The most imposing structure of the Delhi observatory is its lofty Samrāṭ yantra.[5] The instrument is only slightly smaller than the Great Samrāṭ yantra of Jaipur. Its gnomon is 21.3 m tall, measures 34.6 m at its base, and has a width of 3.20 m. Its hypotenuse, i.e., its surface parallel to the earth's axis is 39 meters long, and a flight of steps in the middle leads to its very top. Near the bottom end of the steps there is an iron gate that discourages visitors from climbing up the stairs. The quadrants of the instrument have a radius of 15.09 m and are 2.65 m in width.

The base of the Samrāṭ of Delhi is not built at ground level. The base is 3 to 5 meters below street level in an excavation 38 m long, 36.6 m wide, and lined with concrete and lime. The instrument is built partly below ground level so that the horizon-sun may not be too far below the artificial horizon defined by the line joining the two six o'clock marks on the quadrants.[6]

The general condition of the Samrāṭ yantra is poor now. Seepage water collects at its base.[7] To avoid the collection of seepage water, the lower part of the instrument was filled up many years ago with concrete. This ruined the functional aspect of the instrument for a large part of the day. Moreover, the seepage problem still persists, so much so, that during certain months of the year it is not even possible to approach the west quadrant. One has to wade through standing water to approach the quadrant. The section of the instrument which is above water is not in good shape either. There are deep gouges on the

---

[1] For example, in the building-expense records for the month *Phālguna*, 1791 V.S. (Feb. 1735) a heading translates as "expenses: astronomers on daily wages at the *Jantra*." The heading is followed by listing of payments to a number of astronomers. See Ch. 12. Similarly, in the building records for the year 1794 V.S. (1737-1738), we find monthly expenses on various masonry items listed under the heading *Jantra*. R.S.A., *Tozīs*, Kārkhānā-Jantra, Jaipur State.

[2] It has been suggested that the phrase Jantar Mantar is a corruption of *Yantra Mandira*, or temple of instruments. However, the author has not found the phrase *Yantra Mandira* anywhere in the Jaipur State archives.

[3] De Tassy has also pointed this out in his essay. De Tassy, *Op. Cit.*, p. 537.

[4] The word *Jantar* colloquially means magical diagrams and the word *Mantar* magical words.

[5] For the dimensions of the Samrāṭ, see Kaye, pp. 41-42.

[6] See Ch. 3. The Jaipur Samrāṭ is also built partly below ground level.

[7] When the author revisited the observatory in January 1990, he found for the first time that there was no water standing at the instrument base. He was told by the caretakers of the observatory that the construction of a nearby high-rise building had lowered the surrounding water table. It remains to be seen if this effect is permanent.

Fig. 5-4 The Samrāṭ Yantra of Delhi.

Fig. 5-5 Jaya Prakāśa and Rāma Yantras.

quadrant surfaces where there should have been smooth plaster with scales. In fact, the scale markings on the instrument have disappeared just about everywhere. Consequently, the instrument has lost its value as an observing device.

The Gnomon

According to the visitors' accounts, the gnomon edges of the instrument had been lined up originally with white marble. During the restoration of 1910, however, the scale markings were drawn over a smooth surface plastered with lime instead of marble. Soon after the restoration, the markings began to be effaced because of exposure to weather, as noted by Kaye.[1] Now even the traces of the scale markings are difficult to find except at a few places here and there. The gnomon has been plastered with cement over most of its surface and has thereby lost just about all its scale markings, and from the faint traces that remain, it is not possible to discern the division scheme of its scale.

A few things may still be learned about the instrument, nonetheless. For example, the two surfaces of the gnomon do not lie in one plane. The west surface of the gnomon is at least 2 cm below the plane of the east surface. Further, the horizontal line joining the two six o'clock markings on the quadrants is not tangential with any of the surfaces. The two surfaces are at least 2.5 to 4 cm below the line. As noted earlier, this defect is prevalent in just about every Samrāṭ of Jai Singh's surviving today. It is not clear if the defect at the Delhi Samrāṭ is because of restorations or because of an error during the initial construction.

The Quadrants

The two quadrants of the Samrāṭ were surfaced originally in marble. However, during the restoration of 1910, they were surfaced with lime plaster, of which only traces now remain. Numerous efforts to preserve the masonry base have also erased the scale markings over the surface. The masonry work itself has now become exposed at numerous places. The lower section of the quadrants has been leveled over by pouring concrete on it. As a result, no time-readings are possible for almost six hours of the day, between 9 in the morning and 3 in the afternoon.

Because of the seepage water during most of the year, the east quadrant is accessible only from outside by climbing over the roof of the unfinished Ṣasṭhāṁśa. The west quadrant is not accessible at all unless one wades through water accumulated at the base, as much as one-half meter deep. Besides, there is little left on this quadrant for studying.

---

[1] Kaye, p. 42.

From the little that survives of the scales on the surface of the east quadrant, a number of things can be pieced together. For example, the quadrants had scales both on their surfaces and on their sides. The quadrants were apparently engraved to indicate the modern time-division scheme of hours, minutes and seconds, with the hourly markings drawn all the way across the surface. The hourly markings were divided into one-half hour, 15-minute, 5-min, 1-min, and 1/2 min submarkings. Besides, each one-half minute had been further divided into 15 parts of two seconds each, indicating that the builders or the restorers expected to read time with an accuracy of 2 seconds. The average width of the 2-second divisions that survive today is about 2.2 mm. The divisions vary in width, and divisions as wide as 4 mm are common. The marking of seconds do not come all the way up to the edge, but are recessed by about two centimeters. The upper edge of the quadrants' surface is labeled in English characters and the lower edge in Devanagari. The edge of the quadrants perpendicular to its broad surface has been graduated in degrees, with every sixth division labeled in *ghaṭikās*, the Hindu unit of time. In addition, the degree markings on the edges have been further divided into 20 subparts each, so that the least count of the scale is 3′ of arc. The small divisions are on the average 13 mm in width and have not been divided any further. The author could not check the accuracy of the time measurements because of the dilapidated state of the instrument's gnomon.

The *Yantraprakāra* describes an exercise of collecting data with the Delhi Samrāṭ.[1,2] The exercise enables us to check the accuracy Jai Singh's astronomers obtained with their Delhi Samrāṭ. The *Yantraprakāra* states:

> "On *Saṁvat* 1786, *Vaiśākha Kṛṣṇa* 5 [April 18, 1729, Monday], when 4:42,2 *ghaṭikā* [1 hr, 52 min, 49 sec] night had elapsed, Jupiter's hour angle was [measured as] 9;52,30 W [59;15 W], and *Svāti's* [Arcturus's] hour angle [measured as] 10;4,15 E [60;25,30 E]. Since these two [angles] are on the opposite side [of the meridian], obtain their sum, 19;56,45. The [result thus obtained is] the angular distance between Jupiter and the *nakṣatra Svāti*. This [result] is subtracted from *Svāti's* RA in *ghaṭikās* [which is] 35;9. The remainder [then] is Jupiter's RA in *ghaṭikās*, 15;12,15 [91;13,30]. At the same time, i.e., when 4;42,2 *ghaṭikās* [1 hr, 52 min, 49 sec] had elapsed in the night, the declination of Jupiter was measured to be 23;36 N. . . ."

The exercise described above must have engaged at least four observers to gather the data. In the exercise, the author does not apply any refraction correction to his readings. After applying the refraction correction of 1 arc-

---

[1] *Yantraprakāra* (J) pp. 17-18, and *Yantraprakāra* (S), pp. 84-85.

[2] The *Yantraprakāra* calls the instrument a Nāḍīvalaya.

minute, the declination of Jupiter becomes 23;35. This value is 3 arc-minutes higher than our calculated value of 23;32.[1] Similarly the *Yantraprakāra's* result for the right ascension of Jupiter, 91;13,30, is 0;6,30 higher than our calculated result of 91;7.

The exercise goes on to calculate the true geocentric longitude of the planet and finds it as 91;7,13, which is also higher by 5 arc-minutes, as compared with our computer generated results of 91;1,40. The differences between our computer generated results and the *Yantraprakāra* results indicate that the Delhi Samrāṭ suffered from one or more defects of construction.[2]

Ṣaṣṭhāṁśa Yantra

The Ṣaṣṭhāṁśa yantra, or sixty-degree arc, is located in a chamber 6 m high, 4.6 m long, and only a couple of meters or so wide. The east quadrant of the Samrāṭ rests against this chamber. The instrument is now in ruins, and only its masonry work survives. From the masonry structure, the radius of the arc is estimated to be 8.25 m. At its ceiling, toward the south, the chamber has an opening 1 to 1.25 meters long and about a third of a meter wide. The chamber is now closed from the east-side, but has an arched opening facing the Samrāṭ gnomon to the west.[3] When the author examined the instrument, it could only be approached by wading through water collected around the Samrāṭ base. Bhavan says that the instrument was not restored in 1909 because the engineers encountered a water seepage problem similar to the one now plaguing the Samrāṭ. Suggestions were made, however, to rebuild the instrument above the water table but were never carried out.[4]

There are reasons to believe that, in its original state, the Delhi Ṣaṣṭhāṁśa was an extremely accurate instrument. It approached or surpassed the limit of naked eye observations, which is generally accepted to be one minute of arc.[5] The readings with this instrument have been reported by Jagannātha. He writes: "On *Caitra Kṛṣṇa* 6, 1786 V.S. (March 20, 1729), the zenith distance of the midday sun as measured with the Vṛttaṣaṣṭhāṁsa (Ṣaṣṭhāṁśa yantra) was 28;44,30."[6] A comparison of Jagannātha's readings with computer generated results indicates that the instrument had been accurate within $\pm 1'$ of arc. If the instrument did indeed have a radius of 8.25 m, as we have estimated, its $1'$ divisions must have been only 1.2 mm wide.

---

[1] We used the computer program AstroInfo produced by Zephyr Services, Pittsburgh, Penn., USA.

[2] See Ch. 3.

[3] The instrument is not totally enclosed as Kaye writes. Kaye, p. 42.

[4] Bhavan, p. 73.

[5] This is because the measurements are taken off the pinhole image of the sun produced by the instrument, and thus the resolving power of the eye does not interfere.

[6] SSMC, p. 81, and SSRC, p. 1218.

Claude Boudier, a French Jesuit who visited the observatory at the invitation of the Raja, used the instrument to measure the zenith distance and the diameter of the sun.[1] Boudier was stationed at a mission at Chandernagore, Bengal, and was conducting his own observations of the sun. He came to assist the Raja in his astronomical undertakings. Leaving Chandernagore in January 1734, Boudier and his fellow priest, Father Pons, reached Delhi in late April or early May. They joined the observatory staff in observing a solar eclipse on May 3, 1734.

Boudier stayed on in Delhi for some time before leaving for Jaipur, his final destination. Although Boudier calls the Ṣaṣṭhāṁśa a gnomon, there is little doubt that he is referring to the Ṣaṣṭhāṁśa only and not to the Samrāṭ or any other sundial. He says:

> "The copper plate in which is located the aperture of the gnomon is placed parallel to the polar axis. The rays of the sun fall on the concave surface of the quadrant whose radius is about 26 (Paris) feet (8.44 m).[2] The quadrant is graduated in minutes; the chord of 30 minutes is 522 parts, of which the diameter of the aperture is 32. The image of the sun was without penumbra, at least as far as one could perceive, so that it was easy to measure it exactly."[3]

In Table 5-1, we compare Boudier's data with our computer generated results. Boudier gathered his data regarding the diameter of the sun with a specially engraved plate of copper imported from Europe.[4] Boudier apparently had also used this plate for diameter observations at Chandernagore.[5] "The plate, made by the (French) king's engineer for mathematical instruments, Sr. Bion, had 960 divisions, equivalent to the five inches of a Paris foot," he says in one of his letters to Europe.[6] In this plate one division equalled approximately 0.13 mm.[7,8]

---

[1] For European involvement in the program of the Raja, see Ch. 13.

[2] One Paris foot = 0.3248 m. See *Encyclopedia Britannica*, published yearly.

[3] "Observations Géographiques, faites en 1734 par des pères jésuites pendant leur voyage de Chandernagor à Dely et à Jaëpour." in *Lettres*, p. 776. Also Raymond Mercier, "The Astronomical Tables of Rajah Jai Singh Sawa'i," *Indian J. Hist. Sci.*, **19** (2), pp. 143-171, (1984).

[4] The solar diameter measured by Boudier with the Ṣaṣṭhāṁśa is larger by about one half-minute of arc than its true value. The reason lies both in the penumbra of the image and the diffraction of light at the circular aperture of the instrument. See Ch. 3.

[5] For Boudier's observation see V.N. Sharma, and L. Huberty, "Jesuit Astronomers in Eighteenth Century India," Archives Internationales D'Histoire Des Sciences, pp. 99-107, Vol. 34, June 1984.

[6] *Fonds Brotier*, Vol. 78, f.1, Jesuit Archives, Chantilli, France. Also Pius XII Memorial Library, St. Louis U., St. Louis, MO.

[7] Sharma, V.N., Huberty, L., "Jesuit Astronomers . . . India," p. 103, (see table 2), *Op. Cit.*

[8] It is difficult to see how a small division of width 0.13 mm could be engraved on copper plate. Boudier might have used a Vernier scale to measure such small widths.

Boudier does not state specifically if he applied any refraction correction to his readings. From the analysis of his data it is suspected, however, that he did not do so, although he did have in his possession a copy of de La Hire's tables in which a table for refraction correction is given. The last column in Table 5-1 above indicates deviation from the computer generated values with the refraction correction applied.[1] It is noteworthy that Boudier's readings with the Ṣaṣṭhāṁśa agree within 30 seconds of the computer generated results.

Year 1734	Boudier's Data			Computer-generated Data		Refraction Correction	Deviation with the correction applied†
	Zenith Distance	Diameter of Sun		Zenith dist. without refraction correction	Diameter of Sun		
		Div.*	Angle				
17 May	9;36,16	558	0;32,0,4	9;36,05	0;31,36	9"	2"
18 May	9;22,30	558	0:32,0,4	9;22,38	0;31,35	9"	−17"
19 May	9;09,29	558	0;32,0,4	9;09,34	0;31,35	9"	−14"
21 May	8;44,06	558	0;32,0,4	8;44,12	0;31,35	7"	−13"
25 May	7;57,50	558	0;32,0,4	7;58,01	0;31,33	7"	−18"
26 May	7;47,02	557	0;32,0,1	7;47,20	0;31,33	7"	−25"
27 May	7;36,50	557	0;32,0,1	7;37,00	0;31,33	7"	−17"
28 May	7;26,50	557	0;32,0,1	7;27,03	0;31,33	7"	−17"
21 June	5;24,45	555	0;31,5,4	5;25,02	0;31,28	5"	−22"

* The divisions refer to the markings on the copper plate used by Boudier.
† True zenith distance = apparent zenith distance + refraction correction.

Table 5-1 Ṣaṣṭhāṁśa yantra observations of Boudier compared with the computer generated values. For the diameter of the sun Boudier used a specially engraved copper plate obtained from Paris.

Mercier also has analyzed Boudier's data and determined the latitude of the Delhi observatory from it as 28;37,17.[2] Because this value is only 20" smaller

[1] The refraction correction is based on the *Indian Ephemeris*, published yearly by the Government of India Controller of Publications, Delhi.

[2] Mercier, *Op. Cit.*, pp. 165-166.

than his calculated value from an Archaeological Survey of India map,[1] it is further confirmed that the Delhi Ṣaṣṭhāṁśa was indeed a high-precision instrument capable of measuring angles within one minute of arc.

Jaya Prakāśa

The Jaya Prakāśa of Delhi is located to the south of the Samrāṭ and is built in two complementary halves. The halves consist of two hemispherical bowls of masonry, 8.33 meters in diameter, with entrances from four sides and steps leading down to the bottom. At the bottom of each half is a single drainage hole.

The hemispheres of Jaya Prakāśa consist of a solid section some 8.3 meters wide toward the south end, from which six solid sections emanate joining together once again near the north end. Between these sections there are five walkways. Each walkway has steps for the observer to move around freely. The solid sectors in the two halves do not duplicate but rather complement one another, such that if put together, they will form a complete hemispherical surface. With this arrangement the coordinates of an object can be measured, at any given time, with one hemisphere or the other, as explained earlier. The sections are 53 to 56 cm wide at their base at the south end and about one meter in the middle. The open sections in between are 48 to 53 cm across.

Division Scheme

Although much of the scale markings of the Delhi Jaya Prakāśa have become obliterated, enough of them remain to give some idea of its division scheme. The markings have survived primarily in the sections attached to the rim of the instrument. From these sections, it can be deduced that the instrument once had graduated arcs for measuring the azimuth angle and the zenith distance of celestial objects. In addition, five declination circles including the equator may also be discerned. The markings for azimuth and the equatorial coordinates (right ascension) were three degrees apart. The circle near the rim measuring the azimuth, or the horizon of the instrument, was divided into four quadrants, with its zero markings at the north-south points and the 90-degree markings at the east-west points. Along the rim of the instrument, on the north-east quadrant only, holes have been provided six degrees apart. The reason for these holes is not clear. Another angle-measuring scheme has also been provided with zeroes at the east-west points and 15 principal division marks at the north-south.

The Jaya Prakāśa of Delhi is similar to its sister unit at Jaipur. It once had "circles of the signs of the zodiac, by which a sign on the meridian is indicated

[1] Mercier, *Op. Cit.*, p. 161.

by the position of the sun's shadow," as Kaye reports.[1] No trace of these curves remains now, however. The instrument also had cross wires stretched between the four cardinal points as the theory required. They also disappeared long ago. Kaye, writing in 1917, says: "the cross wires have been discarded, although the pins to which they should be fastened are still there; and iron rods (5 cm diameter galvanized piping) have been fixed at the centers of each Jaya Prakāśa."[2] The author did not find any iron pipes when he examined the instrument in 1981-82.

The instrument shows signs of extensive damage from weathering.[3] There are signs of repairs also; the surface of the instrument has been plastered over and the azimuth lines have been redrawn. But the repair work has been shoddily done and left unfinished for some reason. Besides, the material used in the repair is inferior in quality as the instrument's surface is already cracking and breaking loose in places. For all practical purposes the instrument, as it stands today, is useless for observing.

Rāma Yantra

A few meters to the south of the Jaya Prakāśa is the Rāma yantra. From the outside the yantra appears as a pair of cylindrical buildings with numerous arched windows arranged one above the other in three tiers that enclose a central pillar. The buildings are constructed along a north-south line, and part of their structure is below ground level. In the north building of the yantra, windows exist all around the circumference, giving it the appearance of an open structure. In the south building, however, only the lower section has windows all around its circumference. In the upper two sections, only half the circumference contains windows. In this building a 180-deg-wide upper section extending from the north-east to the south-west has been closed off for some reason. This closed section faces west. The buildings have a single entrance each leading to their inner structure.

The walls of the north-building of the instrument are 1.45 meters wide, whereas the walls of the south-building are 1.68 m. The Rāma yantra of Delhi is almost twice as large as its sister unit at Jaipur. Its inside diameter is 16.65 m, and its inner walls have scales 7.52 m high. Its central pillar is 80.6 cm in diameter and 7.52 m in height.[4] Horizontal radial sectors of white sandstone, 30 in all, spread out from the pillar up to the walls of each yantra. These sectors, engraved with scales, are 6 degrees wide and are separated from each other by open segments of equal width. The sand stone slabs comprising the sectors are raised 1.07 m above the floor and rest on vertical stone slabs or on

---

[1] Kaye, p. 43.

[2] *Ibid.*

[3] The author's last visit to the observatory was in January 1992.

[4] Kaye, p. 38.

arched masonry supports. The radial segments, between the pillar and the walls, are approximately 7.5 meters long. At the circular wall, these segments meet an equal number of evenly spaced vertical columns that comprise the inner surface of the cylindrical structure. The columns are engraved with scales and are, in fact, vertical extensions of the scales of the horizontal radial sectors.

Although the two buildings of the Rāma yantra appear well preserved from outside, their interior condition is poor. There is perpetual dampness on the floor from seepage water. Electrical wiring dangles from sockets. Plaster is gone from several places on the walls.

Division Scheme

The division scheme of the two Rāma yantra buildings is similar. The upper surface of the radial sectors of the instrument are engraved with concentric arcs one degree apart for reading the zenith distance and are labeled in Devanagari. The degree divisions for measuring the zenith distance begin at the base of the central pillar, such that the circumference of the pillar indicates the zero of the scale. The sides of the radial sector are also marked, indicating the altitude. The altitude markings have been subdivided into 10 parts each, and thus the least count of the instrument is 6′ of arc. Near the 45-degree zenith angle, where the radial sectors meet the vertical columns, and where the instrument is theoretically most sensitive, the six-minute divisions are 2.7 cm apart. The author noted an error of 2 mm in the marking of a number of divisions in places. The divisions become progressively smaller in width as one approaches the base of the central column. Consequently, for objects close to zenith, the error in measurement of the altitude or the azimuth could be of several degrees.

The vertical columns forming the circular enclosure are also divided into degree-marks, 0^0 to 45^0, for measuring the zenith distance. The degrees have been further subdivided into 10 parts worth 6′ of arc each. The length of these divisions is smallest near the top end of the wall and gradually increases as one approaches the 45-degree marks at the base. The six-minute divisions on the walls are 2 cm apart between 35 and 36 degrees, and 2.5 cm between 43 and 44 degrees.

Azimuth

The pillar at the center of each building has two circular horizontal scales for reading the angle of azimuth. These scales are divided into degrees and are numbered at every sixth degree, that is, as 6, 12, 18, and so on to 360. The degrees have not been further subdivided.

The arcs on the horizontal radial sectors are used for measuring the angle of azimuth for objects with zenith distance less than 45 degrees. These arcs are divided by radial lines extending all the way from the pillar to the walls. The large divisions of the arcs are once again one degree apart. For objects in

Fig. 5-6 The Rāma Yantras of Delhi.

Fig. 5-7 The Miśra Yantra.

which the zenith distance is greater than 45 degrees, the scales on the vertical columns are used. These are similarly divided.

With a Rāma yantra, the altitude and azimuth of the sun are observed by noting the location of the shadow of the central pillar, or more precisely, the highest point of the shadow of the pillar's edge on the scales of the instrument. The instrument might not have been used when the sun was at the meridian, because other instruments such as the Ṣaṣṭhāṁśa and the Dakṣinottara Bhitti give a far more accurate result for the meridian sun.

Night observations with the instrument are rather cumbersome. For night measurements, the observer uses the space between the solid sectors and columns. For objects with zenith distances up to 45 degrees, the space between the radial sectors is employed. Arc lengths of different radii, fabricated out of brass or thin wood, may have to be used. The observer, crouching between the sectors "places his eye" at the edge of the arc and aligns the edge of the metal or wood arc and the edge of the pillar with the celestial object, say a planet. The angles of altitude and the azimuth are then obtained by interpolation.

If the zenith distance is greater than 45 degrees, the observer has to sit on a plank suitably fitted between the appropriate slots of two adjacent vertical columns and must repeat the process described above. Since the curvature of all the vertical columns is the same, arc-lengths of only a single radius are to be used.

It appears that for zenith distances less than 45 degrees, the readings at night might not have been taken at any arbitrary time because that would have required an inordinate number of arcs of different radii. Instead, they might have been taken at the moment when the object transited the vertical plane defined by the edge of a vertical column, i.e., when the azimuth of the object became just right so that it could be read directly at the edge of the radial or vertical sectors. Although this procedure reduces flexibility somewhat, it eliminates the need of a large number of arc lengths of different radii ranging all the way from less than a meter to 7.52 meters.

Difficulties with the Instrument

While working with the instrument an observer may encounter several difficulties, such as the penumbra of the sun, which make it difficult to repeat observations with accuracy. The author discovered that a small cross-hair held about 2 cm above the surface can eliminate the penumbra problem to a great extent. The cross-hair is held parallel to the coordinate lines engraved on the surface, and moved until the shadow of its cross-point just disappears into the shadow of the pillar's tip. The point where this happens then gives the two coordinates of the sun. It is possible that Jai Singh's astronomers used a similar method to overcome the difficulty caused by the penumbra.

Another difficulty, although not as serious, is to decide with precision where the top of the pillar's shadow is located. This may become particularly

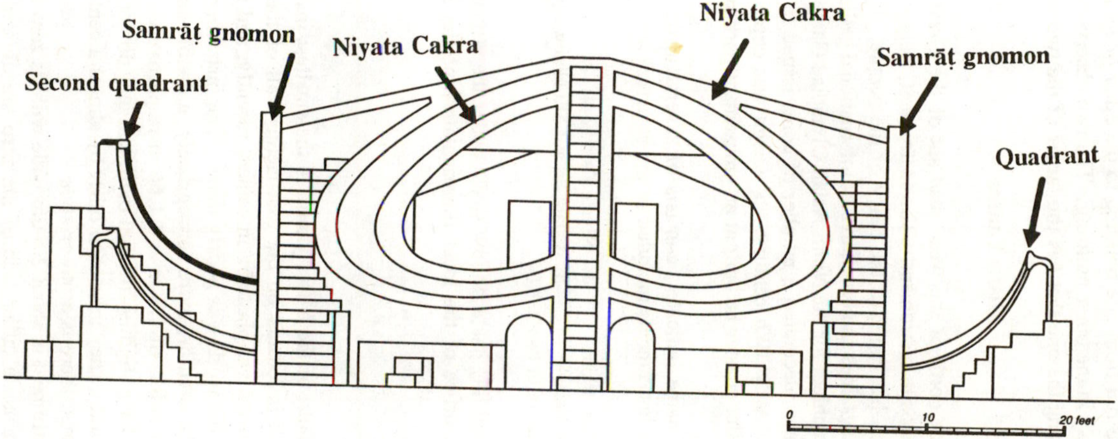

Fig. 5-8 The Miśra Yantra, front elevation.

bothersome when the shadow of the pillar does not fall entirely on a single vertical column of the wall or on a single radial sector. In that case, the observer either has to run from one building to the other or let a helper take a reading simultaneously in the other building. The two observers then compare their readings, retaining the one that has the larger of the two altitude values.

Miśra Yantra

A few meters to the north of the west-quadrant of the Great Samrāṭ is the Miśra yantra or composite instrument. The instrument, being very attractive and unusual in design, has captured the fancy of the people. It has decorated covers of scientific magazines and brochures of international gatherings at Delhi. It was chosen as the emblem for the Ninth Asian Olympiad Games held in Delhi in 1982, and its likeness decorated a number of coins issued to commemorate the occasion. In the early 19th century, its design was copied by a French entrepreneur for a roller coaster he built at an amusement park at Neuilly, near Paris, France.[1]

The yantra, as the name indicates, consists of several instruments within a single structure. The instruments included in the structure are as follows:

1. Dakṣinottara Bhitti
2. Karkarāśi Valaya
3. Samrāṭ yantra in two halves
4. Niyata Cakras (4)
5. Quadrant arc of unknown function

The Miśra yantra building has a hollow space under the structure which has been turned into the office of the local representative of the Archaeological Survey of India.

Dakṣinottara Bhitti

The Dakṣinottara Bhitti of the Miśra yantra is the smallest instrument of the Delhi observatory. It is located on the east-facing wall of the Miśra yantra complex. The instrument was plastered in cement recently, and its scales were redrawn. However, the work was poorly done. The instrument, as it stands today, is a semicircular arch or, more appropriately, a double-arc joined at the top. The arcs have a radius of 1.6 m and 1.66 m respectively. The two arcs of the semicircle thus have slightly different radii, and they do not meet properly at the top as they should. They miss each other by almost a centimeter. Close to the center of the arcs, there is a pin or style.

The scale of the instrument is about 2.5 cm wide with its zero marking at the top. The two quadrants on either side of the zero are divided into 15 main

---

[1] Penelope Chetwode, "Delhi Observatory, The Paradise of an early Cubist," *The Architectural Review*, p. 59, London, 1935.

divisions. As the left quadrant is further graduated with its main divisions subdivided into six parts, a small division of the instrument measures an *aṁśa*, or degree. The divisions have been very poorly marked, however. Although most divisions are 2.8 cm wide, the variations between divisions could be as much as 2.5 mm. The script used for labeling the divisions is Devanagari. In addition to the recent scale markings, faint traces of the 1910 restorations may also be seen on the instrument.

For measuring the zenith distance or altitude of the midday sun with this instrument, one has to hold a pin or a rod above the scale, such that the shadow of the pin falls over the fixed style at the center. The part of the shadow falling on the scale then indicates the zenith distance of the sun. The zenith distance of the meridian moon at night, if bright enough, may also be determined similarly. With this Dakṣinottara Bhitti, the procedure for measuring the altitude of a planet or star, which does not cast any discernible shadow, is cumbersome at best, because the instrument is too close to the ground to facilitate an easy reading. It is debatable, therefore, that this instrument was ever used for any serious work, particularly when another and more accurate Dakṣinottara Bhitti was available to the astronomers of Jai Singh.

Karkarāśi Valaya

On the back wall of the Miśra yantra complex, facing north, is the semicircular scale of Karkarāśi Valaya with its two ends facing upward. The wall on which the instrument is engraved is tilted approximately 5 degrees toward the south and, therefore, is parallel to the plane of the Tropic of Cancer. On June 21, when the sun is at the Tropic of Cancer, its rays at noon fall at grazing incidence on the instrument. It is not known what application the designers had in mind when they built the Karkarāśi Valaya. The instrument is now in ruins. Most of its markings are gone, and deep gouges may be seen in places. According to a photograph in Kaye's *The Astronomical Observatories of Jai Singh*, there used to be a door in the Karkarāśi Valaya wall. The door is now closed with bricks and mortar, and two windows have been added to the wall.

According to Bhavan, the Karkarāśi Valaya scale had zero markings at the two top ends of the scale and the 90 degree marking at the bottom.[1] From what little is left of the instrument, the author concluded that the diameter of its scale had been 11.81 ± 0.03 m and that it was marked according to the Hindu scheme of dividing a circle into quadrants of 15 large divisions and then subdividing them into *aṁśas* or degrees. Next, each degree was divided into 10 parts which were once again parted into six subdivisions of one minute of arc, thus giving the instrument a least count of $1'$ of arc. From the markings still visible

---

[1] Bhavan, p. 70.

at a few places, it is noted that the small divisions indicating minutes were on the average 1.7 mm wide. The markings are done rather poorly, however. For example, a number of divisions of width as small as one millimeter may be noted on the scale. Because the instrument has been designed to measure to the nearest minute of arc, it is the most sensitive device at present in the Delhi observatory.

In theory, the Karkarāśi Valaya may be used once every 24 hours to measure directly the longitude of a celestial object, such as the moon, at the moment when the first point of the sign of Cancer, or the summer solstitial point of the ecliptic is on the meridian. The time when the solstitial point reaches the meridian may be determined from tables or with an astrolabe. For objects such as the planets, which are visible only after dark, observations are possible only after September 22 and before March 21, when the sun is below the horizon at the time the solstitial point approaches the meridian.

Samrāṭ Yantra

The Samrāṭ of the Miśra yantra is constructed in two halves to be used before and after the noon hour for determining the local time. These two halves flank the Niyata Cakras of the Miśra and are identical to a large extent. The gnomon of the east half is 10.97 m long, whereas the gnomon of the west half is 11.00 meters or about 2.5 cm longer. They both are about 46 cm wide and have a quadrant each attached to their outer flanks. The radius of the east quadrant is of 2.63 m, and that of the west 2.87 m ± 4 cm. In addition, the west gnomon has a second quadrant attached to it.

The instrument has suffered from improperly done repairs. Numerous repairs of the surface have all but obliterated the scale markings on the two gnomons. A faint line running parallel to the surface edge near the top of the east gnomon and a few traces here and there on the west gnomon suggest that the instrument was designed to measure declination as well.

The markings from the upper surface of the quadrants are mostly gone, but markings on the sides, approximately 2/3 of the original, may still be seen. The Hindu scheme of dividing a quadrant into 15 parts as well as dividing a quadrant into 90 degrees may be discerned. These markings show that a degree was divided into 20 parts, and that the instrument had a least count of 3 arc-minutes. The width of a small division was approximately 1.5 to 2 mm, and the script used in labeling had been Devanagari.

Niyata Cakra

The Niyata Cakras or "fixed arcs" is the best preserved instrument in the Delhi observatory. The instrument consists of four semicircular scales of marble on either side of a gnomon constructed in the middle. The instrument was originally built in lime plaster, but in 1951 it was surfaced with marble.[1] Stairs are provided between the scales and around the outer scales to approach the top of the structure. Another set of steps is provided in the middle of the gnomon to reach to its very top. The gnomon, or more appropriately, the two edges of the gnomon in the middle, with which the scale ends meet, are the diameters of the semicircular scales. Further, the scales are inclined with the plane of the meridian, or with the vertical surface of the gnomon as Fig. 5-8 indicates.

According to Kaye, who had the angles measured by engineers, the Niyata semicircles lie in the planes inclined with the meridian-plane by 77;18 W, 69;50 W, 69;42 E, 77;22 E respectively.[2,3] The angles are difficult to measure as Kaye points out. The pundits he consulted gave him different answers, however. According to the pundits, "the semicircles are inclined to the plane of the Delhi meridian at angles of 77;16 W, 68;34 W, 68;1 E, and 75;54 E."[4]

The Niyata Cakras scales are 53 cm wide. The Cakras on the outside inclined at about 77 degrees, have a radius of 4.83 m. The radius of the Cakras on the inside, on the other hand, is 4.01 m. At the center of the Cakras, on the gnomon, holes are provided for fixing a pin. The holes, however, have been incorrectly located as they are off from the true center by about 4 cm.

The Niyata Cakras are provided with two scales each running side by side along their inner and outer edges respectively. These scales are not identical, however. The scale on the outer side of a Niyata semicircle measures declination in degrees and minutes. Each degree on this scale is divided into 10 parts, and the parts are once again divided into 3 subparts each. Thus the small divisions of the outer scale measure 2′ of arc. The zero point on this scale is along the east-west line in the middle.

The inner scale, on the other hand, is divided into two quadrants of 15 main divisions each. For this scale the zero marks are on the north-south points. The main divisions here are subdivided into degrees or *aṁśas*, which are once again divided into small divisions of 2′ of arc each. Originally the least count of the instrument had been 1′ of arc, as Hunter reports, but that is changed now. The

---

[1] The repair work was done under the supervision of Kedāra Nātha Śarmā (no relation to the author), the *Rāja Jyotiṣī* of Jaipur at the time.

[2] Kaye, p. 45.

[3] The author attempted to measure the inclination of the outer Niyata Cakra on the western side by observing the time when the stars Castor and Pollux in the constellation of Gemini came into the plane of the instrument. His average value of two readings was 77;12. On the night of the observations, visibility was poor, and therefore his readings may be off by as much 0;45. The observations were conducted on the evening of January 6, 1990.

[4] According to Bhavan, the inclination of the Cakra planes is 77° E, 69° E, 69° W, and 77° W of the Delhi meridian respectively. Bhavan, pp. 68-69.

width of the 2′ divisions, at present, however, is large enough to divide them further into 1′ divisions.

For the Niyata Cakras of larger diameter, the small divisions on their outer scale measure 3.20 ±.05 mm. Similarly the width of the small divisions on their inner scale is 3 mm. The divisions on the two scales of the inner Niyata Cakras (radius, 4.01 m), on the other hand, have widths of about 2.5 mm and 2.35 mm respectively. The divisions of the Niyata scales have been fairly well laid out.

The Purpose of the Instrument

The Niyata Cakras have been constructed for measuring the declination of an object at intervals of a few hours as the object travels from east to west in the sky in the course of a day. According to Bhavan, the Niyata Cakras on the west side of the central gnomon are meant for observing the declination of the sun at 52 minutes, and at 1 hour 24 minutes after sunrise.[1] The procedure, as he illustrates, is to fix a rod into the hole at the center and observe its shadow on the scale at the times just indicated. Similarly, the Niyatas on the eastern side of the gnomon are meant to observe the declination of the sun at 4 hr 36 min, and at 5 hr 8 min after the noon hour.[2] At night, the readings are taken at the moment when the object is at the grazing incidence with the surface of a Niyata scale. For this, the "eye" is placed at the edge of the scale such that the object and the central pin appear in one line. The vantage point on the scale then indicates the declination of the object.

Following the procedure elaborated by Bhavan, we measured the declination of the sun on November 11, 1981, with the Niyata Cakras on the west. We observed that on that day the rays of the sun became tangential to the outer Niyata at 6:57:12 A.M., Indian Standard Time. The sun was extremely weak, casting no appreciable shadows and, therefore, it became necessary to look at it directly. This way the declination was found to be approximately 17 degrees. Next, less than an hour later, at 7:29:26 A.M., the rays of the sun became parallel again to the inner Cakra on the west, and from its scale the declination of the sun was measured to be 16;20. The average of the two readings was 16;40. The ephemeris value of the declination for that day was 17;20,10, and thus there was a discrepancy of about 40′ between the computer generated value and the average reading. Part of the reason for this discrepancy could have been that the hole for inserting the pin was off center by about 4 cm. The Niyata Cakras on the east could not be checked due to the high-rise buildings to the west of the observatory.

---

[1] Bhavan, pp. 67-69. Bhavan's comments regarding the time of observation after sunrise is true only at an equinox and not at any other time.

[2] *Ibid.*

A popular belief is that the Niyatas are meant to duplicate the declination readings for the meridian arcs of four different locations on the globe. According to Bhavan, "the Niyatas are similar to the meridian arcs at Notkey, a small town in Japan; Serichew, a town on the Pic islands east of Russia, in the Pacific; Zurich, Switzerland; and Greenwich, England."[1] All these places except Serichew supposedly have observatories.[2] A check carried out on behalf of the author revealed, that there has never been any observatory at Notkey in Japan.[3] It should be pointed out that the Zurich observatory was built in 1759, 16 years after the death of Jai Singh, and when Madho Singh, his son, was the king, and when anarchy prevailed around Delhi. The observatory of Greenwich was built earlier, in 1675, about 13 years before Jai Singh's birth.

Difficulties with Niyata Cakras

Because the Niyata Cakra are built in planes too close to the horizon, they are cumbersome to use. In the morning, particularly during winter months when the sun is weak and there is a haze over the horizon, the shadow of the pin cannot be discerned easily. It becomes difficult to ascertain the location of the shadow on the scales without looking at the sun directly. Besides, for readings close to the horizon, the refraction correction becomes important and may not be neglected. Neither Hindu nor Muslim astronomers applied refraction corrections to their works. Jai Singh, however, did realize its importance and included a refraction correction table in his *Zīj*.

Although the Niyata Cakras are impressive in appearance, they may not be classified as well-designed instruments. A considerable part of their scales, particularly the section above the horizontal line passing through the center of the arcs, is useless and can not be used. If Bhavan's procedure is followed, the scales above the horizontal line can only be used for measuring the declination of the objects that lie below the horizon, and that can never be possible. One could reverse his vantage point and view the objects from the location of the pin at the center, lining up the pin with the object when the object appears in the

---

[1] Bhavan, p. 69. Also Kaye, p. 45.

[2] The coordinates of these places are reported to be as follows:

Notkey	43;33,0 N.	145;17 E.
Serichew	48;06 N.	153;12 E
Zurich	47;23 N	8;34 E

See Kaye, p. 45.

[3] Profs. Yoshihide Kozai of the Tokyo Astronomical Observatory, Michio Yano and K. Yabuuti of Kyoto Sangyo Univ., and Shigeru Nakayama of Kanagawa University have independently identified Notkey with Notsue, a very remote area of the Hokkaido island of Japan. (Private communication) The place is very desolate, inhospitable, and cold even in summer. An observatory at such a location is highly unlikely. Besides, the author has not found any observatory mentioned at Notsue in literature related to the history of astronomy in Japan. See Ch. 1.

plane of the scales.[1] With this procedure, however, only about one-fourth of the scales of the outer Niyata circles can be used. The two gnomons of the Samrāṭ get in the way for the remaining upper part of the scale. The two inner Niyata Cakras, on the other hand, cannot be worked with this procedure at all. The scales of the outer Cakras obstruct the field of view if this procedure is attempted. The real worth of the Niyata Cakras, or the Miśra yantra as a whole for that matter, lies in its being a good teaching tool.

The Second Quadrant on the West

Attached to the Miśra yantra Samrāṭ's west gnomon is a second quadrant. Kaye erroneously calls it Agrā yantra or amplitude instrument.[2] Hunter does not mention it at all in his paper published in 1799. However, the quadrant can be discerned in Daniell's engravings which were made only a few years after Hunter's visit and were printed in 1816.[3] Bhavan says that he repaired the quadrant in 1910 without altering its shape and that he did not find any special gnomon just for this quadrant.[4] Further, Bhavan is silent about its function. Apparently, there was little left of its gnomon and its markings to give him any clues.

The quadrant has a length of 5.0 m, width 74 cm, and radius 3.18 m. Its radius is thus about 31 cm larger than that of the Samrāṭ's quadrant next to it. The plane of the second quadrant, as the author judged, is tilted by about 5 degrees toward the south, i.e., by about the same angle as the Karkarāśi Valaya. The scale markings etched on its surface in 1910 have more or less disappeared, except for a few traces left here and there. The scale markings on either side of the surface, however, may still be seen intact at places.

From what little is left of the scales, one can see that Bhavan in 1910 divided the scales into 90 degrees with the zero mark at the bottom. The degree divisions had been further subdivided into 10 parts each. Thus the least count of the instrument was 6′ of arc. The Hindu system of marking a circular scale had also been used, and the quadrant was divided into 15 major parts. The least count of this division scheme was also 6′ of arc. The average length of the small divisions was 6 ± 2 mm.

From the five-degree tilt of the quadrant, it appears that the instrument was meant to measure the latitude and longitude of a celestial body directly, once every 24 hours when the summer solstitial point on the ecliptic, or the first point of the sign of Cancer, approached the meridian. However, for these measurements, the section of the gnomon adjacent to the quadrant should also

---

[1] This procedure would require two observers—one viewing from the site of the pin and directing the other with a pin or slit at the scale.

[2] Kaye, p. 45.

[3] Daniell, *Op. Cit.*

[4] Bhavan, p. 66.

be tilted by an additional five degrees to the south. The instrument will then work just like the Karkarāśi Valaya of Jaipur.[1] It is not known if the builders had such an application in mind and if they provided the gnomon with an additional tilt of five degrees to the south as required. By 1910, the gnomon had become so dilapidated that Bhavan, with no clues to its original tilt, simply extended the lower section of the gnomon meant for the west Samrāṭ, as we find it today.

Quadrant on the Floor in Front of the Miśra Yantra

On the masonry floor in front of the Miśra yantra, there used to be a quadrant of 6.1 m radius. Kaye reports seeing traces of this quadrant. According to him, the platform probably was meant "for making measurements when the instruments were being constructed or repaired."[2] Bhavan makes no mention of the quadrant. Strangely, no instrument of the Delhi observatory has a radius of 6.1 m or equal to that of the quadrant. Only the radius of the Karkarāśi Valaya, 5.9 m, comes close to it. The quadrant has become obliterated these days.

Sundial or Agrā

On top of the large Samrāṭ gnomon is a circular column 1.5 m high and almost equal in diameter (1.47 m), which originally must have been the support for either an Agrā or a Palabhā yantra. The top surface of the column has now traces of a European-style sundial of 51 cm in diameter which had been graduated in Roman numerals. This sundial was constructed during the restorations of 1910,[3] but its diagonal plate is gone now.

On the upper surface of the column, where the plaster of the sundial has eroded, traces of a circular scale with radial lines may be seen, indicating that the instrument was once an Agrā yantra of the type found at Ujjain.[4] It is open to question, however, if these radial markings are original or were etched in 1852, when the Samrāṭ was repaired under the supervision of the Archaeological Society of Delhi.

Vertical Columns

To the southwest of the Miśra yantra are located two vertical columns about 5 meters apart and approximately 1¼ m across. The columns do not have a circular cross-section and seem to have no apparent function. They might have

---

[1] See Ch. 7.

[2] Kaye, p. 46.

[3] Bhavan, p. 72.

[4] See Ch. 9.

once supported some heavy metallic equipment of the type found in Jaipur, or supported an instrument, such as *Dhāt al-Thuqbatayn*, with which Jai Singh experimented during the early days of the observatory. The instruments were either dismantled or were vandalized when the observatory was neglected.

Porch

Finally, there is a small covered porch between the Samrāṭ and the Miśra yantra, open on four sides and with a roof on top. It is not certain whether it belongs to the original set of buildings or is a later addition. None of the visitors mentions it in their narrations, and Bhavan says nothing about its origin either. But it has been there at least since the restoration of 1910.

CHAPTER VI

JAIPUR OBSERVATORY I

The Jaipur observatory is located across the street from the City Palace in the business district of the old city. The observatory is popularly known as the Jantar Mantar of Jaipur.

The parameters of the observatory are as follows:

Latitude	26;55,27.4 N
Longitude	75;49,18.7 E
Height above sea level	436 meters
Local Time	I.S.T. +26 min, 43 sec., or Universal Time −5 hr, 3 min, 17 sec.

History

A great deal of uncertainty prevails regarding the construction of Jai Singh's observatories. It is difficult to say with certainty, therefore, just when Jai Singh built the first instruments of the Jaipur observatory. Jai Singh had been interested in observational astronomy since the very beginning of his career as an astronomer.[1] It is possible that he was experimenting with instruments of different designs and observing with them at his Amber palace or at the observatory site only a few miles away, before his observational program was formalized, and observatories were built.

Garrett, citing an inscription on an instrument he restored in 1901-1902, suggests 1718 as the date for the beginning of the construction of the Jaipur observatory.[2] Unfortunately, Garrett does not indicate the instrument on which he saw the inscription, and it is possible that his statement is based on a misinterpretation of the Nāḍīvalaya plaque.[3] The date of construction of the first instruments of the observatory may be uncertain, but there is little doubt that by 1728 a number of instruments had been erected at the present site. This is confirmed by a map and also by the dates of construction given on two masonry structures.[4] The plaque on the Dakṣinottara Bhitti states that its original unit was erected in 1785 V.S. or 1728 A.D. In addition, an inscription

---

[1] In 1716 when Jai Singh was only 20 years old, he acquired two books on the quadrant. See Ch. 13.

[2] Garrett, p. 14.

[3] Garrett, relying on Bhavan, could have interpreted the date on the plaque as 1718. Bhavan, p. 47. For the interpretation of the inscription on the Nāḍīvalaya plaque, see Ch. 7.

[4] These two structures are, (1) a room in the middle of the observatory and (2) the original Dakṣinottara Bhitti yantra which was demolished in 1876.

on the observer's room in the middle of the observatory compound also has this date. The instruments of the Jaipur observatory were built at different times, and thus the observatory's construction was spread over several years and continued on until 1738. The Great Samrāṭ and the Ṣaṣṭhāṁśa were planned around 1732 and completed sometime in 1735.[1] The Jaipur State records of the Rajasthan State Archives indicate unusually large sums expended during these years at the observatory site on some "large instruments," most likely the Great Samrāṭ and the Ṣaṣṭhāṁśa.[2] By 1735, therefore, almost all the important instruments of the Jaipur observatory had been completed. Two years later, in 1737–38, there were again large expenses incurred indicating a major construction project at the site.[3] Thus, the last instrument of the observatory might not have been added until 1738.

Around 1734–1735, when the construction work was at its peak, there were over 23 astronomers and a large assortment of other workers such as masons and engravers employed at daily wages.[4] The work was done under the department of building and construction headed by Vidyādhara, the chief architect of the king.[5]

Old Map of the Observatory

An old map of the observatory drawn in 1728 or earlier shows a number of instruments constructed before this date.[6,7] The map describes a 170-yard-long and 140-yard-wide rectangular area enclosed by a boundary wall of mud with a gate to the north (Fig. 6-1). A 26-yard-wide road leading to the market place and to a government office (*kacaharī*) runs next to its western boundary wall.

---

[1] A French Jesuit, du Bois, writing in 1732, says that Jai Singh was then about to begin work on a large instrument, 108 Roman ft (approx. 35 m) in size. This instrument could be none other than the Great Samrāṭ. Joseph Du Bois, *Op. Cit.*, Ch. 3.

[2] The building records of the Jaipur State for the year 1735 show major expenses at the observatory site. For example, according to one expense report, dated *Pauṣa Śukla* 14, V.S. 1792 (December 28, 1735), a payment of Rs.10,800 was made for lime mortar alone for the work on some instruments in the preceding months. Later, Rāmadāsa Patela (perhaps a contractor) was paid a sum of Rs. 11,335 for various purposes regarding the observatory. R.S.A., Jaipur State, *Tozīs Rojnāmacā Imāratī*, V.S. 1790–1792.

[3] Probably, the Rāśivalayas were built then. Total expenses on the observatory related construction for the financial year of the State in 1794 were about Rs.17,000. R.S.A., Jaipur State, *Tozīs Rojnāmacā Imāratī*, V.S. 1794 (1737-38 A.D.).

[4] *Ibid*. See also Ch. 12.

[5] The expense accounts carry the seal of Vidyādhara. R.S.A., Jaipur State, *Tozīs Rojnāmacā Imāratī*, V.S. 1792 (1735 A.D.).

[6] The Museum, Map No. 15.

[7] The map does not show the astronomer's room, the inscription on which states 1728 as the date of its construction.

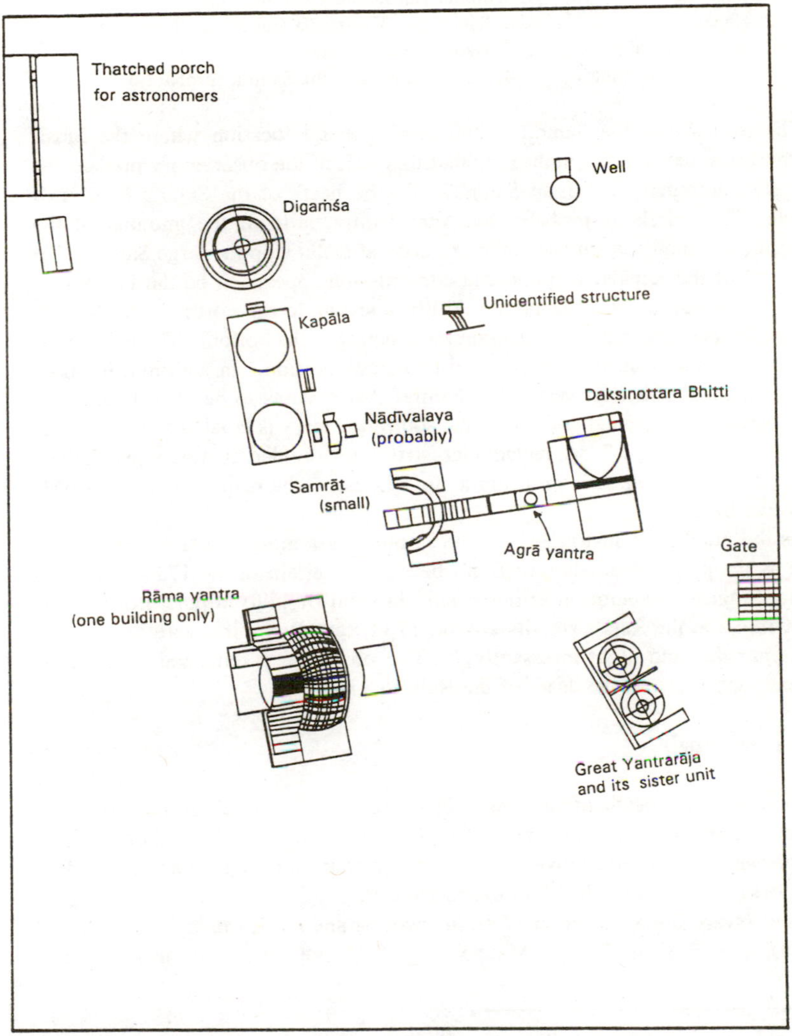

Fig. 6-1 An Old Map of the Jaipur Observatory.
(Courtesy Sawai Man Singh II Museum)

The map shows nine instruments, out of which the following may be identified:

1. Digaṁśa
2. Kapāla
3. Samrāṭ (Small)
4. Dakṣinottara Bhitti
5. Rāma yantra (only one building)
6. The Great Yantrarāja and its sister unit
7. Nāḍīvalaya (probably)
8. Agrā yantra atop the Samrāṭ (probably)

The map shows the Samrāṭ in the same general location where the Small Samrāṭ is situated today, indicating that this unit of the observatory predates its larger counterpart, the Great Samrāṭ. To the north of the Samrāṭ is a small circle. The circle is probably an Agrā yantra built on the gnomon of the Samrāṭ, and thus it is similar to the one seen at Delhi atop the large Samrāṭ. To the east of the Kapāla, a cylindrical construction appears to be the forerunner of a Rāma yantra. A Dakṣinottara Bhitti is shown to the north of the Samrāṭ. The instrument has a double quadrant intersecting at the bottom. The instrument has since been dismantled and moved to a new location. In addition, the map shows a small structure west of the Samrāṭ that remains to be identified.[1]

In the map, a thatched porch near the south boundary is labeled as the "shelter for the astronomers." A rectangular structure, divided in two equal halves adjacent to the shelter, is probably a store house. The map also shows a well and two trees.

As the number of instruments at the Jaipur observatory gradually increased, and as the political situation of Delhi became uncertain in the 1730's, the hub of Jai Singh's astronomical activities shifted from Delhi to Jaipur. Boudier, an eyewitness to the Raja's activity around 1734, says Brahmins were "observing at Jaipur day and night incessantly."[2] The observatory remained a center of intense activity until the death of the Raja in 1743.

After Jai Singh

Soon after the death of Jai Singh, his successor, Īśvarī Singh (1743-1750), found himself engulfed in a war of succession with his half-brother, Madho Singh, and was forced to divert all available resources to raising an army. The support for astronomy therefore evaporated.[3,4]

After Īśvarī Singh's death in 1750, his brother and rival, Madho Singh (1751-1778), ascended the throne. Madho Singh was more interested in astronomy

---

[1] The instrument could be the Śara yantra mentioned in the *Yantraprakāra*. See Ch. 3 and 7. Garrett also mentions a Śara yantra constructed by Jai Singh which was later demolished to make room for a temple and never rebuilt afterwards. Garrett, p. 65. The temple was dedicated to Śrī Ānanda Vihārī.

[2] "Observations . . . " *Lettres*, p. 778.

[3] Jagadish Singh Gahlot, *History of Jaipur*, (in Hindi), pp. 106-107, Jodhpur, 1966.

[4] Andreas Strobl, Letter No. 645, Joseph Stöcklein, *Neuve Weltbott*, Augsburg and Gratz, 1728.

than his predecessor. The Miśra yantra of Delhi was built by him.[1] It is doubtful, however, if the Jaipur observatory ever saw as much activity as it had during Jai Singh's days.

Tieffenthaler

Tieffenthaler visited Jaipur in 1751 when Madho Singh had just ascended the throne.[2] Tieffenthaler has left a brief account of the observatory. Although his account is confusing and inaccurate at times, it nonetheless has some valuable information.

> "Another place that merits more detailed examination is the site where they make astronomic observations. It is a set of works such as has never been seen in these regions and which is astonishing and striking for the novelty and sizes of the instruments.
>
> "This observatory, which is quite large and spacious, adjoins the palace of the King. It is situated on a plain, surrounded by walls, and constructed there specifically to view the stars. As soon as you enter, you notice the 12 signs of the Zodiac, carved in great (arcs of) circles, which are made of the purest lime.[3]
>
> "Then various sections of the astronomic sphere (or of the celestial globe) are presented at elevations corresponding to the altitude of the north star at that place.[4] These sections have diameter of 12 Royal feet or more.[5] Besides these there are large and small equinoctial dials and astrolabes carved into the lime. Additionally, there is a meridian line and a horizontal solar dial carved into a very large rock. "But what really draws one's attention is an *axis mundi* [gnomon] astonishing by virtue of its height of 70 Royal feet and its thickness, and constructed of bricks and lime,[6] situated in the plane of the meridian, with an angle equal to the altitude of the celestial pole.[7] On top of the *axis mundi* there is an observation tower that overlooks the whole town and so tall that one cannot be there without one's head turning.
>
> "The shadow of this *axis* [gnomon] is projected gigantically on a prodigious astronomical quadrant whose tips are turned toward the sky; it is artistically constructed out of the whitest lime or gypsum and

---

[1] Bhavan writes that some of the instruments of the Jaipur observatory were added during the reign of Madho Singh. However, he does not say which ones. See Bhavan, p. 2.

[2] Tieffenthaler, pp. 316-318.

[3] Tieffenthaler is referring to the Rāśivalayas.

[4] Tieffenthaler is again referring to the Rāśivalayas.

[5] Twelve royal feet equal approximately 3.9 m.

[6] Seventy royal feet = 22.7 m.

[7] Tieffenthaler is referring to the Great Samrāṭ of Jaipur.

divided into degrees and minutes. In the morning its [gnomon's] shadow falls on the western quadrant; in the afternoon, on the one [quadrant] facing east. The *axis mundi* passes through the middle of these two quadrants so that at any time one can measure the altitude of the sun in degrees. There is a double dial [Ṣaṣṭhāṣṁśa] near these quadrants. It is enclosed in a kind of chamber, similarly made of plaster, and it [double dial] rises on either side of the room. When the sun passes over the meridian, the rays of this star enter through two holes pierced in a sheet of copper. And when these rays fall exactly on the middle of the quadrants [double dial], they indicate the zenith distance of the sun, [which is] low in winter and higher in summer.

"The instruments that follow are similar to those above: Three astrolabes, cast in copper, hanging on movable iron rings. A circle also cast in copper, equipped with a measuring rod (or alidade) raised according to the height of the pole so as to determine the declination of the sun; whenever you direct this instrument toward the sun, it will indicate to you its declination on the ground.

"I pass over in silence other less important instruments. But one thing that lessens the importance of this observatory is that, situated in a low spot and enclosed within walls, it does not permit observers to see the rising and the setting of the stars; and since the gnomon and other parts of the gnomon are in lime plaster, one cannot draw exact results from the observations."

In his narration, Tieffenthaler clearly mentions the belvedere atop the Great Samrāṭ and the 12 Rāśivalayas that some erroneously allege to have been built after Jai Singh's death.

After Madho Singh

Interest in astronomy declined once again after the death of Sawai Madho Singh, who was succeeded by his son, Pratap (Pratāpa) Singh (1778-1803). The only notable activity during the reign of Pratap Singh was the repair of the Nāḍīvalaya.[1] Garrett writes that a number of instruments such as an Agrā and a Śara yantra, situated in the western part of the observatory compound, were dismantled during this time to make room for a temple and were never rebuilt afterwards.[2]

---

[1] See Ch. 7.

[2] Garrett, p. 65.

Gun Foundry

During Pratap Singh's reign astronomical activity at the observatory ceased, and the observatory itself was turned into a gun factory for casting and boring cannon.[1] A large well with steps down to the water level (*Bāvaḍī*) was dug up and a furnace built immediately to the west of the Great Samrāṭ for melting the gun metal.[2] Between the Jaya Prakāśa and Kapāla, somewhat to the south, another well (*Bāvaḍī*) was dug up and a second furnace erected. Next to the furnace, and to the south of the Kapāla, an elaborate machine for boring gun barrels was constructed.[3] Regarding the gun factory, Lieutenant Boileau visiting the city many years afterwards in 1835 writes:[4]

> "We . . . visited the gun-foundry and observatory, both of which are contained in a very large yard on the N.E. side of the Tripolia (*Tripoliyā*) gate. Though cannon are no longer cast here, the apparatus seems to be in good order, there being a large furnace for melting and running the metal, and very powerful piece of machinery for boring the guns, consisting of an enormous power wheel of wood, well framed and cogged, turned by capstan-bars and communicating its motion to the rest of the machinery by wooden trundles: large compound pulleys are suspended between uprights of strong timber, apparently for the purpose of lowering the gun muzzle downward on to a vertical borer, and steadying it during its descent. The apparatus is contained in a substantial building ornamented with devices of ordnance, and looks as if built by a European."

Boileau's conjecture is correct that a European had been involved in the works of the gun factory. The Rajasthan State archives reveal the name of de Silva, a Portuguese immigrant and amateur astronomer, in charge of the operation.[5] Boileau further notes that instruments of the observatory were suffering from neglect. The plaster, on which the graduations of the instruments had been engraved, had totally peeled off, and from the general appearance it seemed as if the observatory "had never been used since the death of its founder."[6]

---

[1] R.S.A., Jaipur State, records for "*Kārkhānā Topān, Tālukā Jantra.*

[2] The Museum, Map No. 23.

[3] *Ibid.*

[4] A. H. E. Boileau, *Personal Narrative of a Tour through the Western States of Rajwara in 1835*, p. 157. Calcutta, 1835.

[5] The accounts concerning the foundry operation bear de Silva's seal. R.S.A., *Tozīs Kārkhānā Jantra*, V.S. 1849-1851. According to Hunter, de Silva died around 1795, during the reign of Pratap Singh. See Hunter, p. 210. Also see Ch. 13.

[6] A photograph taken sometime before 1900 shows the ruins of a building associated with the gun factory. See Severin Noti, S. J., *Land und Volk Des Königlichen Astronomen Dschaisingh II, Maharadscha von Dschaipur*, p. 71, Berlin, 1911.

The Second Map of the Observatory

A second map of the observatory is from the post-Jai Singh era, after the observatory had been converted into a gun factory, and it shows all the buildings erected during and after Jai Singh's reign.[1] The map, however, does not label some of the instruments by their modern popular names. In the following, the popular names are given in parentheses:

1. Rāma yantra
2. Agrā yantra (Digaṁśa)
3. Kapālas
4. Krāntivṛtta rings (Cakra yantra)
5. Dakṣinottara Nāḍīvalaya yantra (Nāḍīvalaya)
6. Dhruvadarśanārtham Nālikā (Dhruvadarśaka Paṭṭikā)
7. Nāḍīvalaya (Small Samrāṭ yantra)
8. Yāmyottara Bhitti yantra (Dakṣinottara Bhitti)
9. Viṣuvat krāntijñārtham Nāḍīvalaya yantra (Nāḍīvalaya)
10. Jaya Prakāśa
11. Yantrarāja
12. Large Ring of Krāntivṛtta (Unnatāṁśa yantra)[2]
13. Dwādaśa Krāntivṛtta yantra (Twelve Rāśivalayas)
14. Nāḍīvalaya (The Great Samrāṭ)
15. Krānti . . . jñārtham (*sic*), (Incomplete Krāntivṛtta)

In addition, the map shows a number of other features. For example, the map shows a large area to the east of the Great Samrāṭ designated as the mint for the treasury. Then there is a building for a *Jalayantra*, or some water-instrument, where the so-called "ground plan" is situated now. The nature and purpose of the *Jalayantra* is not clear. It very well could have been connected with the mint or with the foundry. The *Ghaḍiyāla Khānā* must have been the place where a time-keeping device was kept. The *Ghaḍiyāla Khānā* could have also kept among other time-keeping devices a sinking bowl type of water clock. Further, the map shows store houses, small rooms, residential quarters for the caretaker, and two wells. Although no traces of the *Jalayantra* and the *Ghaḍiyāla khānā* are left now, the observatory still retains all the instruments except the Nāḍīvalaya to the north of the Kapālas.

---

[1] The Museum, Map No. 23.

[2] This instrument has been incorrectly identified in the map.

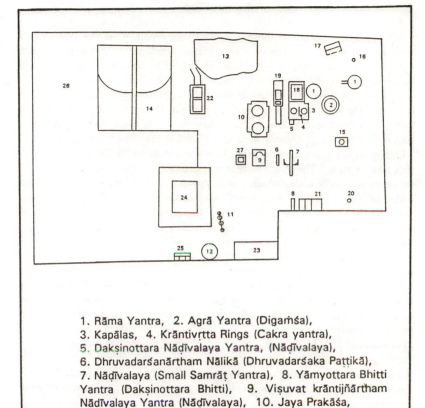

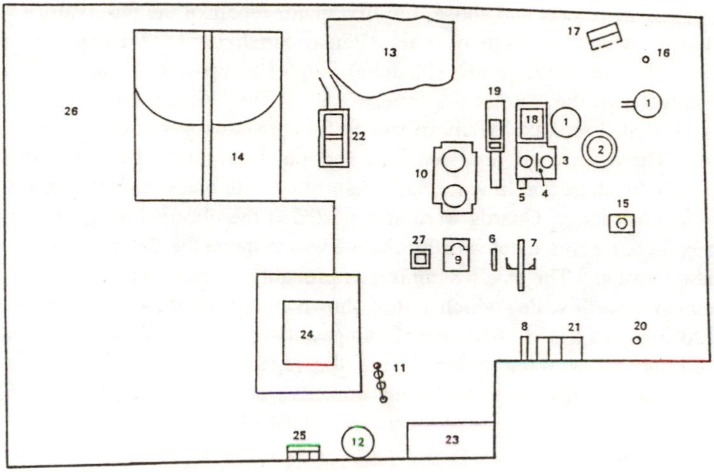

1. Rāma Yantra, 2. Agrā Yantra (Digaṁśa), 3. Kapālas, 4. Krāntivṛtta Rings (Cakra yantra), 5. Dakṣinottara Nāḍīvalaya Yantra, (Nāḍīvalaya), 6. Dhruvadarśanārtham Nālikā (Dhruvadarśaka Paṭṭikā), 7. Nāḍīvalaya (Small Samrāṭ Yantra), 8. Yāmyottara Bhitti Yantra (Dakṣinottara Bhitti), 9. Viṣuvat krāntijñārtham Nāḍīvalaya Yantra (Nāḍīvalaya), 10. Jaya Prakāśa, 11. Yantrarāja, 12. Large Ring of Krāntivṛtta (Unnatāṁśa Yantra),[1] 13. Dwādaśa Krāntivṛtta Yantra (Twelve Rāśivalayas), 14. Nāḍīvalaya (The Great Saṁrāṭ), 15. Krānti . . . jñārtham (*sic*), (Incomplete Krāntivṛtta), 16. Well, 17. Store House, 18. Cannon Boring Machine, 19. *Bāvaḍī*, 20. Well, 21. Store House, 22. *Bāvaḍī* and Large Furnace, 23. Caretaker's Quarters, 24. Building for *Jalayantra*, 25. *Ghaḍiyālakhānā*, 26. Mint for the Treasury, 27. Small Building.

Fig. 6-2 The Second Map of the Jaipur Observatory. (Courtesy, Sawai Man Singh II Museum)

[1] This instrument has been incorrectly identified in the map.

Restorations

After years of neglect and abuse, the first major repair of the observatory was undertaken during the reign of Ram (Rāma) Singh II (1837-1880) and was completed around 1876. Futeh (Phateha) Singh Champawat (Campāvata), the prime minister of the state at the time, writes, "The building had long lain in ruins and wanted repairs. Many of the useful apparatus were out of order and broken. These were all now repaired and put to rights. The lines in the instruments (of stone and lime mortar) were fitted with black lead to ensure that they would last long. Guards were also posted at the observatory gates."[1]

During these repairs, a noteworthy change was to move the Dakṣinottara Bhitti to a new location.[2] The small Samrāṭ was probably surfaced in red stone, and its gnomon's north scale, which is not shown in any of the early maps, was added to it at this time.[3] Although Champawat writes that all the instruments were brought to working order during the repairs, that was not the case. Bhavan comments that only certain instruments were repaired at the time.[4]

From Major Hendley's *Guide to Jeypore*, published in 1876, the year when the repairs were in progress, it appears that there had been no new additions to the observatory since Pratap Singh's reign.[5] Hendley notes little else that is new, other than that the Dakṣinottara Bhitti was being moved to a new location.[6] He identifies the Samrāṭ yantras by the name "Nāḍīvalayas." Nāḍīvalaya is the name reserved now for another instrument which has two circular plates parallel to the plane of the equator. The Samrāṭ yantra indeed was known as "Nāḍīvalaya" at one time or another.[7] According to Hendley, the gun foundry buildings within the observatory compound still existed, although they were in disrepair and were no longer in operation. About two decades later, in 1891, the Rāma Yantra, which had originally been constructed in plaster, was restored in stone.[8]

Restorations of 1901-02

A major restoration of the observatory took place in 1901-1902, under the supervision of Lieutenant A. ff. Garrett.

---

[1] Futeh Singh Champawat, *A Brief History of Jeypore*, Agra, 1899.

[2] See the inscription on the Dakṣinottara Bhitti plaque, Ch. 7.

[3] Garrett erroneously states that the Small Samrāṭ was built in 1876. Garrett, p. 43.

[4] Bhavan, p. 2. Neither Bhavan nor Champawat list the instruments repaired at the time.

[5] Hendley, *Op. Cit.*, Ch. 3.

[6] The names of the instruments and their functions given by him in his book are misleading and often inaccurate.

[7] Jagannātha Samrāṭ in *Samrāṭ Siddhānta*, SSMC, p. 8; also SSRS, p. 1039.

[8] Garrett, p. 38.

Garrett writes:

> "In the beginning of 1901, . . . Maharaja Sawai Madho (Mādho) Singh II of Jaipur, . . ., decided to completely restore the Jaipur Observatory, many of the instruments of which were falling into decay. The necessary funds were allotted, and the work was placed under the supervision of the State Engineer, S.S. Jacob, who handed over the immediate charge of the work to me in May. The work was carried on during the remainder of that year, and finally completed in February 1902, entirely with Jaipur labour, materials and workmanship."[1]

Technical assistance to the project was provided by Bhavan and Chandra Dhar Guleri (Candra Dhara Guleri̅).[2] The repairs of 1901-1902 were the most extensive thus far. The instruments were completely restored, and their scales were redrawn. In certain cases, the restorers even went a step further and altered the instruments.[3] Because the instrument surfaces were done in lime plaster during these repairs, they soon began to erode. As a result in 1945 the scales of the Great Samrāṭ were surfaced with marble.[4] Other instruments, such as the Ṣaṣṭhāṁśa and the Kapāla, were similarly resurfaced at a later date.

The Observatory Today

The Jaipur observatory now occupies a 200 m long and 100 m wide compound of approximately 20,000 m^2 area surrounded on three sides by buildings or high walls. On the south of the compound is an 8 to 10 m high boundary wall. To the west, there is an almost four-story-high building with a temple and residential quarters. To the north, there is a small building adjacent to the boundary, and between this boundary and the City Palace, there is a fairly busy street. To the east, there is a 5 m high boundary wall and a 7 to 7½ m high building next to the wall. One enters the observatory from the north through a small door. As a result of the high buildings on three sides, the observatory is not suited for observing low lying objects.[5]

The observatory is now under the administrative supervision of the Department of Archaeology and Museums of the State of Rajasthan. A supervisor and caretaker staff looks after its upkeep and its instruments. The general appearance of the observatory and its instruments is good.

---

[1] Garrett, p. i.

[2] Bhavan, p. 3.

[3] See Rāśivalaya yantras, Ch. 7.

[4] M. F. Soonawala, *Maharaja Sawai Jai Singh II of Jaipur and His Observatories*, p. 29, Jaipur, 1952.

[5] See Tieffenthaler, pp. 316-318.

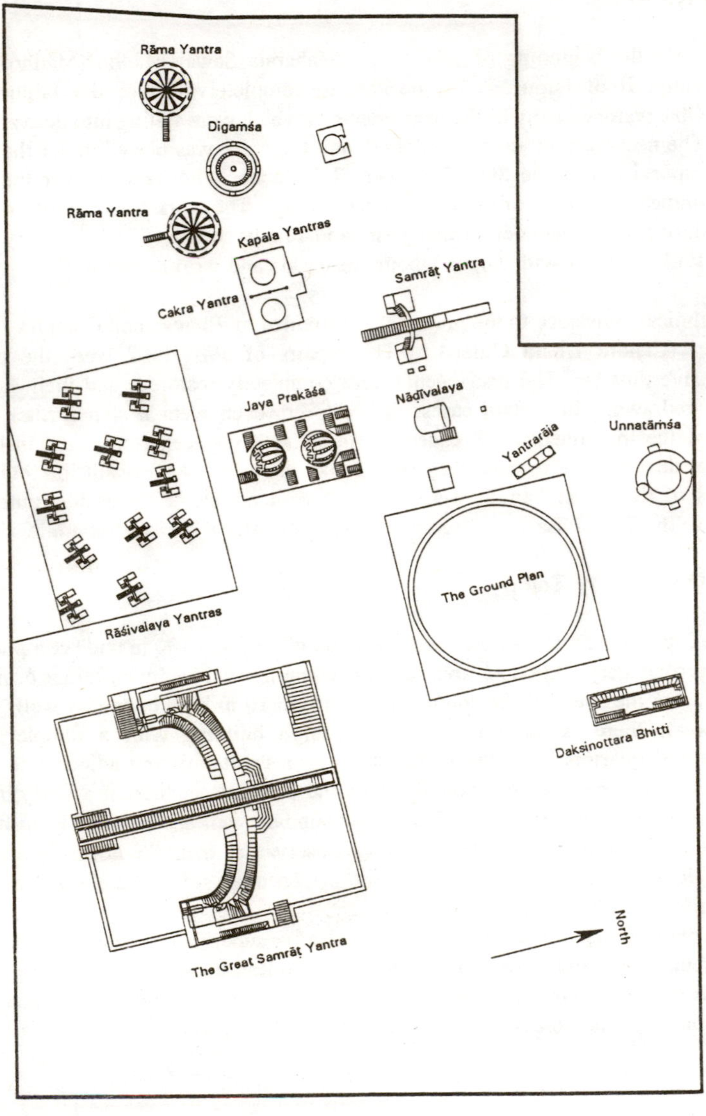

Fig. 6-3 The General Plan of the Jaipur Observatory.[1]

---

[1] based on Kaye, plate XXI.

The instruments are in a good state of repair, and the lawns and the grounds surrounding them are well-kept with flower beds and shrubs. The observatory has the following masonry instruments:

1. Krāntivṛtta
2. Digaṁśa
3. Rāma yantra (2)
4. Kapāla (2)
5. Jaya Prakāśa
6. Nāḍīvalaya
7. Small Samrāṭ
8. Horizontal sundial atop the Naḍīvalaya
9. Dakṣinottara Bhitti (2)
10. Great Samrāṭ
11. Ṣaṣṭhāṁśa (4)
12. Rāśivalaya (12)
13. Horizontal circular dial
14. Dhruvadarśaka Paṭṭika
15. Rāma yantra models
16. Krāntivṛtta II

Metallic Instruments

The instruments of metal at the observatory site on display are as follows:

1. Cakra yantra (2)
2. Yantrarāja (brass)
3. Incomplete Yantrarāja (iron)
4. Unnatāṁśa Yantra
5. Krāntivṛtta
6. Samrāṭ models (2)

In addition there are a number of portable instruments such as astrolabes in storage at the site.

DESCRIPTION OF THE INSTRUMENTS

The Great Samrāṭ Yantra

The Great Samrāṭ is the most imposing structure of the Jaipur observatory. It is the largest sundial in the world. The instrument rises out of a 44.15 m long and 40.03 m wide excavation, 3.5 m below ground level, that is lined with masonry plaster. The instrument is built partly below ground level so that the edges of the quadrants reading six o'clock may not be inordinately above the local horizon. One may still see a number of orthogonal channels, 6.5 cm wide and 5 cm deep, running north-south along the gnomon wall, and east-west a few meters to the south of the quadrants. The channels were originally constructed for leveling the floor. A graceful belvedere atop the gnomon adorns the structure. The instrument is in excellent condition because it has been repaired from time to time. The gnomon edges of the instrument were originally engraved in red stone and the quadrant surfaces in plaster. In 1901, Garrett and his coworkers also finished them as such.[1] Around 1945, however, the

---

[1] Garrett, p. 40.

Fig. 6-4 The Jaipur Observatory.

Fig. 6-5 The Great Samrāṭ of Jaipur

quadrant surfaces and the gnomon edges both were redone in marble, as we find them today.[1]

The Gnomon

The gnomon of the instrument consists of a 2.85 m wide and 22.62 m high triangular wall with a 44.58 m base, and a 50.09 m long hypotenuse.[2] The wall is erected on a north-south axis, in such a way that it divides the excavated enclosure in two equal halves. A number of graceful arches in the wall have been constructed to economize on masonry and to let the wind pass through. At the south end of the gnomon there is a small gate opening to a flight of steps running in the middle of the gnomon and leading to its very top to a small room, 2 m × 1¼ m, on top of which the belvedere stands. There is a second set of steps also built next to the gnomon edges for an observer to sit upon while taking a reading. Garrett checked the angle of the gnomon and found it to be correct within 1′ of the true latitude of the place.[3] He did not check the north-south orientation of the gnomon wall.

Division Scheme

The gnomon edges are engraved to read the declination angles from 61;45 N to 58;15 S. However, because of the belvedere near the top, the section above 58;30 N cannot used. Near the middle section, right above the base of the quadrants, a pair of 11 deg wide secondary scales, one on each side, has also been provided to account for the finite width of the quadrants. The declination angles to the north and to the south of the equator are read by different scales of the quadrants.[4] As the quadrants are 2.84 m apart, the declination scales at the gnomon overlap by exactly the same distance. The secondary scales over the gnomon surface have been provided to take care of this overlapping.

The gnomon is divided into degrees and minutes and labeled in Devanagari. Each degree is divided into 4 major parts, which are further divided into 5 intermediate parts. The intermediate divisions are again subdivided into three parts each. Hence, a small division of the scale reads 1′ of arc, the very limit of naked eye observations. These small divisions are 4 mm wide near the zero mark, and progressively increase to 12 mm where the scale reads 53 deg. Near the zero markings, the divisions may vary by about ½ mm from their true values. Garrett states that, prior to the restoration of 1901-1902, there were two scales on the gnomon superimposed one on top of the other. He writes: "It appeared as if some errors were made and subsequently corrected without

---

[1] Soonawala, pp. 29-30, *Op. Cit.*

[2] The height is measured from ground level and not from the excavated floor.

[3] Garrett, p. 42.

[4] See Ch. 3.

removing the graduations first engraved. The result is very confusing, so new scales have now been engraved inside the old ones. Some of the old graduations are absolutely correct, others are in error by amounts varying up to a maximum of 3/4 inch, corresponding to an error of declination 3 arc-minute."[1] The instrument graduations Garrett talks about could have been inscribed at a later date during the reign of Madho Singh, the second son of Jai Singh. A number of instruments were repaired in 1873-76, during the reign of Ram Singh II, but it is not known if the Great Samrāṭ was one of them.[2] The present graduations were inscribed around 1945.

The edges of the gnomon should be straight lines. Any deviation from the linearity requirement may lead to an erroneous reading of time or angle.[3] We checked the linearity of the gnomon with a taut string (with its obvious limitations) and noted that both edges of the gnomon vary at places about ±3 mm perpendicularly and up to ±3 mm laterally.

A larger error is expected from the fact that the zero markings of the gnomon do not line up with the six o'clock markings on the quadrants on either side of the gnomon.[4] A taut string held between the two six o'clock markings revealed that the zero markings of the two inner scales are inscribed to the south of the line. The inner scale of the eastern edge of the gnomon is shifted 5 cm to the south, and the scale near the western edge is shifted 3 cm, also to the south. The outer scale, that is, the main scale zero-markings, were found to the south by 5 cm for the eastern edge and 3.5 cm for the western. It is unclear whether this difference is due to an improper marking of the gnomon or due to the quadrants' deviation from the plane of the equator.[5] If improper markings are causing this difference, then the declination readings for the objects near the celestial equator could be off by as much as 10′ to 12′. For large angles of declination the error would not be that great, however. When using the taut string procedure, the author noted that the string grazed the gnomon surface, as required by the theory of the instrument.[6,7]

The Quadrants

The quadrants on either side of the gnomon have a radius of 15.15 ±0.01 m. The quadrants are 2.84 m wide and are flanked by stairs on both sides of their scales. The combined width of the wall with the stairs and the quadrants is

---

[1] Garrett, p. 42.

[2] See Champawat, *Op. Cit.*

[3] See Ch. 3.

[4] *Ibid.*

[5] *Ibid.*

[6] *Ibid.*

[7] With the string, it could not be ascertained if the gnomon surface was above the line joining the two six o'clock marks.

about 5.23 m. A number of archways connect the courtyard on both sides of the quadrants. The quadrants are surfaced with 10-11 cm thick white marble slabs on which the scales have been engraved. The scales have been engraved parallel to one another near the two edges and are 2.76 m apart.

Division Scheme

A quadrant of the Great Samrāṭ of Jaipur reads time according to the modern time-reckoning scheme. Its divisions are hours, one-half hours, minutes and seconds, and not *ghaṭikās* and *palas*. The original markings on the quadrants must have been in *ghaṭikās* and *palas*. During Jai Singh's era, time was measured in *ghaṭikās* and *palas*. The use of hours and minutes had not come into vogue yet. Apparently the scales were changed during the repairs of 1901-1902 or earlier. Each minute-division of the scale has been further divided into 10 intermediate parts, which in turn have been subdivided into 3 small divisions each. The result of this division scheme is that the least count of the scale is 2 sec. The small divisions, on the average, are 2.2 mm wide. However, divisions with widths as little as 2 mm and as large as 2.5 mm may also be noted. In contrast to the Samrāṭs of Delhi and Varanasi, the Great Samrāṭ of Jaipur does not have scale markings on the narrow sides or edges of the quadrants. The labeling is done in Devanagari.

How the Instrument was Erected

The erection of a large instrument such as the Great Samrāṭ of Jaipur, requiring tolerances in millimeters, is undoubtedly an impressive feat. The question arises how an instrument of such precision was erected. In this regard Garrett's remarks are worth examining.[1] He writes:

> "Although there is no detailed record of this, yet the general method is apparent from a close inspection of the building and from passages in the *Yantrādhyāya*.[2] First of all then, the site was carefully levelled, the levels being tested by constructing narrow masonry channels and filling them with water. On the level platform thus prepared, a north and south line was laid down by means of observations of the shadow of a vertical rod cast by the sun. The next step was to determine the latitude, if not already known, and this was doubtless found by the method described . . . under the Dakṣinottara Bhitti Yantra.[3] The gnomon could now be erected accurately in the north-south line, and

---

[1] Garrett, pp. 41-42.

[2] The *Yantrādhyāya* referred to is a chapter in the *Samrāṭ Siddhānta* of Jagannātha Samrāṭ.

[3] See Ch. 3.

with its hypotenuse inclined at an angle equal to that of the latitude or 26;56. This angle was probably set out by means of its cotangents. Thus, supposing the height of the gnomon to be 90 feet, then the base must be 90 cot-latitude. A truly level base was obtained from the surface of the water in the masonry channels, and true perpendicular by means of a plumb line. Thus the accurate orientation of the gnomon would seem to have presented little practical difficulty. But it was otherwise with the quadrants. The planes . . ., in which these quadrants lie, are at right angles to the hypotenuse of the gnomon; in other words, the gnomon points towards the pole, and the quadrants lie in the plane of the equator. Jai Singh, therefore, marked out the plane of the equator first, and then built his quadrants touching this plane. On either side of the gnomon and parallel to it at a distance of about 50 feet, lofty walls were built. From the points . . ., on the hypotenuse of the gnomon representing the centers of the upper edges of the quadrants, lines . . . were drawn on both sides of the gnomon at right angles to the hypotenuse. Similar lines were drawn on the inner faces of the parallel walls, and from points on these lines, horizontal threads or wires were stretched and attached to corresponding points on the lines on the sides of the gnomon. It is evident that these threads all lay in the plane of the equator, and the iron nails to which they were attached may still be seen on the walls of the Samrāṭ.

"A thread 49 feet 10 inches long being now attached at one end to a center [point D in Fig. 3-3] of a quadrant, and being revolved about this point touching the series of horizontal wires, the other end [of the thread] traced out the upper edge of the quadrant arc, and it was then a comparatively simple matter to build the masonry up to this. The result was eminently satisfactory."

Although Garrett's remarks display considerable insight, he is incorrect in stating that the work on the quadrants began only when the gnomon had been completed. An old "observatory progress report" suggests that the work on the Samrāṭ's quadrants began when its gnomon had been half completed. The report indicates the progress of the work with a drawing, in which the gnomon is shown partly constructed. The report states, "when the construction (on the gnomon wall) reaches the center point, the work on the quadrants would commence."[1] Apparently, the builders did not wait for the gnomon to be completed before starting their work on the quadrants.

[1] The Museum, Map No. 24.

Measurements

The function of a Samrāṭ is primarily three-fold as discussed earlier: one, to tell time during day or night; two, to measure the hour angle; and three, to determine the declination of a celestial object. The time during the daylight hours is measured with the sun and at night with the stars. The main difficulty in the day is that of the penumbra. The difficulty may be eliminated by using a short of piece of string as explained in Ch. 3.[1,2] The difficulty of the penumbra becomes compounded, however, when the gnomon-shadow falls on the grey areas of the marble surface of the quadrants of the Great Samrāṭ. The procedure for measuring the hour angle is similar to the measuring of time at night as explained in Ch. 3.

Declination

On 25 December 1981, the author measured the declination of the sun around 1:37 P.M. by directly looking at it through a suitable filter.[3] He found it to be 23;23 ± 1′, which after a refraction correction of ± 1′ , became 23;24 ± 1′. The computed value for the declination for the day was 23;24. The author repeated his readings the same day, two hours later, around 3:31, and found that the declination readings after refraction correction came out to be 23;22 ± 1′. When the author revisited the observatory on 1 January 1990, he took another set of readings of the sun's declination. The declination measured between 1:30 and 1:37 P.M. came out to be 23;2,30 ±75 arc-sec. The expected value for the day was 23;3,41. Hence, at least certain sections of the Great Samrāṭ's declination scale still function with the intended accuracy of ± 1′ of arc.

---

[1] By the time Garrett restored the instrument, there was no one to show him how to eliminate the penumbra problem. In frustration he writes, ". . . the quadrants are graduated to read seconds of time. But unfortunately it is hopeless to try and observe to seconds, owing to the shadow being so ill-defined that it is difficult to fix its edge with any exactitude. . . . Still, a skilled observer would probably obtain his apparent time correct to 15 seconds, and possibly even nearer." See Garrett, p. 40.

[2] Unable to realize that the penumbra problem can be easily eliminated, Blanpied has erroneously concluded that the instrument could not be read with a precision better than ±12 seconds. He has also made an error in assuming that the finest of the divisions correspond to 4 seconds of time. Blanpied (1975), pp. 1025-1035, *Op. Cit.*, Ch. 3.

[3] For a filter, the author used a strip of silver-based, exposed black-and-white photographic film. One should be careful in selecting a filter. Lightly smoked glass and ordinary black-and-white exposed films do not sufficiently filter out the infrared radiation of the sun and are thus dangerous to the eye.

Declination at Night

The measurement of declination at night requires two observers and a torch bearer. Following the procedure elaborated in Ch. 3, the author measured the declination of a number of winter stars in January 1982. We give some of these readings in Table 6-1.

Star	Measured declination	Declination from Ephemeris
Sirius	−16;44 ± 0;3	−16;41
Rigel	−8;19 ± 0;1	−8;13
Capella	45;57 ± 0;1	45;59

Table 6-1 Declination of three winter stars. The measured values represent an average of five readings taken on the evening of January 22, 1982. The readings are without refraction correction, which is within the observational error.

The above readings should be taken as an illustration of a technique only and not a reflection on the capability of the instrument. With experience, a team of observers may obtain readings closer to the accepted results than those obtained by the author and his assistant.

A Morning and an Afternoon at the Samrāṭ

During the winter months of 1981-82, on several consecutive days, the author took time-readings with the Great Samrāṭ in the morning and afternoon hours. He took these readings at random intervals, as little as 2 to 3 minutes apart. The readings were compared with a digital watch. On a typical morning, such as on January 22, no shadow of the gnomon edge could be seen on the west quadrant before 7:36 Indian Standard Time. A tree to the east of the Samrāṭ had been blocking the rays of the sun. Even without the obstruction of the tree, it would not have been possible to take any readings because the sun was too weak at the time to cast any appreciable shadow. At 7:40 A.M. some sort of shadow gradually did begin to appear. However, on the white marble surface, the shadow of the gnomon edge cast by the weak orange-yellow rays of the morning sun was not distinct. Because of this, one could not take readings by the "method of string."[1] By looking directly at the weak sun, and with the north edge of the quadrant as the vantage point, however, a reading could be

[1] See Ch. 3.

possible.[1] Slowly, the sun rose higher, and a very diffused shadow of the gnomon began to be noticeable on the quadrant surface. With the diffused shadow, the author could not read the quadrant scale with an accuracy better than ±30 sec until 7:46 A.M. It was only around 7:58 A.M., or almost at 8 A.M., that the "method of string" could be used allowing the accuracies of ±3 sec or better. Up until the noon hour on the Samrāṭ, the difference between the digital watch and the time readings on a Samrāṭ scale was consistently within ±3 seconds. A few seconds before the noon-hour, as indicated by the instrument, the shadow disappeared from the west quadrant but did not appear on the east quadrant on the other side of the gnomon wall until a minute and a half later. The cause of this delay in the appearance of the shadow is inherent in the observable penumbra and has already been explained earlier.[2]

The time readings on a typical winter afternoon, such as January 21, 1982, were similar to those in the morning hours of other days. Also the difference between the time readings and the digital watch also was consistently within 3 seconds. After 4:45 P.M., on January 21, as the sun again began to get weak, and as the ground-level haze developed, the readings began to be less reliable. The shadow of the gnomon edge became somewhat diffused at that time, and it became extremely difficult to determine its edge with any degree of certainty. Besides, at this time the penumbra was estimated to be 6-7 cm wide. A few minutes before 5 P.M., the readings had to be abandoned altogether due to the shadow of a tree and then that of a building falling on the instrument.

Hence, the Great Samrāṭ may only be used after 8 in the morning and before 5 or so in the afternoon, or for only about 9 hours out of 11 hours of daylight of a January day.

Accuracy of Readings

The accuracy of time readings with the instrument, as the instrument exists today, depends on the day of the year, time of day, and the scale used at the time. The south scale of the west quadrant, read in the morning hours, is somewhat more accurate. Its readings are 30-35 seconds ahead of the true local time. The readings with the north scale of the same quadrant, on the other hand, could be ahead by as much as 40-50 seconds of the true local time. Thus the readings of the north scale of the west quadrant are always higher than its south counterpart by about 10 seconds. The "Samrāṭ clock" thus runs faster during the morning hours on both of its morning scales, and the noon arrives about 35 seconds earlier according to the West quadrant.

---

[1] Because of the danger involved to the eye in directly looking at the sun, this method is not advised. Observers should use a filter that eliminates most of the harmful rays striking the eye and yet allows the sun to be seen.

[2] See Ch. 3.

The "Samrāṭ clock" runs slower in the afternoon. The south edge of the east quadrant, used during the afternoon hours, tells its time 40-50 seconds behind the true local time. True noon arrives at this scale 40-50 seconds later. Part of this discrepancy can be explained, as elaborated earlier, due to the visible penumbra. The string used in these measurements did not indicate the center of the sun but indicated a point that is away from the center by an angle of 9′.[1] And the Sun takes about 35 seconds to cover an angle of 9′.

The late arrival of the noon-hour on the east quadrant is, in part, due to another factor—the width of the gnomon wall near the quadrants. We noticed that the gnomon wall is not uniform. The gnomon wall near its bottom, where the quadrants join it and where the two 12 o'clock markings are, is 2.85 m wide. On the other hand, the edges of the scales over the gnomon, near 20^0 S, are 2.865 m apart. They thus differ by 1.5 cm.[2] This explains an additional discrepancy of 6−7 seconds.

The Small Samrāṭ Yantra

The small Samrāṭ yantra is located in the north-west section of the observatory, near the north boundary railing. According to Garrett, and also according to Soonawala, the instrument was built in 1876.[3] However, the instrument is depicted on a map drawn around 1728.[4] It must have been, therefore, built much earlier. In 1876, the instrument was restored and, most likely, the "north gnomon" was added to it. The author believes that at that time it was also surfaced with red and white stones that we see there today.

The instrument differs considerably in appearance from the other Samrāṭs of Jai Singh. Its unique shape may be viewed as two gnomon walls of uneven height joined together back-to-back. With this design the instrument acquires the capability of measuring—in theory at least—the declination angles of objects in the sky above the northern horizon. None of the other Samrāṭs of Jai Singh have this capability. In actual practice, however, the buildings surrounding the instrument, in particular the City Palace to the north, limit the observing close to the horizon.

The triangular meridian wall of the Samrāṭ is 1.29 m wide, 6.9 m tall and 19.75 m long at its base. The south-facing gnomon of the instrument is 14.725 ± 0.005 m long and is inclined at an angle of approximately 27;30 with the horizontal. Its counterpart, the north-facing gnomon, is somewhat smaller, only

---

[1] See Ch. 3 for explanation.

[2] In December-January, it is the gnomon edges near the 20 deg S that cast their shadow on the quadrants below.

[3] Garrett, p. 43 and Soonawala, p. 30, *Op. Cit.*

[4] The Museum, Map no. 15.

Fig. 6-6 The Small Samrāṭ Yantra.

Fig. 6-7 The Twelve Rāśivalayas.

8.58±0.02 m long, and makes an angle of about 41 degrees with the horizontal.[1] Because the two gnomons differ in height, a 1.22 m long vertical section of the wall separates their top ends. Archways, three in all, have been cut into the wall for easy passage from one side to the other. A flight of stairs runs in the middle of the gnomon diagonals.

Conforming to the usual design of a Samrāṭ, the gnomon wall of the small Samrāṭ is flanked on either side by quadrants, which in turn are supported by masonry platforms. The radius of the quadrant on the east flank is 2.780 m everywhere, however, the radius of the west quadrant varies between 2.780 m and 2.785 m. The quadrants are 98 cm wide and have their lower ends about 71 cm above the floor. The masonry platforms supporting the quadrants are 10.74 m long and 4 m wide. Steps have been provided on the platforms for reaching the upper sections of the quadrants for a reading.

Division Scheme

The gnomon edges of the Samrāṭ are of white marble on which declination scales have been engraved. The declination graduations start out at about 2 m from the lower end of the south gnomon, where the marking indicates 63 deg south and continue on to the north gnomon. On the north gnomon the markings have been engraved all the way down to a distance of 3 m from the bottom end, where it indicates 70 deg north. The upper end of the north gnomon terminates at 76 deg N. The 90 deg mark lies about 3/4 of the way up from the bottom. The upper end of the south gnomon terminates at 71;44 N. The vertical section of the wall joining the two ends thus has division markings between 71;44 N and 76 N. All these scales are divided into degrees and minutes. Every degree division of these scales has been divided into 6 parts, and each of these parts has been further subdivided into 5 subparts. The least count of the scale, therefore, is 2′ of arc. The degree divisions near the zero mark are about 2 mm wide and those near the 70 deg mark on the south gnomon are 14 mm wide. The divisions have been filled with lead and thus provide agreeable contrast with the marble base. The script used in marking the divisions is Devanagari.

Quadrants

The quadrants of the Samrāṭ have two scales along their edges. The division scheme is in hours, minutes and seconds. The hours have been divided into 60 divisions of one minute each, which, in turn, have been subdivided into three parts. The smallest division of the quadrant scales thus measures 20 seconds. The length of these divisions varies between 3.5 and 5.0 mm, with an average

---

[1] The significance of 41^{0} angle is not clear. In fact any angle of inclination would be equally satisfactory.

length being little over 4 mm. The quadrant markings have not been inlaid with lead as the gnomon divisions are. The script used in marking once again is Devanagari. The small Samrāṭ does not have the Hindu time-reckoning scale of *ghaṭikās* and *palas*.

Defects of Construction

The small Samrāṭ suffers from a number of defects of construction. The two edges of its south gnomon deviate from the linearity requirement. The western edge of this gnomon deviates up to 10 mm laterally and 5 mm perpendicularly. The eastern edge similarly deviates up to 5 mm in both directions. Further, the zero markings on this gnomon's scale do not fall on the line joining the two six o'clock graduations of the quadrants on either side. The zero markings are shifted by 1.1 cm to the north from the line joining the six o'clock graduations at the quadrants. In addition, the plane of the eastern scale of the gnomon lies about 2 mm below the line.

The north gnomon is no better than its counterpart on the south and suffers from similar defects of construction. Its eastern edge deviates 10 mm laterally and 4 mm perpendicularly from the linearity requirement. The western edge is slightly better, however. It deviates no more than 3 mm in either direction.

As a result of these errors of construction, there should be an error of nearly 6 deg in the measurement of declination for the objects near the celestial equator. The time measurement error due to the nonlinearity would be of the order of one minute. On a typical winter afternoon, the penumbra of the shadow is about 1 cm wide. The uncertainty due to the penumbra may be eliminated by superimposing the shadow of a string held a centimeter above the shadow edge of the gnomon.

Just as at any other Samrāṭ yantra of Jai Singh, the noon hour at the small Samrāṭ also does not arrive simultaneously on the two quadrants on either side of the gnomon. The shadow of the gnomon, as it turns out, approaches about a minute later on the east quadrant, as the author checked it on a December day. This is not due to a defect of construction but is inherent in the "visible penumbra" as discussed earlier.

Rāśivalayas

The Rāśivalayas are located to the west of the Great Samrāṭ, adjacent to the south boundary wall of the observatory, on a 41 m long and 39 m wide masonry floor. Although Jagannātha does not describe the Rāśivalayas in his *Samrāṭ Siddhānta*, the instruments were indeed built during the reign of Jai Singh.

Their prototypes are described in the *Yantraprakāra*,[1] and they are also noted by Tieffenthaler.[2] The Rāśivalayas are a set of 12 instruments, based on the principle of the Samrāṭ yantra, and are designed for directly measuring the latitude and longitude of a celestial object. They are, in fact, the immobile masonry version of the torquetum, which is a cumbersome device to work with.

Each Rāśivalaya unit refers to a particular sign of the zodiac, and a small painting pertaining to its lore adores its gnomon. A Rāśivalaya measures the longitude and latitude of a celestial object at the time when the first point of the sign approaches the meridian. The moment the first point of a particular sign approaches the meridian, the gnomon of that sign points towards the pole of the ecliptic, and the plane of its quadrants becomes parallel to the plane of the ecliptic. As the 12 signs approach the meridian in succession, the set permits 12 different readings of the object in a period of one sidereal day.[3] By this method, one eliminates the need for transforming coordinates from the horizon or the equator system into the longitude-latitude system or ecliptic system.

The Rāśivalayas occupy floor areas anywhere from 5.75×4.03 m^2 to 7.50×5.60 m^2. The lengths of their gnomons vary from 4.22 m for Gemini to 6.21 m for Aries. The radius of the quadrants of the Rāśivalayas is either 1.24 m or 1.68 m. The orientation and the inclination of their gnomons differ from one unit to the other. This is because the pole of the ecliptic is not a fixed point in the sky, but a point that revolves around an hour-circle 23½ degrees from the north celestial pole, approximately once in 24 hours. As a result, the inclination and the orientation of the gnomons, which point toward the pole when readings are to be taken, also change accordingly.

The surface of the gnomons and of the quadrants both, where the scales have been inscribed, are covered with white marble slabs, which are turning yellow with age. The gnomon scales are divided into degrees and minutes and may read latitude up to 60 degrees on either side of the ecliptic. The degree divisions on these scales are divided into six parts, such that the least count of the scales is 10′ of arc. The small division on the gnomon for the sign of Gemini, near its 60-degree mark, are about 14 mm wide and the divisions near the zero mark are 3.5 mm. On the other hand, the division length for the sign of Aries, near the 60-degree mark, is 19 mm. The script of the markings is Devanagari.

The scales on the quadrants, as explained earlier, read directly the longitude of a celestial object. The scales on the two quadrants of a Rāśivalaya unit put together cover a 180-degree wide section of the ecliptic and, consequently, they overlap considerably from one unit to the other. However, no two quadrant scales start out with the same longitude markings. The scales are divided into

---

[1] *Yantraprakāra* (J), p. 16; and *Yantraprakāra* (S), pp. 24, 70.

[2] Tieffenthaler, p. 316.

[3] The earth rotates 360 degrees once in 23.934469 hours, the length of a sidereal day.

degrees and minutes, with their small divisions reading 10′ of arc each. In addition to a scale on the upper surface, a quadrant also has a scale on the sides. This scale has divisions according to the sexagesimal scheme. Here each quadrant is divided into 15 major divisions which are further partitioned into 6 parts. The parts are again divided into 6 small divisions. Thus, once again, the least count of the scale is 10′ of arc. The small divisions on the Gemini quadrants are almost 4 mm wide, whereas the same divisions on the Aquarius quadrants are 5 mm.

Sign	Upper End of east quadrant	Upper End of west quadrant
Capricornus	180;00	360;00
Aquarius	222;50	42;50
Pisces	255;30	75;40
Aries	281;25	101;30
Taurus	305;30	125;30
Gemini	331;30	151;25
Cancer	360;00	180;00
Leo	29;30	208;30
Virgo	54;20	234;20
Libra	78;30	258;35
Scorpio	104;30	284;30
Sagittarius	137;10	317;10

Table 6-2 The scales on the quadrants of the Rāśivalayas. The markings on the scales increase from west to east.

In order to observe with a Rāśivalaya unit, one has to know the precise moment when the first point of the sign represented by the unit would transit the meridian. In fact, all one has to know is the exact time of transit of one of the 12 signs—say the first point of Capricornus—only. Once the time for one sign is known, the time for the rest of the signs may be calculated from Table 6-3. The astronomers of Jai Singh had similar tables for their work.[1] The time for

[1] *Yantraprakāra* (J), pp. 24-25, and *Yantraprakāra* (S), pp. 85-86.

the first point of Aries or the vernal equinox in particular may also be obtained from an ephemeris.[1]

Sign	Longitude	Time elapsed since equinox transit			Time elapsed since the last sign		
		hr.	min.	sec.	hr.	min.	sec.
Aries	00	00	00	00	00	00	00
Taurus	30	1	51	20	1	51	20
Gemini	60	3	53	2	1	59	18
Cancer	90	5	59	1	2	8	23
Leo	120	8	7	24	2	8	23
Virgo	150	10	6	42	1	59	18
Libra	180	11	58	2	1	51	20
Scorpio	210	13	49	22	1	51	20
Sagittarius	240	15	48	40	1	59	18
Capricornus	270	17	57	3	2	8	23
Aquarius	300	20	5	25	2	8	23
Pisces	330	22	4	44	1	59	18
Aries	360	23	56	4	1	51	20

Table 6-3 The timings for the Rāśivalayas instruments. The timings have been calculated to the nearest second from the moment the first point of Aries or the vernal equinox is on the meridian.[2] After knowing the precise time of observation for any one of twelve Rāśivalayas, the timing for any other may be calculated from this table.[3]

---

[1] The procedure is as follows. First, find the universal time of the equinox transit at the Greenwich meridian from an ephemeris. Then subtracting 5 h, 3 min, 17 sec, determine the local time of the equinox transit at the Jaipur meridian. The earth takes 5 h, 3 min, 17 sec to rotate 75;49,18.7, the angular difference between the longitudes of Greenwich and Jaipur. Now if Indian Standard Time is to be used, then add 26 min, 43 sec to the local time.

[2] Although the earth rotates at a fixed rate, the signs of the zodiac, or the 30-degree sectors of the ecliptic do not transit the meridian at the same rate.

[3] Similar tables have been given by Prahlad Singh and also by Bhavan. Unfortunately, both of these tables have been prepared by taking the earth's sidereal day of 24 hours and are only approximately true. See Singh, p. 145, *Op. Cit.*, Ch. 3; and Bhavan, p. 30.

Although the above table gives the time of observing to the nearest second, in actual practice one does not have to be that precise. The Rāśivalayas are designed to have a least count of 10′ of arc on both their quadrant scales and on their gnomon scales. Now it takes a minimum of 37 seconds for the longitude of an object at the meridian to change by an angle of 10′ of arc. Thus, if a reading is taken at a quadrant within 1/2 minute of the required timing, no appreciable error would occur in the longitude measurements.[1] Further, for the latitude readings one does not have to be even that precise, because the declination or the tilt of the pole of the ecliptic changes at a much slower pace. A 10′ change in the latitude value near the equinox will imply a deviation in time of about 100 seconds at worst.[2] Therefore, a reading taken within 1½ min of the true time will be well within the accuracy of the instrument.

On a clear day one may also obtain the timings for a reading from the Jaya Prakāśa. The Jaya Prakāśa has a set of curves indicating the approach of the signs on the meridian. When a sign approaches the meridian, the shadow of a cross-hair cast by the sun falls on the curve belonging to that sign.[3]

Measuring the longitude of the sun does not require knowledge of the exact time when an observation should be taken. For the sun, all one has to know is which Rāśivalaya unit to use next. Because the latitude of the sun is zero, the shadow of a rod held at the zero marking of the gnomon of that unit will fall on a quadrant edge the moment that quadrant is operative. At that moment, as explained earlier, the quadrant becomes parallel to the plane of the ecliptic, and its gnomon points toward the pole of the ecliptic. The observer thus waits with a rod held at the zero marking of the gnomon and directs his assistant to watch the shadow of the gnomon on the quadrant below. When the rod's shadow coincides with the edge of the quadrant, his assistant takes the reading. Now if the time on the watch is also noted down when the rod's shadow approached the quadrant edge, the timings for subsequent readings may be easily calculated from Table 6-3.

Restorations

A photograph of the observatory taken before 1900 and published by Noti shows the Rāśivalayas in ruins.[4] But Garrett found the Rāśivalayas in 1901 "in fair state of preservation."[5] The instruments had been, evidently, repaired during the reign of Sawai Ram Singh in 1870's along with the other instruments

---

[1] When one of the solstitial points is at the meridian, the time rate of change of longitude has a maximum value of 10 arc-second per 37 sec.

[2] At equinox, the time rate of change of latitude is maximum and equals 10 arc-seconds per 104 sec.

[3] See Ch. 4.

[4] Noti, p. 71, *Op. Cit.*

[5] Garrett, p. 71.

repaired at that time.[1] The repair work was improperly done, however, and Garrett had considerable difficulty in deciphering the Rāśivalayas' true purpose. During 1901-02, while restoring the instruments, Garrett altered the Rāśivalayas once again. Explaining his reasons for altering the instruments, Garrett says:

> "At the time the restoration of the Observatory was begun, they (Rāśivalayas) were found in a fair state of preservation, so that the angles representing the azimuths and altitudes of the gnomons, as actually measured at this time, probably represented fairly accurately the angles as originally constructed. The origin of these instruments is unknown, and none of the pundits or astronomers of Jaipur could give reliable information concerning them, though it was at once seen that they were intended for the direct measurements of celestial latitude and longitude, and being twelve in number, it seemed probable that one was connected with each of the signs of the zodiac. Assuming this, there were two theories to be tested; first, that one instrument out of the twelve was to be used as each sign of the zodiac rose on the eastern horizon; and second, that instruments were to be used in succession as each sign arrived at its culmination on the meridian. The former theory seemed the more probable, as the Hindus attach great importance to the position of the ecliptic on the horizon. The necessary computations were, however, made for both hypotheses. . . . (As) neither theory fully accounts for the angles as measured, but the second theory gives nearly correct results for the azimuths. . . . Further computations were therefore made to ascertain whether the gnomons, as actually existing, could ever point in this direction, and it was found that in the majority of cases they could not do so. It was therefore clear that the Rāśivalayas had been wrongly calculated, and as the azimuths calculated on the second hypothesis agreed so nearly with existing azimuths, it was decided to make the necessary alteration in the altitudes only in conformity with this hypothesis, so that one Rāśivalayas could be used as each sign culminated on the meridian."[2]

The Table 6-4 indicates the angles of the instruments as Garrett found them and as he changed them.[3]

---

[1] Garrett is apparently unaware of the repairs.

[2] Garrett, p. 71 ff.

[3] Garrett, pp. 71, and 73.

Ṣaṣṭhāṁśa Yantra

A Ṣaṣṭhāṁśa yantra is a meridian dial in a dark chamber over which a pinhole image of the midday sun indicates its declination and zenith distance. At Jaipur there are altogether four units of the Ṣaṣṭhāṁśa, and they are located within the lofty chambers on both sides of the Great Samrāṭ. The quadrants of the Samrāṭ in fact rest against these chambers. The instruments within the chambers are accessible through small doors from the north and from the south. The chambers

Rāśivalaya Sign	Original Gnomon Angles (approximate)		Gnomon Angles as Altered in 1901-02		Longitude of gnomon's plane[1]
	Azimuth	Altitude	Azimuth	Altitude	
Aries	−26;0	27;0	−25;56	24;32	11;26
Taurus	−21;30	15;30	−21;17	14;25	35;33
Gemini	−12;30	7;0	−12;19	6;36	61;27
Cancer	0;0	3;30	0;0	3;28	90;00
Leo	12;30	7;0	12;19	6;36	118;33
Virgo	21;30	15;30	21;17	14;25	144;27
Libra	26;0	27;0	25;56	24;32	168;34
Scorpio	26;0	38;0	25;37	35;33	194;28
Sagittarius	18;0	46;30	17;40	45;42	227;09
Capricornus	0;0	50;30	0;0	50;22	270;00
Aquarius	−18;0	46;30	−7;40	45;42	312;51
Pisces	−26;0	38;0	−25;37	35;33	345;32

Table 6-4 The azimuth and altitude of the gnomons of the Rāśivalayas before and after the restorations of 1901-02.

are not totally dark as one might suppose, but have small stone grill-work or *jālī* high above that lets some daylight in. This light does not interfere with observations, however. In fact, a little light is useful to enable the observers to move around freely without bumping into one another or into the chamber walls.

The chambers enclose two units of the Ṣaṣṭhāṁśa or sixty-degree scales each along their two east and west walls. These scales are separated by a 2 m wide

[1] Garrett, p. 73.

aisle. High above a scale's south end, some 9 m from the ground, is embedded its metal plate with a pinhole. Initially, the scales had been inscribed on a smooth surface of lime plaster. The restoration work of 1901-02 was also done in lime plaster. Later on, however, the scales were surfaced with marble. The four units of the Ṣaṣṭhāṁśa are identical. The account that follows below is based on the unit built adjacent to the western wall of the western chamber.

The Ṣaṣṭhāṁśa yantra next to the western wall, which we studied in detail, is just like its sister units and has two parallel scales separated by a distance of 7.5 cm. The scales have a radius of about 8.64 m. A single pinhole of diameter 3.8 mm in a metal plate high above the two serves them both.

Division Scheme

The declination scale of the Ṣaṣṭhāṁśa has been inscribed for measuring angles between 23;30 N to 23;30 S only, whereas the zenith distance scale has been engraved for measuring angles from 0 to 60 degrees. The scales are divided into degrees and minutes. A degree division is divided into 10 major parts which are further divided into six subparts, thereby giving the instrument a least count of 1' of arc. The average length of the small divisions measuring 1' of arc is 2.51 mm. Although the divisions have been carefully laid out, a few divisions ranging in length from 2.75 to 2.25 mm may also be observed. The divisions are labeled in Devanagari.

Observations

The Ṣaṣṭhāṁśa of Jaipur is by far the most sensitive and easy-to-use instrument of Jai Singh's observatories. The pinhole image of the sun falling on the dial of the instrument is usually sharp. One can easily discern sunspots on it, if any are present. During the winter months of 1981-82, the author took a number of readings with the instrument. During this period, the author observed sunspots just about every day he took a reading. After developing a certain degree of familiarity with the instrument, the author noted that the readings do fall within ±1' of arc, the intended accuracy of the instrument. For example, on 22 December 1981, the declination for the center of the sun, after applying the refraction correction, was noted as 23;26±0;1 S. This reading compares favorably with the Ephemeris value of 23;26,21 S. Similarly, the solar diameter was observed as 0;33, whereas the Ephemeris value for the day was 0;32,34. The image of the sun that day was 8.35 cm wide as measured on the instrument's surface. The two scales, namely, the declination and the zenith distance, have apparently the same degree of accuracy. Around June 21, when the sun is nearly overhead, one reads the scale directly below the pinhole. At that time the error is suspected to be about 2' of arc. The lower edge of the

Fig. 6-8 The Ṣaṣṭhāṁśa chamber. The West quadrant of the Great Samrāṭ rests against it.

Fig. 6-9 The Dakṣinottara Bhitti Yantra.

zenith distance scale does not lie exactly below the pinhole.[1] Although the instrument, or more appropriately certain parts of it, measures angles with a high degree of accuracy, it does not indicate the meridian transit of the sun accurately for setting a clock without applying a correction. The pinhole of the Ṣaṣṭhāṁśa yantra studied in detail does not lie right above any of its two scales below, that is, it does not lie in the meridian plane defined by either of the two scales on the surface below. The pinhole of the yantra examined lies 15 mm due east of the west edge of the declination scale. It is possible that the builders did not consider this application of the instrument.

Garrett writes that he had found a square aperture in the plate when he restored the instrument in 1902 which he replaced with a small round hole as we find there today. It is not certain, however, when the square aperture had been put in. Jagannātha does not specifically prescribe a square aperture.[2] It is possible that the aperture plate with a square hole had been put in at a later date. If the square aperture had been there since the very beginning, the image could not have been very sharp when Jai Singh carried out his observations.

Dakṣinottara Bhitti Yantra

The Dakṣinottara Bhitti Yantra is located to the north of the Great Samrāṭ. The present instrument was built in 1876, during the reign of Sawai Ram Singh, as a replacement for the dilapidated one constructed by Jai Singh in 1728. The original instrument had its divisions inscribed in lime plaster and not in marble as the instrument has them now. According to the plaque on the instrument, and also according to a map of the observatory, the old instrument had been located to the north of the smaller Samrāṭ.[3] It had double quadrants on either sides of a north-south wall. The area where the original instrument had been located before has now been turned into a sidewalk of the street adjacent to the north boundary of the observatory.

The present instrument actually has two separate Dakṣinottara Bhitti units inscribed in white marble on two 15.41 m long and 7.22 m high parallel walls of a narrow north-south chamber. The building, including the chamber and the stairs on its sides leading to the roof, is 4.68 m wide. The east-facing wall of the building has two quadrants intersecting at the bottom, and two 1.5 cm wide and 21 cm long iron bars near the upper end where the center of the arcs are located. The radius of the quadrants is 6.06 m each. On the west-facing wall,

---

[1] We disagree with the claims made by Garrett regarding the accuracy of the Ṣaṣṭhāṁśa. Garrett claims that the instrument can measure the angular diameter of the sun within 15." Garrett, p. 37. We also disagree with Singh that the planet Venus can be seen on the instrument's surface. Venus appears on the meridian in daytime only, between 9 am and 3 pm and thus cannot be seen. Prahlad Singh, p. 110, *Op. Cit.*, Ch. 3.

[2] SSMC, pp. 7-8; also, SSRS, pp. 1038-1039.

[3] The Museum, Map no. 23.

on the other hand, there is a semicircle with its center near the top end of the wall. At the center an iron style 22 cm in length and 1.5 cm in width has been fixed. The diameter of the semicircle as measured at the top is 12.13 meters. Around the semicircle, steps have been provided to facilitate a reading.

The instrument is designed to measure the altitude of an object and thus has the zero markings of its scales near the top ends and the 90-degree marking at the bottom. Its degree markings have been divided into 10 main divisions which, in turn, have been further subdivided into 3 small subdivisions each, so that the least count of the instrument is 2′ of arc. The small divisions have an average width of 3.55 ± 0.5 mm, and their markings have been filled with lead. The instrument is sturdily constructed as, even after more than 100 years of existence, it is in excellent shape.

The location of the instrument is well chosen as it can measure the zenith distance of objects within 3 degrees of the horizon. Although the instrument is said to have been designed, according to the plaque, to measure a large assortment of astronomical parameters, it is most suitable for measuring the meridian altitude or zenith distance of the sun. For objects other than the sun, particularly those near the zenith, the instrument is difficult to work with. The clearance between the scales of the units on the east side and the floor below is only 72 cm. An observer may have to crouch to take a reading. The west side of the instrument, in this regard, is even more cumbersome to work with. There the clearance is only 36 cm. Further, as the scales are flush with the wall, and do not project out outward, it is difficult to measure the time of the meridian transit of an object, such as a planet, with accuracy.

In December 1981 and January 1982, the author took several readings of the midday sun with the instrument and compared them with the values calculated with a computer. The readings with the instrument on the east facing wall were invariably lower, anywhere from 17′ to 24′ of arc.[1] The difference was too large to be accounted for by the uncertainty of the penumbra alone, which will be about 10′ of arc due to the finite width (1.5 cm) of the rod casting the shadow. The west-facing unit of the instrument did slightly better. There the dial read approximately 10′ to 15′ higher.[2]

[1] The program used was AstroCalc Plus, produced by Zephyr Services, Pittsburgh, 1988.

[2] The rod on the west facing wall is indeed off-center by about 0.5 cm.

CHAPTER VII

JAIPUR OBSERVATORY II

Jaya Prakāśa

The Jaya Prakāśa is constructed on a centrally located masonry platform, 20.5 m long, 11.22 m wide and 30 to 90 cm above the surrounding ground level. The instrument consists of two complementary hemispheric bowls, concave upwards, 5.44 m in diameter, laid out in a north-south direction. The hemispheres are surfaced in marble and have entryways on the east and west sides of the platform. At the bottom of each bowl, there is a drainage hole.

The rim or upper edge of the hemispheres of Jaya Prakāśa represents the local horizon and the lowest point on its concave surface, the zenith.[1] Between the east and west points on the horizon and passing through the zenith point on the instrument is the great circle of the prime vertical. Similarly, another great circle, passing through the north and south points and through the zenith, represents the meridian. The meridian and the prime vertical both are divided into degrees and minutes such that their small divisions measure 10′ of arc each. The average length of these divisions is 8 mm. The rim of the instrument, representing the horizon, is also divided into degrees and minutes. The small divisions on this scale, measuring 6′ of arc each, are 4.75 mm wide.

The north celestial pole is marked at a distance of 27 degrees below the south point on the meridian. Radiating from this pole and reaching up to the horizon, 24 hour-circles are engraved. The surface area between two alternate hour circles in both units of the instrument has been eliminated and steps provided in its place for an observer to move around freely for his readings.[2] Because the two hemispherical units of the instrument complement each other, the same two sections from the units have not been removed. In fact, the sections have been removed in such a way that the remaining sections, if put together, would form a complete hemisphere with some overlap. It should be pointed out, however, that only a portion of the space between two adjacent hour circles of a unit has been eliminated. A large section around the pole, about 28 degrees wide, toward the southern end has been left intact in both units, and that is where the overlap would mostly occur.

Cross wires were originally stretched between the cardinal points marked on the horizon. These days the cross wires have been replaced by a ring that casts

---

[1] See Ch. 4.

[2] It should be pointed out that Jagannātha recommends only a *ghaṭī*-wide section to be removed. See Ch. 4.

Fig. 7-1 Jaya Prakāśa Yantra. The Nāḍīvalaya is in the background.

an elliptic image of the sun over the surface of the instrument below. The center of the image indicates the coordinates of the sun. At night the coordinates of a planet or star are read by observing its passage through the ring as viewed from an appropriate vantage point at the edge of a section left intact.

The instrument is designed primarily to measure the coordinates of a celestial body in the horizon and the equatorial systems and to indicate the local time and the approach of a sign of the zodiac on the meridian.

Horizon System of Coordinates

For the horizon system, the bottommost point of the instrument represents the zenith and the rim, the horizon. Equal-azimuth lines, spaced 6 deg apart, radiate out from the zenith point. Equal zenith-distance circles intersect these equal-azimuth lines at right angles. The circles have their centers on the vertical axis passing through the zenith. They are parallel to the horizon and are also drawn 6 deg apart.

While measuring the coordinates in the horizon system, an accomplished observer, with proper interpolation, can obtain an accuracy of $\pm 3'$ of arc for the zenith distance for all his readings, because this accuracy does not depend on the location of the object in the sky.[1] For the azimuth readings, however, the

---

[1] See Ch. 4.

situation is different. The accuracy of these observations depends on the zenith distance. The azimuth readings for an object near the horizon have the accuracy of ±3′ of arc, because at this location the small divisions, worth 6′ of arc, are 4.75 mm wide and may be interpolated easily to ±3′. It should be pointed out, however, that the instrument, because of its low elevation and because it is practically surrounded on all sides by other structures, is incapable of taking readings of the objects near the horizon. Hence the accuracies of ±3′ are of theoretical interest only. As the altitude of the object increases, the accuracy of the measurements gradually deteriorates. In fact, near the zenith point where the equal-azimuth lines merge into a point, the readings are practically useless. Jai Singh's astronomers, therefore, must have avoided observations of overhead objects with the Jaya Prakāśa. The Rāma yantra also suffers from this shortcoming.

The Equatorial System of Coordinates

For the equatorial system of coordinates, a point on the meridian 27 degrees below the south end of the horizon, represents the North Celestial Pole and 90 degrees further down the line the great circle of the equator. The equator is divided into degrees, which in turn are further subdivided into four parts of 15′ of arc each. The 15′ divisions are on average 11.5 mm wide. On both sides of the equator, parallel to it, the diurnal circles, eight in all, are drawn at an angular distance equal to the declination of the first point of the signs of the zodiac; and at 66;30, the angular distance of the pole of the ecliptic. Considerable deviation from the true values of the declination of these curves may be noted as shown in Table 7-1.

The equatorial coordinates of an object are measured following a procedure similar to the one for the horizon system of coordinates. For the equatorial measurements, the instrument has accuracy limitations similar to the horizon ones. Just like the altitude measurements, the declination readings do not depend on the location of the body in the sky, and they could have an uncertainty of ±3′. The hour angle readings, on the other hand, are not independent of the location of the object in the sky. For the hour angle readings, the best results are obtained near the equator, where the division length is the largest. At the equator, the uncertainties are of the order of ±4′ to ±5′ of arc. This accuracy falls off rapidly as one approaches the pole, where the hour circles merge into a point and the readings, as a result, become practically meaningless.

Saṅkrānti

The entry of the sun into a sign is indicated by the shadow of the cross hair falling on the surface of one of the "first point" diurnal circles as described above.[1] This indication is according to the *sāyana* reckoning of signs, in which

Sign	Declination	True Value
Capricornus	23;22,30	23;26,24
Sagittarius-Aquarius	20;00	20;09,04
Scorpio-Pisces	11;20,25	11;28,20
Equator	00;00	00;00,00
Taurus-Virgo	11;30	11;28,20
Gemini-Leo	20;15	20;09,04
Cancer	23;30	23;26,24

Table 7-1 The declinations of the diurnal circles on the Jaya Prakāśa and their true values.

the first point of Aries is always at the vernal equinox and not at a point fixed on the ecliptic by Hindu astronomers some 1700 years ago.[2] If the popular *nirayana* system is to be used, then a correction for the precession of the equinox has to be applied to the time of the *sāyana-saṅkrānti*.[3] Because the "first point" curves deviate from their true value by as much as 4′ of arc—particularly near the solstices—the time of the sun's entry into the signs of Cancer or Capricorn could be off by as many as four to five days and is approximate at best.[4] Near Taurus-Virgo or Scorpio-Pisces the error is only about ±4 hr.[5]

---

[1] Because a number of Hindu religious observances are tied to the entrance of the sun into a sign or *Saṅkrānti*, the knowledge of the passage is of importance to the Hindu public at large.

[2] In Hindu astronomy the first point of the Aries is the same as the vernal equinox of 285 A.D.

[3] For March 21, 1990, this correction was +37 minutes and 2.7 seconds. After 1990, the correction will be (37 m, 2.7 sec) + (3.07 sec/yr) × *n*, where *n* is the number of years elapsed since 1990.

[4] Near a solstice the declination of the sun changes rather slowly; therefore, the determination of the exact moment of the solstice would be difficult with an instrument such as the Jaya Prakāśa. A change in the angle of declination by 4′ of arc would take about 4 to 5 days at solstice.

[5] The error is less because near an equinox point the sun travels at a rate of approximately 1400″ of arc perpendicular to the equator per day.

Local Time

The local time is read from the shadow of the cross-wire on one or the other unit of the Jaya Prakāśa. The units indicate the time alternately, in segments of one hour. As it turns out, the instrument is most accurate at equinox when the sun travels along the equator of the instrument, where the small divisions are wide enough to be read with the accuracy of one-half minute. The accuracy at other times could be only as much as $\pm 1\frac{1}{2}$ minutes. At other times when the accuracy is not the best, the main difficulty arises from the fact that there are no subdivisions between any two hour circles, and one has to measure or estimate the angular distance of the cross-wire shadow from the nearest hour circle and then by interpolation determine the time. At night the procedure would be even more cumbersome when there are no shadows to fall on the instrument's surface. Because the procedure for reading time off the Jaya Prakāśa is cumbersome, it is doubtful if the instrument was used much for this purpose, particularly when simpler devices such as the Samrāṭ were available nearby.

The Sign of the Zodiac on the Meridian

The approach of a sign on the meridian is indicated by a set of 12 curves, designated as the *madhyalagnavṛttam*, inscribed on the instrument's surface. This information is useful to an astrologer for his trade. The information also alerts an observer measuring the longitude of the sun with the Rāśivalayas as to which Rāśivalaya unit to choose for his next reading. In daytime when the sky is clear, the shadow of the cross-wire falling on one of the *madhyalagnavṛttam* curves indicates the approach of the first point of that sign on the meridian. The *madhyalagnavṛttams* have been drawn according to the *sāyana* system. An astronomer interested in the *nirayana* reckoning of the signs must apply a correction for the precession.[1]

A *madhyalagnavṛttam* or arc is a locus of the coordinates of the sun from one day to the other during the course of a year when a sign approaches the meridian. In other words, it represents the orientation of the ecliptic when a sign is at the meridian. These locus points were either calculated directly from the formulae of spherical astronomy or read directly off an astrolabe made for the latitude of Jaipur. Another method of drawing these curves is to appropriately divide the diurnal circle of the pole of the ecliptic into 12 equal parts. Then with the points of division as centers, draw arcs of radius of 90^0, or of radius equal to the angular distance of the ecliptic from its pole, on the instrument's surface. These arcs then represent the path of the sun's shadow on the instrument.

---

[1] Ref. 2, p. 160.

Difficulties with the Instrument

Although the procedure for daytime measurements, that is, measuring the coordinates of the sun from the shadow of the cross-wire, is straightforward, nighttime readings are rather cumbersome if they are to be taken at an arbitrary moment. In order to take readings at an arbitrary moment, metallic or wooden arc-lengths of various radii may have to be used. These arc-lengths have to be placed in between the open sections. In the other method, the specially prepared arc lengths become unnecessary if the edge of an uncut section is used as the vantage point. The observer lines up the object, the cross-wire, and a ring placed at the observer's vantage point. However, with this method, nighttime readings can only be taken at one-hour intervals when the rotation of the earth brings the object to an observable position and not arbitrarily at any given moment. It is reasonable to believe that Jai Singh's astronomers might have preferred the second method.

Kapāla Yantras

To the west of the Jaya Prakāśa, the two Kapālas are located on a 13.96 m long and 7.88 m wide, and one-half meter high east-west platform. The Kapālas are two concave non-complementary hemispherical bowls with a diameter of 3.46 m each. The instruments differ considerably with each other in engraving and in purpose. Originally, their surfaces had been finished in lime plaster, and only the rims were of marble,[1] but now they are finished entirely in white marble. Drainage holes, one each, are provided at the bottom.

Kapāla A

On the Kapāla platform, the instrument Kapāla A is located to the west. Kapāla A is designed to measure the coordinates of the sun in the horizon and the equator systems, and to indicate the local time. In addition, it also tells the ascendant,[2] and the *Saṅkrānti* or entry of the sun into a sign. The coordinates, local time, ascendants, and *Saṅkrāntis* are all determined by observing the shadow of a cross-wire stretched between the cardinal points marked on the rim. In recent times, the cross wires have been replaced by a small circular plate with a hole in the middle that casts an elliptic image of the sun on the surface below. The center of the image serves the same purpose as the shadow of a cross-wire. Because the instrument has been designed to work with the shadow of a cross-wire cast by the sun, it may not be used at nighttime. The instrument has two sets of coordinate systems inscribed on its surface, namely, the altitude-azimuth

---

[1] Garrett, p. 49.

[2] The instrument, in fact, tells the emergence of the first point of a sign on the horizon.

and the equatorial, enabling the observer to record his data in the system of his choice.

The Altitude-Azimuth or the Horizon System of Coordinates

In the altitude-azimuth or horizon system, the lowest point on the instrument's surface represents the zenith. With the zenith point as the center, parallels of altitude, or more appropriately, the equal zenith distance circles, have been drawn up to the very rim of the instrument. The rim then becomes the local horizon. These circles are 6 degrees apart. The horizon circle on the rim is divided into degrees and minutes, such that its small divisions measure 10′ of arc. Radiating in all directions from the zenith point on the instrument's surface, there are 20 equal azimuth lines drawn up to the horizon or rim. For some reason, these lines have not been spaced evenly and are anywhere from 3;40 to 45;23 apart. The parallels of altitude circles and the equal azimuth lines have not been divided into degrees or minutes.

Equatorial Coordinates

In the equatorial system of coordinates, a point on the meridian, 27 degrees below the south point of the rim, represents the North Celestial Pole. A great circle, at a distance of 90 degrees from the pole, is the equator labeled as the *Meṣa-Tulā Horā Vṛttam* or Aries-Libra diurnal circle. Both sides of the equator contain six diurnal circles corresponding to the declination of the first point of the signs of the zodiac. An additional diurnal circle indicates the path of the pole of the ecliptic, and it is drawn at a distance of 23½ degrees around the pole. These curves have been labeled in Devanagari, such as "*Mithuna Siṁha Horā Vṛttam*" (Gemini-Leo diurnal circle). Next, arcs representing the hour circles, spaced at 6-degree intervals, emanate from the pole in all directions. The 6-degree interval between two adjacent hour circles may have been chosen because a point on the equator takes exactly one *ghaṭikā* or one Hindu unit of time to travel a 6-degree distance. Here again there are no subdivisions of any kind.

Because there are no subdivisions on the arcs, and because it is inadvisable to step down to the marble surface of the instrument for better reading, it is cumbersome to take a reading with precision. Besides, the precision of a reading itself depends on the coordinate being measured. For example, both the zenith distance and the declination may be measured, with some interpolation, with a precision of ± 5 arc minutes or better. For both the azimuth and the hour angle, the accuracies vary and, in fact, depend on the zenith distance and the declination of the object respectively. For large zenith distances, such as for an object near the horizon, the accuracies in the readings of the azimuth are of the order of $\pm 3'$ of arc.[1] For small zenith distances, where the parallels of altitude shrink in circumference, an error of several degrees may easily occur. When the sun is near the zenith (around noon during the summer months) the azimuth readings become practically meaningless. Similarly, near the north celestial pole, where the hour circles converge, and the diurnal circles progressively shrink in size, the readings may be highly imprecise. The readings are most sensitive at the equator, however, where the separation between the hour angle divisions is the greatest. There the uncertainties are of the order of $\pm$ 3 arc minutes.

The instrument has a set of 12 curves inscribed on its surface, labeled in Devanagari, for determining an ascendant. The elliptical image of the sun falling on one of the curves indicates the sign emerging at the horizon at that very moment. The signs indicated by this method are according to the *sāyana* system, in which the first point of Aries is always at the vernal equinox. Having read a *sāyana* sign off the instrument, one may obtain its *nirayana* counterpart by applying a correction for the precession of the equinoxes.

Theoretically, the curves indicating the rising sign, or ascendant, are the loci of the sun's image on the instrument's surface as a sign rises on the horizon from one day to another during the course of a year. The curves may also be considered as the projection of the ecliptic on the instrument's surface when a sign appears on the horizon. The loci, or projection points, on the instrument might have been calculated either with equations of spherical trigonometry or read directly off an astrolabe made for the latitude of Jaipur.

Another method of drawing these curves is much simpler and does not require any calculations. It exploits the fact that the first point of Aries always rises at the east-point of the horizon, and the pole of the ecliptic at that moment is exactly 90^0 away from the east-point on its diurnal circle. With the location of the pole as a starting point, divide the diurnal circle of the pole into 12 equal parts. Then with these points of division as centers, draw arcs of radius 90^0 on the surface. These arcs then represent the path of the sun's shadow on the instrument's surface. A spot check revealed that some of the curves have been improperly drawn on the surface of Kapāla A.

Local Time

Local time is read from the shadow of the cross-wire on the instrument's surface. In this respect the instrument is similar to any other hemispherical sundial, or similar to the Jaya Prakāśa. As it turns out, an instrument of this type is most accurate at equinox when the sun travels along the equator of the instrument, and where the distance between the hour circles is the widest. At

---

[1] Because of its low location, the instrument is unsuitable for observing objects near the horizon.

equinox the time may be read with the Kapāla A with an accuracy of about ±½ minute.

Saṅkrānti

The *Saṅkrānti* or entry of the sun into a sign is indicated by the shadow of the cross-wire falling on one of the first-point diurnal circles described above. The indication is, of course, according to the *sāyana* system. If the *nirayana* system is to be used, a correction for the precession of the equinox has to be applied to the time of the *sāyana-saṅkrānti*. The error in *saṅkrānti* may be anywhere from a few hours near an equinox to several days at the solstices. In this regard also, the instrument is similar to the Jaya Prakāśa.

The instrument Kapāla A is a good teaching tool because it shows a number of elements of interest to an astronomer, such as the relationship between the horizon and the equator system of coordinates. The instrument is also quite useful to an astrologer for casting a horoscope.

Kapāla B

Kapāla B is situated to the east of its sister unit, Kapāla A, on the same platform. Kapāla B is the only instrument at the Jaipur observatory which is not meant for observing. Instead, its object is to transform graphically the horizon system of coordinates into the equatorial system and vice versa for the latitude of Jaipur. The transformation implies converting the zenith distance and azimuth of an object into its corresponding declination and hour angle respectively. These transformations require the application of the formulae of spherical trigonometry, which may involve lengthy calculations.[1] Kapāla B, in theory at least, is an ingenious device as it eliminates these calculations.[2]

The surface of Kapāla B has arcs representing the two sets of coordinate systems, namely, the horizon and the equatorial. For the horizon system of coordinates, the rim of the instrument represents the meridian, and the north point of the rim represents the zenith. See Fig. 7-2. Another point, almost 27 degrees to the east of this zenith on the rim, designates the north celestial pole. The north point of the "original horizon" is located at the east point of the rim.

---

[1] The equations for transforming the azimuth and the zenith distance to declination and to hour angle are as follows:

$$\text{Cos } z = \text{Sin } \delta \text{ Sin } \lambda + \text{Cos } \delta \text{ Cos } \lambda \text{ Cos H}.$$

$$\text{Sin } \delta = \text{Sin } \lambda \text{ Cos } z + \text{Cos } \lambda \text{ Sin } z \text{ Cos A},$$ where δ = declination, z = zenith distance, λ = latitude of the place, H = hour angle measured from the observer's meridian, and A = Azimuth.

[2] Tycho Brahe also constructed a device for graphically transforming coordinates. His device was a sphere of wood covered with brass, about 1½ m in diameter. Later, he used the sphere to depict the stars whose coordinates he had measured. See Thoren (1973), pp. 25-45, *Op. Cit.*, Ch. 1.

The "original horizon" lies in the vertical plane passing through the east-west points on the rim. A great circle passing through the north and south points of the rim and the bottom-most point of the instrument, represents the prime vertical.

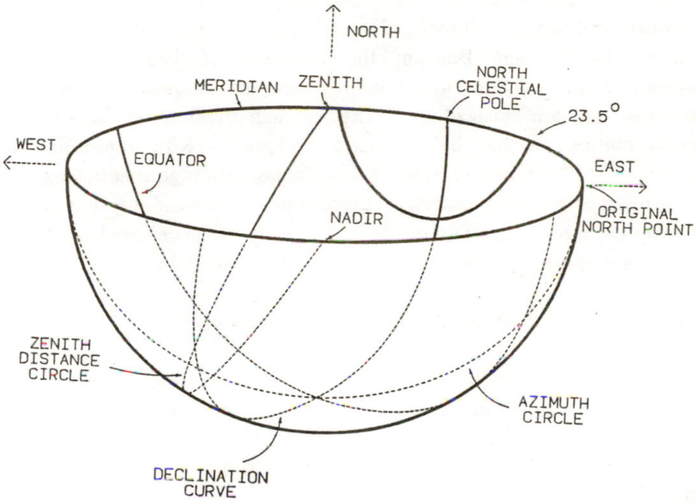

Fig. 7-2 The Principle of Kapāla B.

In order to comprehend this peculiar arrangement of the celestial points on the rim and the orientation of the great circles, such as the original horizon and the prime vertical, it is helpful to think of the instrument's surface as representing the celestial sphere, which has undergone two rotations as follows: First, rotate the celestial sphere along a horizontal east-west diameter until the zenith point is at the north point of the rim. Next, rotate the sphere again, this time about a north-south diameter, counterclockwise looking northward by 90 deg. When this is done, the original meridian turns into the rim of the instrument, and the plane of the horizon becomes the vertical plane passing through the east and west points on the rim. Representing this orientation, the bowl of the instrument may be considered to be one-half of the celestial sphere to the west of the meridian plane of the place. The rim of the instrument, representing the meridian, has been divided into degrees and minutes, such that its small divisions measure 10′ of arc, and they are labeled in Devanagari.

In literature, the instrument has been described inaccurately. For example, Garrett, who was in charge of the engineering of the restoration of 1901-02, states that there are latitude and longitude lines engraved on the instrument's surface and that the instrument is capable of converting declination and right ascension into these two coordinate angles.[1] In addition, he calls the rim of the instrument the solstitial colure, which is also somewhat misleading.[2] In order that the rim may represent the solstitial colure, the pole of the ecliptic must be a fixed point on the rim, at a distance of 23½ degrees from the north celestial pole. The pole of the ecliptic, however, revolves on an hour circle inscribed on the instrument's surface, 23½ degrees from the celestial pole. Kaye, Soonawala and Singh, following Garrett's lead, commit similar errors.[3]

Because Garrett and Bhavan, the restorers of 1901-02, did not fully comprehend the function of the instrument, the plaques they erected at the instrument site are misleading. The English version of the plaque states: "Representation of the half celestial sphere. Rim represents solstitial colure. . . ." The Hindi version of the plaque, although identifying the rim as the meridian correctly, is otherwise inaccurate. The Hindi version identifies the instrument as a representation of the eastern half of the celestial sphere, whereas it is a representation of the western half of the celestial sphere.[4]

Horizon or Altitude-Azimuth System of Coordinates

Both the zenith and nadir of the horizon system of coordinates are located on the rim. They are 180 degrees apart, along a north-south horizontal line. See Fig. 7-2. The zenith distance is measured from the zenith point on the rim along the great circles drawn 6 degrees apart from the zenith point to the nadir point.

A number of great circles, 6 degrees apart, inscribed from the east point to the west point of the rim represent the curves along which the azimuth is measured. The great circle between the east and west points and passing through the bottom-most point of the instrument represents the horizon or, more appropriately, the western half of the horizon. The great circle of the horizon divides the hemisphere into two equal halves.

Equatorial System

The north celestial pole is marked on the rim, at a distance of about 27 degrees to the east of the zenith (the north point of the rim). The south celestial pole is, similarly, at a distance of 27 degrees to the west of the nadir (the south point on the rim). Great circles or hour circles spaced six degrees apart are drawn from

---

[1] Garrett, pp. 48-49.

[2] The solstitial colure is a great circle on the celestial sphere passing through its poles, the poles of the ecliptic and the solstitial points.

[3] Garrett, p. 48; Kaye, pp. 52-53; Soonawala, p. 37, *Op. Cit.*, Ch. 6; Singh, p. 164, *Op. Cit.*, Ch. 3. Kaye, confusing the Kapāla B with its counterpart A, makes another mistake in stating that Kapāla B indicates the rising signs.

[4] Bhavan repeats the error of the Hindi plaque in his book as well. Bhavan, p. 11.

one pole to the other. Intersecting these hour circles are seven diurnal circles, with the one in the middle representing the equator, which divides the hemisphere into two equal halves. The diurnal circles have been drawn on either side of the equator at distances of approximately 11;20, 23;25 and 66;30, respectively. These arcs represent the declinations of the first points of the signs of the zodiac and the pole of the ecliptic. Some of these curves have not been drawn very accurately and are off by as much as 8′ of arc from their true values.

In principle the instrument works as follows. If one wishes to convert the azimuth and zenith distance of a body into its hour angle and declination respectively, one should first plot a point on the instrument's surface according to the given coordinates. Next, read the angular distance of this plotted point from the equator along the hour circles. The angular distance is then the declination of the body. Similarly, the angular distance from the rim or the meridian provides the hour angle. Finally, by adding or subtracting from the hour angle the angular distance between the vernal equinox and the meridian when the observation was made, one obtains the right ascension.

The angular distance between any two stars may be determined by plotting two points on the instrument according to their coordinates and spanning the points with a divider. Next, placing the ends of the divider on the graduated scale on the rim, the separation is converted into degrees and minutes.

In principle, Kapāla B is quite elegant. However, its accuracy varies. Its scales have been divided at intervals of only 6 degrees, and their largest separation is about 18.26 cm. With careful interpolation, where the separation is largest, accuracies of the order of 3′ of arc may be achieved. The accuracies deteriorate if the object is close to the zenith point or near the poles where the great circles of equal azimuth and those of the hour circles merge into points respectively. In this aspect, the instrument has the same limitations as the Jaya Prakāśa or the other Kapāla. Besides, every time a conversion is to be made, the observer may have to climb down to the surface of the instrument for better interpolation, a practice which could be detrimental to the scale markings, and thus is inadvisable.

Rāma Yantra

The Rāma yantra of Jaipur is located near the south-west corner of the observatory compound. The instrument measures the azimuth and altitude of a celestial object and comprises two complementary cylindrical buildings 4.58 m tall and with an inside diameter 6.95 m. A Rāma yantra building may be viewed as a set of equally-spaced columns standing in a circle and joined together with a horizontal masonry ring at the top. A flight of steps leads up to the top of the ring.

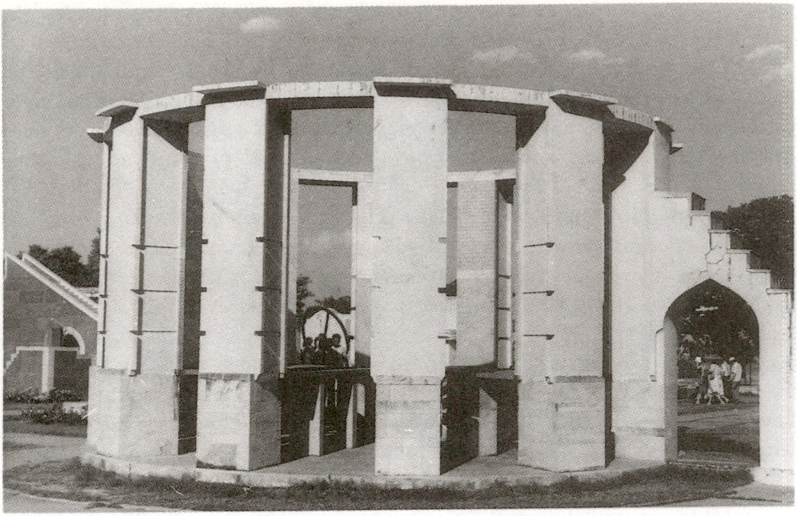

Fig. 7-3 The Rāma Yantra of Jaipur.

Fig. 7-4 The Nāḍīvalaya of Jaipur.

At the center of each yantra, stands a vertical pole approximately 8 cm in diameter and of roughly the same height as the columns surrounding it.[1] From the base of this column horizontal sectors radiate out to the vertical masonry columns just described. The sectors consist of white sandstone slabs supported by vertical slabs resting on a masonry floor. The upper surface of the sectors, where the azimuth and altitude scales are inscribed, is about 1.12 m above the floor. The interior walls of the vertical columns are surfaced with marble and engraved with vertical and horizontal scales. The differing coloration of the marble slabs on these surfaces suggests repairs undertaken at different times. The section of the columns that is engraved with scales is 3.45 m tall as measured from the radial sectors. The side walls of the columns carry three slots each for inserting wooden planks for the observer to sit on while taking nighttime readings. The scales are labeled in Devanagari.

The inner surface of the connecting masonry ring atop the columns is engraved with a circular scale divided into 360 degrees and has its zero marking at the north point. The degree markings of this scale have been divided again into 5 parts each, thereby giving the scale a least count of 12′ of arc. The width of the small divisions of this scale is 12 mm. At the cardinal points of this scale, pins have been provided for stretching cross wires in between. Because cross wires are not prescribed for the instrument, their purpose apparently is to check the alignment of the vertical pole located at the center.[2]

The radial sectors of the two Rāma yantra units, although equal in number, that is, 12 each, are of unequal width. The sectors of the unit to the east are 12 degrees wide and separated by 18-degree open sections in between. The sectors of the unit to the west, on the other hand, are 18 degrees in width and separated by 12 degree wide open sections. The two units are, however, complementary, such that if put together, they would form a complete cylindrical enclosure with an open top.

The radial sectors have degree markings, engraved both radially and concentrically on their upper surface for measuring the altitude and azimuth, respectively. The degree divisions have not been divided any further, however. The two vertical sides of every sector, on the other hand, are subdivided into minutes, such that a small division measures 6′ of arc. The division-lengths of the scales on the vertical sides increase progressively as one approaches the base of the vertical columns where the altitude markings terminate at 45-degree marks. Near the 45-degree marks, the width of a small division measuring 6′ of arc, is about 13 mm, whereas the same width near the 85-degree marks, close

---

[1] The poles were loosely fixed to the base at the time when the author examined the instrument. Also they were of unequal length. The pole in the western yantra was about 3 cm shorter than required by theory, whereas the one in the eastern yantra was found to be longer by the same amount.

[2] It should be pointed out that cross wires can also be used in place of the central pillar. It will be hard to detect their shadow, however.

to the pole, is approximately 6 mm. The procedure for measuring coordinates in daytime and at nighttime has been described earlier.[1] While measuring the coordinates of the sun with the Rāma yantra, the main difficulty arises due to the penumbra of the pole's shadow, which may be eliminated with a cross-hair.[2]

Digaṁśa Yantra

The Digaṁśa yantra is located a few meters to the west of the Kapāla platform. The instrument is situated in a poor location of the observatory; it is too close to one of the Rāma yantra buildings. Because of the close proximity of the Rāma yantra, a part of the sky is always blocked out from the instrument and may never be observed. The Digaṁśa predates the Rāma yantra buildings.[3]

The function of the instrument is to measure the angle of azimuth of a celestial body as described in Ch. 4. The instrument consists of a 96 cm high pillar surrounded by two coaxial cylindrical walls. The height of the inner wall is 96 cm or the same as the height of the pillar. The height of the outer wall is 1.95 m, or nearly twice that of the inner wall. The pillar is 1.04 m in diameter. The inner diameter of the two walls is 4.47 m and 7.77 m respectively. The inner wall is 43 cm in width, whereas the outer wall is 61 cm. Passageways have been provided at the cardinal points of the wall for the observer to move in and out of the instrument freely. The upper surfaces of the pillar as well as those of the two circular walls are constructed with marble slabs and engraved with scales.

The circular scale over the pillar is inscribed around the rim and graduated into four quadrants of 90 degrees each. The zero markings of this scale are on the north and the south points and the 90-degree markings are along the east-west line. The degree divisions of the scale are further divided into three parts each such that the least count of the pillar scale is 20′ of arc. The width of these divisions varies between 2.75 and 3.0 mm. At the center of the scale is a brass knob for securing the strings used for measuring the azimuth of a celestial object.

The inner wall surrounding the pillar has two concentric scales on its top surface. The scales have been divided into four quadrants like the one on the central pillar. The degree divisions on these scales have been subdivided into 10 parts each, thereby giving the scales a least count of 6′ of arc. The small divisions of the inside scale on this wall are on the average 4 mm wide, but divisions of width 3.5 mm may also be noted here and there. Similarly, the average division length of the outer scale is 5 mm, but divisions of width 4.25 mm may also be noted at places.

---

[1] See Ch. 4 and 5.

[2] See Ch. 5.

[3] See map no. 15, The Museum.

The outer wall also carries its two scales on its vertical edges near the top surface. The division scheme of these scales is similar to the one on the inner wall. The small divisions of the inner scale on the outer wall are on the average 6.5 mm wide, and the divisions on the outer edge are 7.5 mm with little variations noted. The cardinal points on the outer scale once carried pins, of which only three now survive. Cross wires were stretched between these points. The procedure of observing with the Digaṁśa has already been described in Ch. 4.

Nāḍīvalaya

The Nāḍīvalaya yantra is located to the east of the small Samrāṭ and consists of two large circular discs or plates of stone embedded in masonry, facing north and south respectively. The diameter of the north-facing plate is 3.70 m, whereas that of the south-facing plate is 3.74 m. The top end of the plates stands almost 4 m above the floor. The plates are separated by a distance of 5.37 meters and are oriented parallel to the plane of the equator. The space between the plates has been turned into a storage chamber measuring 5.34×6.10 m^2 at the outside. The instrument may thus be viewed as two huge discs fixed to the north and south walls of a small room.

The two plates differ in the number of scales engraved over their surfaces, as well as in the base material on which the scales have been inscribed. The north plate has one scale only, engraved around its circumference on a marble base, whereas the south plate has three concentric scales inscribed in red stone and marble. The north scale is 3.62 m in diameter and is divided into two halves with 12 major divisions each. The divisions are 15 degrees wide, measuring one hour of time each. The zero markings of this scale are at the top and at the bottom, thereby enabling the observer to measure time from noon or from midnight. The large divisions on the north scale are further subdivided into four parts, so that the least count of the scale is one minute of time or 15′ of arc. The small divisions measuring one minute of time are well laid out and are 8 mm wide. The script of the scale is Devanagari. At the center of the scale is a rod 15 cm long and about 1 cm wide pointing toward the north celestial pole.

On the south plate, two scales are engraved in limestone and the third, between the two, is engraved in red stone. The innermost scale of the three has a diameter of 84 cm and is divided into four quadrants of 15 *ghaṭīs* each. The *ghaṭīs* in turn have been divided into six parts each, thereby giving the scale a least count of 10 *palas* or 4 min. The zero markings of this scale are, once again, at the top and at the bottom, and the 15-*ghaṭī* markings on the east-west points are along a horizontal line. The script of this scale is Devanagari. The scale in the middle, engraved on red stone, has a diameter of 1.09 m and is divided into hours and minutes just as the north facing plate is. Its large divisions, representing hours, have been further divided into 12 parts, such that one part measures 5 minutes. The small divisions on this scale are on the average 12.3 mm wide. The scale is labeled in English characters.[1] This is the only place in the observatory of Jaipur where any Western characters have been used on a scale. The outermost scale of the south plate is similar to its counterpart on the north side. It has a diameter of 3.63 m and has a least count of one minute. Its small division length is 8 mm.

With the Nāḍīvalaya inst.ument one can easily ascertain the arrival of the sun at equinox with an uncertainty of a few hours if the event occurs in daylight hours. The exact moment when the center of the sun crosses the equator is difficult to observe, unless one looks directly at the sun with a sun-filter from a vantage point at the instrument's rim. With this technique, the author, on September 22, 1985, observed that the sun was still on the north side of the instrument's surface until the noon hour.[2] In the afternoon the clouds moved in. On the following day, September 23, in the afternoon when the sky cleared up, the author observed the sun on the south side of the instrument. The sun had crossed the equator by then. The sun arrived at the autumnal equinox that day a few seconds after 7:27 A.M., Indian Standard Time.

Nāḍīvalaya Plaque

The south plate of the Nāḍīvalaya carries a plaque with seven verses in Sanskrit, which, according to Bhavan, were restored in 1901-02 without any corrections or changes. We reproduce these verses in Appendix II. The inscription on the plaque states the date of the "second restoration" of the Nāḍīvalaya. Unfortunately, the inscription is in code, and no single interpretation can be obtained from it.

Dvivedī in his *Upapattīnduśekhara*, interprets the code as 1640 S.E., or 1718 A.D.[3] Bhavan also interprets the date on the plaque to be 1718 but points out that the inscription is from the period of Sawai Pratap Singh who ruled much later, between 1778 and 1803.[4,5]

At the author's request, several scholars of Hindu astronomy examined the inscription and noted that the verses are ambiguous and may be interpreted in

---

[1] The English characters are of recent origin. In Jai Singh's time no such characters were used.

[2] At 10:04 A.M., on September 22, 1985, the southwest edge of the sun touched the rim of the north plate of the instrument. The sun was not shining on the south side at this time.

[3] The date according to Dvivedī is *Vaiśākha Śukla* 9, and the *nakṣatra Kṛttikā*, 1640 S.E., which is April 9, 1718 (Gregorian). The nakṣatra that day was *Puṣya* and not *Kṛttikā*, however. Durgā Prasāda Dvivedī, *Upapattīnduśekhara,* pp. 65-66, Ahemadabad, 1936.

[4] The day and date according to Bhavan is Friday, *Śrāvaṇa Kṛṣṇa* 9, 1640 S.E., which translates into Saturday, August 20, 1718 (Gregorian). That day the *nakṣatra* was *Rohiṇī*, however. The day, according to Bhavan also does not match with the calendar.

[5] Garrett, perhaps taking a clue from this inscription, concludes that construction of the observatory began in 1718.

At the author's request, several scholars of Hindu astronomy examined the inscription and noted that the verses are ambiguous and may be interpreted in many different ways. David Pingree has interpreted the date as Friday, January 25, 1771.[1] We reproduce his interpretation also in Appendix II.

Because no single unambiguous date could be deduced from the verses, the author searched the records of the Rajasthan State Archives for *Dharmādhikārī* Deva Kṛṣṇa, under whose guidance, according to the inscription, the restoration had been carried out. In the *Dastūr Kaumvār* records, the most likely place to find the name in the archives, there was no mention of any Deva Kṛṣṇa for the period of Sawai Jai Singh.[2] However, for the post-Jai Singh period, between 1753 and 1806 A.D., the records listed five different people under the name Deva Kṛṣṇa. Of these five, a Deva Kṛṣṇa, son of Phakīra Dāsa, was recorded to have received state recognition in 1806 for something related to the *Darogāship* of the *Kārkhānā* or department of *Puṇya*. The *Darogāship* of the department of *Puṇya* (religious affairs) comes closest to being a *Dharmādhikārī* or officer in charge of religious affairs. This Deva Kṛṣṇa appears to be the person the plaque refers to. Moreover, Pratap Singh's name on the plaque strongly suggests that the restoration occurred during his reign.[3] The year 1806, during which Deva Kṛṣṇa was honored, falls close to the reign of Pratap Singh (1778-1803).

Hence, if the identification of the man in charge is correct, the second restoration of the instrument took place any time between 1771 and 1803, just before or during the reign of Pratap Singh, the grandson of Jai Singh. The first restoration might have taken place during the reign of Madho Singh (1751-1767), who also patronized astronomy and constructed instruments.[4] However, no record to substantiate this conjecture has come to light.

Krāntivṛtta Yantra

Near the north-west corner of the observatory compound, close to the observatory office, is a masonry structure, which Garrett identifies as the

---

[1] Appendix 2. However, 1771 does not fall within the ruling period of Sawai Pratap Singh.

[2] The phrase "*Dastūr Kaumvār*" literally translates into "caste-wise listings of ceremonial gifts." Originally, the *Dastūr Kaumvār* records were kept on 13 cm × 17 cm, loose leaves of paper, known as *aḍasaṭṭās*. In the later years of the 19th century, the *aḍasaṭṭās* were copied into large-size books of almost 1000 pages each, as we find them today. The books have almost 100,000 entries arranged alphabetically according to the caste of the recipients. However, exceptions to the rule are numerous and scattered throughout the *Dastūr Kaumvār* volumes. For example, the *Dastūr Kaumvār* lists Jesuits, such as Manuel Figueredo, under the caste "*Musalamān*" (Muslim), and not under *Firaṅgī* (European), as one would expect.

[3] Pratap Singh ruled Jaipur between 1778 and 1803. It is possible that the restoration took place in 1771 a few years before Pratap Singh ascended the throne, but the plaque was erected during his reign.

[4] See Ch. 5.

unfinished Krāntivṛtta yantra. The yantra, or its unfinished structure, consists of a circular plate of red stone, 3.39 m in diameter, resting against a masonry support and oriented in a plane parallel to the equator. The plate has a 5 cm thick and 60 cm long rod at the center.

The instrument has two circular scales concentric with the plate, of which the inner scale has a diameter of 40 cm, whereas the outer one, engraved around the rim, has a diameter of 3.39 m. The inner scale is divided into 60 parts with each part measuring a 6 degree wide angle. The outer scale, on the other hand, is divided both into degrees and minutes and has a least count of 6′ of arc.

In order for the instrument to function as a Krāntivṛtta or torquetum, there should be a mobile superstructure, preferably of metal, mounted on the central rod. However, the superstructure was not built during the restoration of 1902 because it was thought too heavy for the support. Garrett believes that the instrument's metallic part was never completed.[1] This instrument could be the Śara yantra described in the *Yantraprakāra*.[2,3]

Palabhā Yantra or Horizontal Sundial

On top of the Nāḍīvalaya building, accessible via a number of rather steep steps on either side of the north-facing plate is a Palabhā or horizontal sundial. The sundial has been constructed on a 76.3 cm diameter red stone slab on a 40 cm high circular platform. The sundial has a 13 cm high right triangle gnomon fashioned out of a brass plate. The plate has a 24.2 cm long base, and a 27.5 cm long diagonal at an angle of 28;15 with the horizontal. Because the latitude of Jaipur is only 26;55, it is not clear why the gnomon is off by more than a degree. The gnomon-angle is, however, close to the latitude of Delhi, which is 28;38, and it is possible that the sundial was fabricated for the latitude of Delhi. On either side of the gnomon, on the flat surface, are marked straight lines for measuring time in the units of *ghaṭikās*.

Near its rim the circular slab is engraved with a scale divided into 10′ divisions. The scale is not labeled, however. A horizontal sundial with a triangular gnomon does not require an evenly divided circular scale. It is possible that the circular dial was inscribed so that the instrument may also function as an Agrā yantra. A conical projection atop the vertical edge of the triangular gnomon might have also been provided for this purpose. The *Yantraprakāra* describes construction of a Palabhā yantra.[4] Jai Singh could have built his Palabhās according to these instructions.

---

[1] Garrett, p. 45.

[2] The author's conjecture regarding the Śara yantra is in contradiction with Garrett's statement that the yantra was dismantled to make room for a temple. See Garrett, p. 65. The author has not been able to locate any evidence in support of Garrett's statement.

[3] *Yantraprakāra* (J), p. 15, and *Yantraprakāra* (S), pp. 67-68.

[4] *Yantraprakāra* (J), p. 26, and *Yantraprakāra* (S), pp. 76-77.

Dhruvadarśaka Paṭṭikā

The Dhruvadarśaka Paṭṭikā is perhaps the simplest of all instruments found at Jai Singh's observatories. The instrument is nothing else but a small trapezoidal structure whose upper surface points toward the pole star. The purpose of the instrument is, apparently, to show the north star to a lay person. Viewing of the north star by a bride and her groom is a part of the wedding ceremony in certain parts of India. It is possible that the instrument was erected at the time of a prince's wedding. The date of its construction is uncertain.

The Dhruvadarśaka Paṭṭikā is erected on a 3.07 m long and 54 cm wide masonry base. Its lower end is about 76 cm above the ground and upper end 2.32 m. A person glancing along the slanted upper surface of the instrument can easily recognize the pole star on a clear night.

The Ground Plan or Seat of Jai Singh

The so-called Ground Plan or Seat of Jai Singh is inscribed on a level platform 15 to 35 cm above ground level to the north of the Great Samrāṭ. The plan is, in fact, a large circular engraving on 30 cm wide red stone slabs neatly arranged in a circle of approximately 30 m diameter. The origin of the plan is uncertain. Neither of the two maps of the observatory discussed earlier, shows any circular scale labeled as the Ground Plan or the Seat of Jai Singh.[1] The second map shows an area of just about the same size as the plan with a building in the middle. The building is labeled as *Jalajantra ko baṅgalo bado*—"large building for the water machine," and has long since been dismantled. It is possible that the circular engraving is of recent origin and that it was constructed in the 1870's, when some of the instruments of the observatory were renovated. However, Garrett says that it had been there even before that and that "the platform and its graduations were formerly made of plaster, and were restored in stone in 1876."[2] The Ground Plan consists of five concentric circles running along its entire circular length. The diameter of the outermost circle as measured along north-south is 30.35 m and along the east-west, 30.38. The circles do not have scale divisions all around but only in their north-east and south-west quadrants. These quadrants have been divided into 24 large divisions each, which, in turn, have been subdivided into 15 intermediate parts. The intermediate parts have been divided into 10 parts, each of which is divided into three parts. Because of this division scheme, the smallest division of the scale represents 30″ of arc. At the center of the plan there is embedded a 8×8 cm^2 plate of brass.

---

[1] Map Nos. 15 and 23, The Museum.

[2] Garrett, p. 65.

The purpose of the instrument is not clear. One may gain a clue to its purpose from the similarity of the dimensions of the plan and its division scheme to those of the quadrants of the Great Samrāṭ. It is likely, as Garrett suggests, that the division scheme of the Great Samrāṭ was first laid out on a flat surface before transferring them to the Samrāṭ quadrants. Bhavan also states that the circular scale was meant to draw the divisions of "hours, minutes, and seconds of the Great Samrāṭ."[1] Bhavan further adds that "now the instrument is another Digaṁśa for measuring the azimuth of the sun."

If the instrument was indeed originally meant for dividing the scales of the quadrants of the Samrāṭ, its division scheme must have been different than its current scheme. In the days of Jai Singh, the units of time were *ghaṭikā, pala*, and *vipala*, and not hours, minutes and seconds in which the Ground Plan scale is seen divided at present. Obviously, the present division scheme is the result of renovations.

Cakra Yantra

The Jaipur observatory has two units of the Cakra yantras or ring instruments.[2] They are mounted between the Kapālas on stone posts along a north-south line. The instrument are made of heavy, molded brass and pivoted to rotate freely about a diameter parallel to the earth's axis. The object of the instruments is to measure the declination and the hour angle of a celestial object. The units have been made in duplicate so that two observers working side by side may compare their readings.

The Cakra yantra unit which is located to the north has a diameter of 1.84 m and the other unit, to the south, has a diameter of 1.79 m. The circular scale of the north Cakra yantra is divided into 60 major divisions that are further divided into 6 subdivisions. Each subdivision, in turn, is partitioned into 10 small divisions, so that the least count of the scale is 6′ of arc. The small divisions on the average are 1.6 mm wide, however, divisions with width as little as 1 mm and as large as 2 mm may also be seen. The south yantra is divided into degrees at a few places of its circular scale only. It has not been further subdivided into minutes.

At the south pivot end of the instruments, there is a circular brass plate each, attached to the post and parallel to the plane of the equator, for measuring hour angles. The plates have a diameter of about 51 cm. The least count of the hour angle scale on the plate is 1 deg. The divisions on this scale are on the average 4 mm wide.

In order to measure the declination and the hour angle of an object, a sighting tube is mounted at the center of the instrument. The tube, with a pointer

---

[1] Bhavan, pp. 43-44.

[2] The Varanasi observatory also has a Cakra yantra.

attached to it, rotates about a perpendicular axis passing through the center of a Cakra ring. The observer, rotating the Cakra about its polar axis and the tube about the center, obtains the object in sight, and then reads the declination off the circular scale on the Cakra, and the hour angle off the plate at the post.

Today, because of wear and tear of the stones at the pivot points, the polar axis of the instruments has become off center. It should be pointed out, however, that Jai Singh did not rely on the metal instruments, such as the Cakra, because their axes soon wore out, leading to inaccuracies.[1]

Unnatāṁśa Yantra

The Unnatāṁśa yantra is located next to the north boundary wall of the observatory. No other example of this instrument exists at Jai Singh's observatories. The instrument is essentially a large circular ring suspended from a wooden beam resting over two huge masonry pillars, about 7 m tall and 1.5 m wide. The pillars stand along a line which is inclined at an angle of 12 degrees to the north-south direction. The beam, which originally had been plastered over, is now exposed and is also supported by a graceful arch between the pillars. The pillars, themselves are erected over a somewhat oval shaped platform 1.7 m above ground level. Between the pillars, the platform has a shallow masonry pit, circular in shape and having a diameter of 5.68 m. Steps have been provided all around the pit down to its very bottom for an observer to move around freely for his observations. The instrument is suspended in such a way that its lower section stays within the pit, below the platform. The instrument can rotate freely about a vertical axis.

The ring of the Unnatāṁśa is fabricated out of molded sections of brass of cross section 6×4.5 cm^2. Its inner diameter is 5.27 m, and the outer diameter is 5.35 m. The ring has two cross beams, one vertical and the other horizontal. At the center of the beams, where the center of the ring also lies, there is a hole for mounting a sighting tube. The total mass of the instrument is estimated over 600 kg.

The circumference of the instrument is divided into 60 major divisions following the sexagesimal scheme preferred by Jai Singh's astronomers. The major divisions have been divided into 6 parts of 1 degree each, which have been, once again, partitioned into 10 small divisions of 6′ of arc each. Although the degree divisions of the instrument have been well laid out, the small divisions may vary in width anywhere from 4 to 5 mm, and have an average

---

[1] See Ch. 2.

Fig. 7-5 The Unnatāṁśa Yantra of Jaipur.

Fig. 7-6 The Great Astrolabe and Its Sister Unit.

length of 4.63 mm. The instrument does not have any characters marking the divisions.

The instrument works as follows. The observer mounts the sighting tube at the center, and then rotating the tube about its horizontal axis and also turning the instrument about its vertical axis, obtains the object in his sight. The altitude of the object, indicated by the pointer attached to the tube, is then read off the circular scale. The steps around the pit facilitate this operation by eliminating the need for a stepladder.

The Unnatāṁśa is quite a heavy piece of equipment, and yet it has only two cross beams to support its structure. It is suspected that the limited support provided by a pair of beams is not sufficient, and its sections are likely to sag, causing an error.[1] Besides, the divisions on the top half of the instrument are difficult to read because there are no labels anywhere. However, the divisions are not essential for a reading. The divisions were probably provided to check from time to time if the instrument's horizontal axis had shifted from its position.

The Great Astrolabe

The Great Astrolabe is suspended from a massive wooden beam supported by 3.3 m tall masonry pillars to the west of the Ground Plan. The orientation of the pillars is such that the line joining them makes an angle of about 23^0 with the plane of the meridian. The astrolabe is the largest instrument of its kind in the world, measuring 2.43 m vertically, including the crown; 2.115 m horizontally; and weighing over 400 kg. The metal rod, or pin from which the instrument hangs down, juts out to the west, i.e., the instrument does not hang directly under the beam. This positioning prevents the beam from obstructing observations around the zenith point of the sky. The instrument is in good shape except for a large hole approximately 3×1 cm^2 in area and two small ones 3 to 4 mm in width. The instrument has been fabricated from a single plate of molded brass patched up in places with lead. The plate varies in thickness anywhere from to 1½ cm around the middle and about ¼ to 1¼ cm near the edges. The plate has been engraved as a *Safīhā* for 27^0 latitude. Its back side is unfinished and rough. The instrument does not have a rete as such, instead it has an ecliptic ring with the signs of the zodiac marked on it and a sighting tube that mounts at the center.

Division Scheme

The rim of the instrument is divided according to the sexagesimal scheme, i.e., into four quadrants of 15 major divisions each, or a total of 60 divisions of one-

---

[1] The cross beams themselves are estimated to be of 240 kg mass.

Under this division scheme, the small divisions measure 10′ of arc each or 1.67 *palas* of time. The average width of the small divisions is 3 mm. For the purpose of easy reckoning, the degree marks have been displayed with dotted lines. Next to the circular scale near the rim, the usual circle of the tropic of Capricorn is inscribed. The circle is a 10 cm wide ring with its inner diameter equal to 1.94 m. The ring of Capricorn is followed by a ring of equator with an inner diameter of 1.268 m. Next, the ring of the tropic of Cancer has, similarly, a diameter of 0.83 m.

The instrument is built for the latitude of Jaipur as there are 27 degree markings between the zenith and the pole. The almucantars, or altitude circles are drawn one degree apart, with every sixth almucantar indicated with a dotted line. The equal azimuth lines are also 1 deg apart and have every sixth member indicated by a dotted line for clarity. The equal azimuth lines, for some reason, do not terminate at the horizon as is the usual practice with ordinary astrolabes, but continue on below the horizon down to the very rim. This feature is also noted in other astrolabes of the same general period in India. Strangely, the tympan (plate surface) of the instrument has a permanently engraved ecliptic indicated by a 6 mm thick ring, which has a diameter of 1.38 m and touches the tropic of Capricorn over the south point at the meridian, indicating its own orientation in the sky when the winter solstitial point is on the local meridian and the equinoxes are at the horizon. The pole of the ecliptic, as expected, is then situated on the meridian. The ecliptic ring is divided into degrees. Further, the tympan has permanently marked positions of some 50 stars. The locations of these stars have been indicated by pin holes inlaid with silver. The star names have been described in Sanskrit using Devanagari script. The orientation of the Big Dipper indicates its position in the evening sky when the winter solstitial point is on the meridian.

Although the instrument is unique in certain aspects, its workmanship is not of the highest quality as compared with the workmanship of the fine astrolabes in the collection of Jai Singh. At several places, double lines drawn by mistake may be seen on the instrument's surface.

How the Instrument Works

The chief function of the instrument is telling time. In order to determine time, the observer mounts the ecliptic ring and the sighting tube. He consults an almanac or some other astrolabe to find the position of the sun on the ecliptic for the day.[1] The position of the sun is marked with a piece of chalk on the ecliptic ring. Next, with the help of the sighting tube the observer measures the zenith distance or altitude of the sun at the moment. The chalk mark,

---

[1] If a Hindu almanac constructed according to the *nirayana* system is consulted, correction for the shift of the first point of the Aries would have to be applied. Ref. 2, p. 160.

representing the location of the sun on the ring is then coincided with the appropriate altitude circle or almucantar on the surface (tympan). As the almucantars for the morning and the afternoon hours are different, care is taken in choosing them. As pointed out earlier the circumference of the instrument is divided into *ghaṭīs*. The angular distance between the meridian and the line joining the mark of the sun on the ring with the pole, indicates the time from midday. The angular distance is read directly from the circumference of the instrument. If the time is desired in hours, minutes, and seconds, the angle between the meridian and the sun-pole line may be converted into these time units, recalling that 15 degrees correspond to an hour.

The instrument also enables the observer to determine, for a given moment, the rising and the setting signs of the zodiac on the horizon, as well as those on the meridian above and below the horizon. In daytime the procedure is similar to that of finding the time. Once the ring of the ecliptic has been properly located over the surface as described above, its intercepts with the horizon and with the meridian indicate the signs at desired locations. The signs thus indicated are *sāyana* and not *nirayana*, and a correction may have to be applied to change them to *nirayana*.[1]

At night both the time and the signs are found with the stars. The instrument has positions of 50 or so stars, covering all seasons, inlaid on the instrument's surface. The procedure involves orienting the ring of the ecliptic in a proper position by observing a prominent star visible in the sky at the time. First, rotate the ecliptic ring such that it coincides with the ecliptic engraved on the *Safīhā*. When this is done, the solstitial-point points to the top of the instrument. Next, rotate the sighting tube or alidade so that it is over the star's inlaid position, and place a mark on the alidade over that position. Also put a mark on the ecliptic ring where the alidade intercepts it. Having done this, next determine the altitude of the star as usual and mark off the appropriate almucantar—east or west—representing that value. Now rotate the alidade so that the point marked on it indicating the star coincides with the almucantar just marked. Finally, rotate the ecliptic once again so that the mark on it lines up with the alidade. These series of operations set the ecliptic in position for telling the time, as well as the *lagna* or rising sign.

The time is found from the angular distance of the vernal equinox, or first point of Aries, from the meridian. However, with this procedure the time indicated is sidereal. For example, if the angular distance of the vernal equinox from the meridian is zero, the sidereal time is 00:00 hr. The local time, which is reckoned with the sun, is determined by applying a correction based on the angular distance of the sun from the vernal equinox or first point of Aries on the ecliptic that day. Tables such as the one used for the Rāśivalayas may be used for this purpose. The sign on the ecliptic intercepted by the eastern horizon is

---

[1] Ref. 2, p. 160.

the *lagna*. The other signs of interest may be determined from the intercepts of the ecliptic with the western horizon and the meridian.

Iron Plate

Next to the Great Astrolabe, sharing with it a common support pillar, is a circular plate of iron of 2.1 m in diameter, or roughly the same dimensions as the astrolabe. This circular plate may be called the sister unit of the Great Astrolabe. The sister unit is made of 55-60 sheets of rusting iron, ¼ to ¾ cm thick and riveted together. On one side of this unit, along its rim, runs an iron strip, 5-6 cm wide and ½ cm thick. The strip does not cover the entire circumference, however, but has a 90 cm long section missing. At the center of the unit is a 1.5 cm wide hole, perhaps for attaching a sighting tube. Near the hole, there is another similar hole, possibly drilled by mistake. On the whole, the instrument, if it may be called one at all, is a rather crude device.

Markings

The sister unit of the Great Astrolabe has very few markings. The strip running along the rim has a faint scale which is difficult to read. It can be discerned, however, that the scale had been divided into six-degree divisions. These divisions were subdivided into 10 parts which were further divided into six each. The least count of the scale thus had been 6′ of arc. The scale, in other words, has a division scheme similar to the periphery of the Great Astrolabe. Finally, above the center of the instrument there is a 17 cm long vertical line.

The purpose of the instrument is not clear. One thing, however, is certain—that in itself, the instrument cannot be an astrolabe, and that it predates the Great Samrāṭ. In an early map of the observatory that does not show the Great Samrāṭ, the Great Astrolabe and its sister unit may be identified appearing as a pair of round objects.[1] Possibly, the unit had been meant to be the back-support for a brass astrolabe that was never completed.

---

[1] Map no. 15., The Museum.

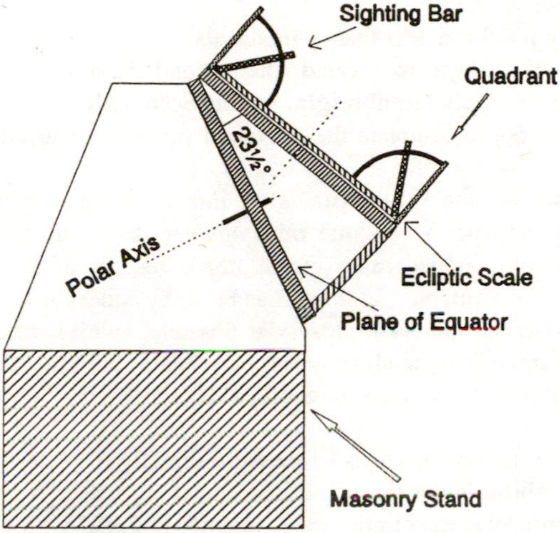

Fig. 7-7 The Principle of Krāntivṛtta II.

Fig. 7-8 Krāntivṛtta II.

Krāntivṛtta Yantra II

On a masonry platform, less than a meter above the surrounding turf, to the north of the Nāḍīvalaya is located the second Krāntivṛtta yantra. The instrument is of relatively recent origin, having been built in 1901-02 by Garrett when he decided not to complete the "original yantra."[1] Garrett writes:

> "As, however, the Krāntivṛtta is an interesting instrument, it was decided to construct a new one on a smaller scale, utilizing for this purpose an old divided brass circle, three feet in diameter, which happened to be available. In this it has been the endeavor to follow the instructions of Jagannātha as closely as possible, and it is therefore that the new Krāntivṛtta is, in all essential particulars, the exact counterpart of the instrument as used in Jai Singh's days."[2]

The purpose of the instrument is to measure directly the celestial latitude and longitude of an object in the sky. The instrument consists of a graduated circle or, more appropriately, an ecliptic scale, inclined at 23;27 with the plane of the equator as illustrated in the Fig. 7-7. As the scale rotates about the earth's polar axis, it can be made parallel to the plane of the ecliptic at any desired moment. A bar with a quadrant each at its two ends is mounted over the scale such that the bar rotates about a perpendicular axis through the center of the ecliptic scale. The quadrants attached to the bar have sighting strips. The least count of the equator scale fixed to the masonry is ⅓ deg, and that of the ecliptic scale 1/6 deg.

For measuring the coordinates of a celestial object, the ecliptic scale is aligned with the ecliptic in the sky. This may be done by sighting a "guide star" first whose longitude is already known. Line up the sighting bar with the star's longitude on the ecliptic scale. Next rotate the entire ecliptic plate along with the sighting bar such that the guide star is visible through one or the other sighting slits. This procedure orients the ecliptic plate of the instrument in proper position. For observing the object in the sky, rotate the sighting bar around so that its sighting strip and the slit mounted on one of the quadrants line up with the object under observation. The longitude of the object is then read off the ecliptic scale and its latitude from the quadrant scale. Because the orientation of the ecliptic continuously changes with the rotation of the earth, the instrument is somewhat cumbersome to work with. The instrument may also be used for indicating the signs of the Zodiac at the horizon and at the meridian by merely aligning the ecliptic scale in proper position with the sun or with a guide

---

[1] Garrett pp. 45-47.

[2] Garrett, p. 45.

star whose longitude is known, and then reading off the signs on the scale intercepted by the prime vertical and the meridian on the instrument.

MODELS OF INSTRUMENTS

The observatory has the following models of the instruments.

1. Krāntivṛtta yantra (1)
2. Samrāṭ yantras (3)
3. Rāma yantra (2)

Krāntivṛtta Yantra

An incomplete model of Krāntivṛtta, with its sighting bar missing, is displayed on a portable stand of wrought iron some 1½ m high. The model has only two plates, with upper surfaces finished smooth for scales. The plates, about 82.5 cm in diameter, have been fabricated from molded brass about a centimeter thick and inclined with each other approximately at an angle of 23;30, the obliquity of the ecliptic. The plates represent the planes of the equator and the ecliptic respectively. The lower plate representing the plane of the equator makes a 28;47 angle with the vertical, suggesting that the instrument was built for Delhi, where the latitude is 28;38.[1] The upper plate, representing the plane of the ecliptic, rotates about an axis—the polar axis—in such a way that one edge is always in contact with the circumference of the lower plate.

Both plates have scales around their circumference, which have been divided into 12 sections each, similar to the face of a clock. These scales have also been divided into 60 main divisions according to the sexagesimal scheme. The divisions have been further subdivided into 3 parts. Thus the least count of the scale is 20′ of arc. The upper plate representing the plane of the ecliptic is attached to a ring that slides over the lower plate. This ring also has a scale around its periphery divided into 60 main divisions. It also has four pointers spaced 15 main divisions apart. A few of these main divisions have been further divided into small divisions, 0.71 cm wide, and measuring a degree each. A few *nakṣatra* names such as *Aśleṣā* and *Ārdrā* also appear on it. The script of the inscription is Devanagari. The workmanship of the instrument is of the same order as that of the Great Astrolabe.

---

[1] The *Yantraprakāra* describes a Krāntivṛtta of one cubit (40-50 cm) in diameter. *Yantraprakāra* (S), p. 69; and *Yantraprakāra* (J), p. 14.

Samrāṭ Yantras

In the observatory compound there are three models of the Samrāṭ yantra. Two of these models are of metal and bolted to portable stands of iron bars about 3/4 meter high, and the third is made out of wood.

Model A

The model A of the Samrāṭ appears to be of earlier origin than its sister unit model B. For this model, instead of nuts and bolts to hold the different parts together, iron rivets have been used. The nuts and bolts have been used only to secure it to the stand. The gnomon of the model is 87.25 cm tall, 191.1 cm long, and fabricated out of bar iron. Its diagonal makes an angle of 27;09,57 with the horizontal. The upper surface of the gnomon is 1 cm thick and has a scale divided into degrees.

The west and the east quadrants of the instrument have been fabricated out of a 2 cm thick and 13.5 cm wide sheet of brass and have radii of 42.9 and 43.0 cm respectively. The quadrant scales have been divided into 15 *ghaṭikās* each, according to the Hindu scheme of time division. The *ghaṭikās* have been further subdivided into 6 subparts, thereby giving the instrument a capability of measuring time down to 10 *palas* or 4 minutes. The four-minute divisions are engraved, however, on the edge of the quadrants and not on the face.

Model B

Model B has a 92.8 cm high and 204.5 cm long gnomon fabricated out of iron bars. Its quadrants are made out of brass sheets 15.6 cm wide. The gnomon makes a 26;59 angle with the horizontal. The quadrant on the west of the gnomon has a radius of 49.7 cm and the one on the east, a radius of 52.3 cm. The quadrants are divided into hours, one-half hours, and minutes, with the small divisions measuring 5 minutes. From the nuts and bolts used in the model, and from its general appearance, the instrument appears to have been built in modern times. The division of the quadrants into hours and minutes also suggests that the instrument is of recent origin.

Model C

The third model of the Samrāṭ is made out of wood on a roughly 2×2 m^2 base, and it is of more recent origin as its construction suggests. Most likely, the model was once displayed in the former Prince Albert Museum along with other instruments of astronomy. It has a 1.12 m tall gnomon and quadrants of appropriate radii.

Rāma Yantra

The Rāma yantra models are, essentially, two complementary cylindrical structures of masonry located to the south of the Kapāla platform. The units have been built on circular platforms about ½ m above the surrounding turf. The cylinders are 1.28 m high and have an outer diameter of 2.12 m and a wall thickness of 18 cm. They have 12 radial sectors and an equal number of columns each. The columns and the radial sectors both are 15 degrees wide. They are not duplicates of their larger counterparts in the observatory. The inner surfaces of the models are of white stone and profusely graduated. There are no inscriptions on the scale, however.

One may conclude that the Rāma yantra models are of recent origin as they are not shown in any of the old observatory maps.

PORTABLE INSTRUMENTS

At the observatory site:

1. Geocentric or Ptolemaic model of the universe (1)
2. Terrestrial globe of wood (1)
3. Zarqālī astrolabe (1)
4. Persian astrolabes (3)
5. Devanagari astrolabes
   a. complete (3)
   b. incomplete (8)
6. Dhruvabhrama and Turīya yantra (night dial and daytime dial) (1)
7. Cūḍā yantra (ring dial) (2)
8. Soṭā yantra (time measuring device) (3)
9. Ghoṭā yantra (time measuring device) (2)
10. Yantrādhipati (time measuring device) (1)
11. Telescope (1)

Geocentric or Ptolemaic model

The geocentric or Ptolemaic model of the solar system consists of a number of circular brass plates. An inscription on the model says: "Francois le Maire Paris." The author believes that the Jesuits in the service of the Raja brought the model from Europe. In the inventory of the Jantar Mantar the model is listed as the "*Model of brass yantrarāja.*" The model is broken into two pieces now, and the pieces are stored separately.

The base plate of the model, with the signs of the zodiac inscribed near its periphery, has a diameter of 32.8 cm. Over this base plate, there are 6 large circular plates depicting planetary orbits. These plates rotate about different axes and have small circular plates attached to them representing the planets' epicycles. The diameter of the largest plate, representing Saturn's orbit, is 21.6 cm and that of its epicycle, 1.8 cm.

Terrestrial globe[1]

The terrestrial globe is made out of wood. It has a circumference of 119.2 cm and a diameter of 37.9 cm. The globe is of Jai Singh-Madho Singh era and illustrates the geographical knowledge of the times in India. On the globe, the land mass is shown in brown color and the seas in deep olive green. It has a number of latitude and longitude lines inscribed on its surface. The globe displays continents, islands, and countries. The language of the globe is Rajasthani Hindi. The globe maker has made some obvious errors. For example, he shows the sea port of Goa on the east coast of the Indian peninsula.

Zarqālī astrolabe[2]

The Zarqālī astrolabe consists of one plate of brass 1.54 cm thick and 55.2 cm in diameter. It is inscribed in Persian, but it also has Sanskrit terms engraved here and there. Its tag says in Devanagari, *Yantra Jarakālī Sarvadeśī*. The instrument was constructed in 1680 by an astrolabe maker of Lahore.

Persian astrolabes[3]

At Jaipur there are 3 astrolabes engraved in Persian script. Their diameters range from 16.4 to 34 cm, and they have 1 to 7 plates. Their workmanship is good to excellent.

Devanagari astrolabes

The Jantar Mantar stores 11 Devanagari astrolabes, of which only 3 are complete, and the others are in various stages of completion. Their diameters range from 11 to 35 cm. It appears that the astrolabes of the same design were being mass produced, but the production was abruptly halted for some reason.[4] The instruments have anywhere from a single to 7 plates. Their retes vary from a very simple ecliptic circle and a pointer to artistically chiseled brass-work. The workmanship of these instruments ranges from mediocre to excellent.

---

[1] This globe should not be confused with the European globe Jai Singh obtained from Surat. See Ch. 13.

[2] For a detailed description of this instrument, see Kaye, pp. 27-30.

[3] See Kaye, pp. 16-26.

[4] It is possible that Jai Singh or Madho Singh started an astrolabe workshop which was closed down after their death.

Dhruvabhrama and Turīya yantra

Dhruvabhrama yantra is a time measuring instrument for nighttime hours.[1] The instrument is of the Jai Singh-Madho Singh period. It consists of a 1 mm thick and 17.9×16.3 cm^2 brass plate. The instrument is based on the principle that the line joining the stars Kochab (β Ursae Minoris) and Polaris (α Ursae Minoris) revolves around the celestial pole approximately once every 23 hr and 56 min.[2] The instrument essentially measures sidereal time. It has a four-sided pointer which indicates the sidereal time, *lagna* or ascendant, the sign on the meridian, and the sign and the *nakṣatra* setting in the west. Working with the instrument, the observer sights the two stars through a slit in the plate and reads the parameters off the scales. From the knowledge of the sun on the ecliptic and from the sidereal time, the local time may be calculated. On the backside of the Dhruvabhrama is a Turīya yantra, or quadrant, for measuring time with the sun during daytime.[3]

Cūḍā yantra[4]

A Cūḍā yantra is a ring dial for telling time during daytime with the sun. At the observatory there are two units of this yantra. They are hollow cylinders of brass, open at both ends, and with a number of holes around their curved surfaces. The larger of the two yantras is 12.2 cm in diameter and 9.1 cm long. The smaller yantra is 5 cm long and has a diameter of 9.9 cm. The instrument might have been fabricated during Madho Singh's reign.

Soṭā and Ghoṭā yantras

The Soṭā and Ghoṭā yantras, as their tags label them, are time measuring devices. Their total number is five: three Soṭās and two Ghoṭās. They are based on Gaṇeśa Daivajña's (b. 1507) *Pratoda* yantra.[5] The instruments are 19.6 cm to 86.6 cm long rods of metal and wood. These instruments are most likely of the Madho Singh period.

Yantrādhipati

Yantrādhipati is circular brass plate of 34.7 cm in diameter. Both sides of the Yantrādhipati plate are engraved with a number of concentric scales. They also have alidades to attach at their center. The purpose of the instrument is to

---

[1] Padmanābha (*fl. ca.* 1400) wrote a text on Dhruvabhrama yantra.

[2] For the working of the instrument, see Garrett, pp. 62-64.

[3] *Ibid.*

[4] For the working of the Cūḍā yantra, see *Yantraprakāra* (S), pp. 79-82.

[5] For a description of the *Pratoda* yantra, see Shakti Dhara Sharma, *Op. Cit.*, Ch. 1.

measure time. According to the inscription, the instrument is of Madho Singh's period.

Telescope

The telescope at the observatory consists of two brass tubes about 19 cm long each and of about 11 cm diameter. The length of the instrument may be changed from 29 cm to 37 cm by sliding one tube inside the other. The tubes have a converging lens each of approximately 23 cm focal length. The telescope is a crude device and unsuitable for astronomical work. It does not reflect the art of telescope making practiced in India during the time of Jai Singh.

Celestial globes

The Hawa Mahal Museum of Jaipur, has a 30.6 cm diameter globe of brass. The globe has a number of mythological figures depicting major constellations engraved on its surface. The constellations are labeled in Devanagari. A second celestial globe is preserved at the Sawai Man Singh II Museum. It has a diameter of 17 cm, and it is also made out of brass. The constellations on this globe are inscribed in Persian. The workmanship of both of these globes is good.

Armillary sphere

The armillary sphere is preserved at the Hawa Mahal Museum. It is 53 cm in diameter and has 8 rings.[1]

Models of the Observatory Instruments

These instruments have been transferred from the former Prince Albert Museum of Jaipur to the observatory storage. They are models of the observatory instruments prepared for display at the museum.

1. Azimuth-altitude measuring circles (1)
2. Cakra yantra (1)
3. Krāntivṛtta (1)
4. Plaster of Paris, wood, and stone models of observatory instruments such as the Jaya Prakāśa.

[1] For further details see Ch. 2.

CHAPTER VIII

VARANASI OBSERVATORY

Jai Singh built his Varanasi observatory on the roof of Māna Mahala, a building overlooking the Ganges River.[1] The Māna Mahala is surrounded by a neighborhood called Māna Mandira. The observatory may be reached either via a narrow winding alley from the main street leading to the famous Daśāśvamedha Ghāṭa, or from the river after climbing up a good number of steps and then walking a short distance through the alley.

The parameters of the observatory are as follows:[2]

Latitude	25;18,24.9 N
Longitude	83;0,46.1 E
Elevation	110 m (approx.)
Local Time	I.S.T. +2 min, 4 sec, or Universal Time +5 hr, 32 min, 4 sec

History

Varanasi has been a center of Hindu learning since the time of Buddha, over 2,500 years ago. The city has attracted scholars and students from all over India. Jai Singh's forefathers sent their children to Varanasi for education. The French traveler, Jean Baptiste Tavernier, during his visit to Varanasi around 1650, reports meeting two sons of Mirzā Raja Jai Singh, who were studying at a "college" founded by the Raja.[3,4] The college was in the same building—or one close to it—on the roof of which Sawai Jai Singh built his observatory almost a century afterwards.

The building where Tavernier met the two princes is said to have been constructed by the princes' grandfather, Māna Singh, in 1614. However, Prinsep, after studying the architectural details of the building around 1830, believes that the building must have been much older. He writes:

> "This specimen of architectural effect can with difficulty be ascribed to so recent a period as the age even of Māna Singh, from whom this observatory derives its name and existence; for the stone is quite worn away in many places by the weather. It may have belonged to a more

---

[1] The other names of the metropolis of Varanasi are Benares and Kashi (Kāśī).

[2] Kaye, p. 61.

[3] Mirzā Raja Jai Singh had two sons, Rāma Singh and Kīrata Singh.

[4] Tavernier, *Op. Cit.*, pp. 234-235, Ch. 1.

> ancient building before being set up in its present position by that chief; but whether executed or borrowed by him, it bears away the palm of antiquity in the town, and is a *chef d'oeuvre* of its kind. . . ."[1]

Scholars have argued that an old observatory already existed at the site and that Jai Singh merely renovated it. The old observatory is said to have been built during the reign of the Mughal Emperor Akbar (1542-1605). However, scholars have found no evidence of this assertion, other than some hearsay accounts of European travelers such as Robert Barker in the late 18th century. Barker writes:

> "This observatory at Benares is said to have been built by the order of the emperor Ackbar [Akbar]; for as this wise prince endeavored to improve the arts, so he wished also to recover the sciences of Hindostan, and therefore directed that three such places should be erected; one at Delhi, another at Agra, and the third at Benares."

Barker's belief that the Emperor Akbar built the observatory 200 years before Jai Singh, merely represents the opinions of his local guides and are open to serious questioning. Evidence has yet to surface that Akbar took any interest in astronomy and that he built observatories at the three places enumerated by Barker's sources.

Most likely, the Brahmins who received endowments from some ancestor of Jai Singh, from Raja Māna Singh, perhaps, used the open terrace of the building for routine observations. This led to the belief in later years that there had already been an observatory at the place for hundreds of years before Jai Singh built his own of masonry and stone.[2,3]

The date when Jai Singh erected his observatory is uncertain. Scholars have suggested several dates ranging from 1693 to 1737. The early date of 1693 is due mainly to a footnote in the English version of the travelogue of Tavernier, edited by V. Ball and published in 1889.[4] A plaque at the observatory site, erected by Bhavan, gives 1710 as the date.[5] John Lloyd Williams, who investigated the observatory in 1792, thinks it was 1737. Kaye also argues in

---

[1] James Prinsep, *Benares Illustrated in a Series of Drawings*, Calcutta, 1830.

[2] For example, see Krishnan D. Mathur, "Indian Astronomy in the Era of Copernicus," *Nature*, pp. 283-285, Vol. 251, Sept. 27, (1974).

[3] Prior to Jai Singh, the Hindu astronomers generally preferred portable instruments such as the astrolabe and armillary sphere, which did not require any fixed moorings and could be taken to any place where the astronomer wanted to observe.

[4] V. Ball, commenting on the travels of Tavernier, erroneously writes that the observatory of Varanasi was built in 1693 by Jai Singh. The Mirzā Raja Jai Singh died in 1667 and Sawai Jai Singh was only a child of five in 1693. Therefore neither of them could have erected an observatory. See Tavernier, *Op. Cit.*, p. 234, Ch. 1.

[5] The plaque was erected during the restoration of 1911. The date is therefore unreliable.

favor of 1737.[1] However, the observatory is mentioned in the *Samrāṭ Siddhānta*, which had been completed by 1730.[2] The observatory therefore must have been constructed before 1730. The most likely date for the observatory's construction is between 1724 and 1730.

Travelers' Accounts

Varanasi, or Benares, is a principal city of north India and is conveniently located on a main route of travel. Thus many Europeans visited it during the 18th and 19th centuries, and several of them have left detailed accounts of their visits to the observatory. The accounts of these travelers, although unreliable at times, are nonetheless informative.

Robert Barker

The earliest account of the Varanasi observatory is by Robert Barker. Barker came to India in 1749 and for a time was the commander-in-chief of the army of the East India Company in Bengal.[3] Barker was curious about the astronomical knowledge of the Brahmins. In particular, he wanted to know how they predicted solar and lunar eclipses. He narrates his experience as follows:[4,5]

> ". . . I made inquiry, when at that place in the year 1772, among the principal Brahmins, to endeavor to get some information relative to the manner in which they were acquainted of an approaching eclipse. The most intelligent that I could meet with, however, gave me but little satisfaction. I was told, that these matters were confined to a few, who were in possession of certain books and records; some containing the mysteries of their religion, and others the tables of astronomical observations, written in the Sanskrit language, which few understood but themselves: that they would take me to a place which had been constructed for the purpose of making such observations as I was

---

[1] Kaye, p. 64.

[2] SSRS, p. 1162; and SSMC, p. 38. According to Bahura, the *Samrāṭ Siddhānta* manuscript of the Sawai Man Singh II Museum was completed in 1728. Bahura II, p. 58. The author has not been able to consult this manuscript of the Museum.

[3] *The Dictionary of National Biography*, ed., Leslie Stephen and Sidney Lee, pp. 1128-1129, Oxford Univ. Press, 1964.

[4] Robert Barker, "An Account of the Brahmin's Observatory at Benares," *Philosophical Transactions of the Royal Society of London*, Vol. 67, Pt. 2, pp. 598-607, (1777).

[5] Barker's report also appeared in its German translation in *Le Gentils Reisen in den Indischen Meeren in den Jahren 1761 bis 1769, und Chappe d'Auteroche Reise nach Mexiko und Californien im Jahre 1769 aus dem franzosischen, nebst Karl Millers Nachricht von Sumatra und Franziscus Masons Beschreibung der Insel St. Miguel aus dem Englischen*, published by Carl Ernst Bohn, Hamburg, 1781.

inquiring after, and from whence they supposed the learned Brahmins made theirs.

"I was then conducted to an ancient building of stone, the lower part of which, in its present situation, was converted into a stable for horses, and receptacle for lumber; but, by the number of court-yards and apartments, it appeared that it must once have been an edifice for the use of some public body of people. We entered this building, and went up a staircase to the top of a part of it, near the river Ganges, that led to a large terrace, where, to my surprise and satisfaction, I saw a number of instruments yet remaining, in the greatest preservation, stupendously large, immovable from the spot, and built of stone, some of them being upwards of twenty feet in height; and, although they are said to have been erected two hundred years ago, the graduations and divisions on the several arcs appeared as well cut, and as accurately divided, as if they had been the performance of a modern artist. The execution in construction of these instruments exhibited a mathematical exactness in the fixing, bearing and fitting of the several parts, in the necessary and sufficient supports to the very large stones that composed them and in the joining and fastening each into the other by means of lead and iron."

Barker goes on to give detailed descriptions of the instruments. He describes six instruments in all, namely: the Dakṣinottara Bhitti, large Samrāṭ, Nāḍīvalaya, Cakra yantra, Digaṁśa, and small Samrāṭ. Barker's description agrees fairly well with the appearance of the instruments today. Barker did not comprehend the function of the Digaṁśa; nonetheless, his description of the instrument is accurate. He measured the dimensions of the large Samrāṭ. He judged the radius of a Dakṣinottara Bhitti located on a separate roof to be about 20 feet (6 m). The small divisions of this instrument were 5 mm wide and measured 3′ of arc. Barker missed in his report the second Dakṣinottara Bhitti yantra, which was located on the east-facing wall of the large Samrāṭ's gnomon.

The sketches accompanying Barker's report were made by Archibald Campbell, an engineer in the service of the East India Company. Campbell's drawings are accurate in their overall appearance but they may not be relied upon for details. There are obvious errors in them which were pointed out by later visitors. For instance, Campbell depicts the small Samrāṭ a little distance away from the Digaṁśa. Additionally, he leaves out the scale markings on the instruments.

John Lloyd Williams

Barker's report created a certain degree of excitement in the members of the British Royal Society, and they requested John Lloyd Williams, a resident of Varanasi, to send additional information on the observatory.[1] Williams took precise measurements of the instruments illustrated in Barker's article and added two more to the list that Barker had prepared. One of these instruments was a pair of meridian arcs on the eastern wall of the Samrāṭ that Barker had missed. Williams writes:

> "On the outer wall of the style [gnomon of Samrāṭ], fronting the east, at the height of 10 feet and 10 inches from the base, are fixed two iron pins, each forming a centre, from which circular lines are drawn, intersecting each other as in the annexed representation with a parallel line drawn underneath, which has the hour, *gurry* and *pull* lines marked on it." He also describes two pairs of iron rings fixed along the outer edge of the gnomon of the large Samrāṭ.[2,3]

About the second instrument he says:

> "In addition, . . ., on the south-east quarter of the building is a large black stone, 6 feet 2 inches in diameter, fronting the west; it stands on an inclined plane. I could not learn the use of this instrument; but was informed that it never had been completed."

Williams does not describe the Dakṣinottara Bhitti located on a separate terrace. This instrument had been reported by Barker but left out by Campbell in his drawings. Williams also leaves out the small Samrāṭ. In describing the history of the place, Williams quotes a local resident, Nabāb ᶜAli Ibrāhīm Khān:

> "The area, or the space comprising the whole of the buildings and instruments, is called in Hindoo, *maun-mundel*; the cells, and all the lower part of the area were built many years ago, of which there remains no chronological account, by Rajah Maunsing [Māna Singh],

---

[1] John Lloyd Williams, "Further particulars respecting the Observatory at Benares, of which an Account, with Plates, is given by Sir Robert Barker, in the LXVIIth Vol. of the Philosophical Transactions," *Philosophical Transactions of the Royal Society of London*, Vol. 88, Part 1, pp. 45-49, (1793).

[2] The rings apparently were used for sighting the North Star. At Jaipur there is a special device just for this purpose. See. Ch. 7.

[3] The purpose of the parallel lines under the arcs, as depicted by Williams in his drawing, is not clear.

for the repose of holy men, and pilgrims, who come to perform their ablutions in the Ganges, on banks of which the building stands.

"On the top of this the observatory was built, by the Rajah Jeysing, for observing the stars, and other heavenly bodies; it was begun in 1794 *Sambut* (1737 A.D.), and, it is said, was finished in two years. The Rajah died in 1800 Sambut (1743 A.D.).

"The design was drawn by Jaggernaut [Jagannātha], and executed under the direction of Sadashu Mahajin [Sadāśiva Mahājana]; but the head workman was Mahon [Mohana], the son of Mahon a pot-maker of Jeypoor. The Pundit's pay was five rupees per day; the workmen's two rupees, besides, presents; some got lands, or villages, worth 3 or 400 rupees yearly value; others money."[1]

Williams adds that his Brahmin sources, "all agreed that this observatory was never used, nor did they think it capable of being used, for any nice observations; and believe that it was built more for ostentation, than, the promotion of useful knowledge."

William Hunter

A few years after Williams's report appeared in the *Transactions of the Royal Society of London*, Hunter published his paper on Jai Singh in 1799, based on his notes taken in 1786. Hunter correctly explained the function of the Digaṁśa, and he also corrected William's report. On the south-east quarter, the 6 ft. 2 inch diameter black stone was not facing west as Williams had reported but was in the plane of the equator. Although he compared it with the Nāḍīvalaya, he did not observe any graduations on the circle and thought that the instrument had never been completed. The instrument could have been another Nāḍīvalaya, as Hunter suggested, or a Krāntivṛtta of the type at Jaipur.

Bāpūdeva Śāstrī

Bāpūdeva Śāstrī (1820-1889) was a professor of astronomy at Benares Sanskrit College. In 1866, he wrote a monograph in Sanskrit, *Kāśī-Mānamandira-vedhālaya-varṇanam*, describing the observatory, its instruments, and their function.[2] Śāstrī describes 8 instruments in all, including the one that he could not identify.

---

[1] Considering the wages of the astronomers at Jaipur, only Rs. 1 per day, the workmen's wages of Rs.2 per day seems a little high. See Ch. 12.

[2] *Mānamandira Observatory of Kāśī by Bapudeva Shastri*, ed. and tr. by Shakti Dhara Sharma, pp. 21-28, Kurli, 1982.

He writes:

"In Varanasi on the banks of Ganges, not far away from *Maṇikarṇikā Ghāṭa* there is a building, named Mānamandira in the northwest direction, which was constructed by the king named Māna Singh of Amber city in the province of Rājapūtānā.[1] Approximately 150 years ago, his (Māna Singh's) famous descendant Jai Singh constructed astronomical instruments in that building for observing planets and stars."

Śāstrī then goes on to describe the instruments and their application. He describes the Dakṣinottara Bhitti yantra first. Since this instrument was modified by Bhavan at a later date in 1911, and also because Śāstrī's description differs from that of Barker, we reproduce it here in full. Śāstrī writes:

"There is a Yāmyottara yantra in the north-south direction in the form of a wall made up of lime-brick-stones. It is 7⅓ hands in height, 6 hands, and 1⅔ aṅgulas in breadth and 16½ aṅgulas in thickness. One of its sides, facing east is whitewashed and smooth. On this side at the points near top corners there are fixed two gnomons (Śaṅkus) made of iron, which are separated by 5 and 5/24 hand from each other and are fixed at a height of 7 hands less 2 aṅgulas from the ground. With the gnomons as centers and radii equal to the distance between them, there are drawn two intersecting quadrants. Below both of these quadrants are concentric quadrants which are uniformly divided consecutively in 15 parts (6^0 each) 90 parts (1^0 each) and 900 parts ($6'$ each). This instrument can be seen by the visitors at first sight while entering the Mandira."

Śāstrī also describes the unknown instrument as follows:

"Toward east of this north-south wall (Dakṣinottara Bhitti yantra) there is an adjacent plane with breadth equal to that of the wall and length equal to seven hands. Although it was initially leveled like water [surface] but now it has become uneven. On this plane there were fixed two pegs of iron with holes at the top. These are fixed along eastern direction determined by the two gnomons in the wall. At present only one of them in the east exists there.

"Quite near this plane, there is a leveled circular plane made of lime bricks and with diameter equal to one hand, 9¼ aṅgulas. There is also another one (second) leveled circular plane made up of stones and

[1] The modern state of Rajasthan.

> having diameter equal to two hands and 7 aṅgulas. Nearby this plane there is a leveled square with side equal to one hand and 11 aṅgulas. The graduations on these two circular platforms and the square are erased now. But it seems that earlier these were made to determine the gnomonic shadow and the azimuth."

The two circles described in Śāstrī's narration could have been an Agrā yantra and a Palabhā yantra. The observatories of Ujjain and Mathura also had a pair of similar instruments. Śāstrī notes that the graduations on the instruments in general had become erased and that some of the instruments had either tilted or had broken down.

Joseph Dalton Hooker

Joseph Hooker, having visited Varanasi in March 1848, wrote the following report: "The building is square, with a central court and flat roof, round which the astrolabes, &c. are arranged."[1] Hooker drew sketches of three instruments, namely, the large Samrāṭ, Nāḍīvalaya, and Krāntivṛtta. According to his drawings the large Samrāṭ still had some graduations left on its quadrants. The Nāḍīvalaya had apparently lost most of its markings. In his drawings, Hooker shows a post to the east of the Nāḍīvalaya. These days the post is to the south of the instrument as Campbell shows in his drawings made in 1772.

When Hooker visited the observatory, it was in a state of neglect. He says that the observatory was ill kept, "dirty and ruinous, and the great stone instruments were rapidly crumbling away. . . ."

> "The observatory is nominally supported by the Rajah of Jeypore. . ., who doles out a too scanty pittance to his scientific corps [employees at the observatory]. . . . The observatory at Benares, and those at Delhi, Matra [Mathura] on the Jumna, and Oujein, were built by Jey Sing, Rajah of Jayanagar, upwards of 200 years ago. . . ."

Obviously, Hooker confuses Sawai Jai Singh with the great grandfather of the Raja, the Mirzā Raja Jai Singh.

Claude Boudier, a French Jesuit at Chandernagore, passed through Varanasi in March 1734 to meet Jai Singh at Jaipur and measured the latitude and longitude of Varanasi.[2] However, he is silent about the observatory. Tieffenthaler, who certainly must have visited the city of Varanasi during his wanderings within the country, also does not mention the observatory. Perhaps

---

[1] Joseph Dalton Hooker, *Himalayan Journals*, Vol. 1, pp. 67-70, reprint, New Delhi, 1969.

[2] "Observations . . . ," *Lettres*, pp. 775-776.

the observatory, located in an inconspicuous place, skipped the attention of these two travelers. Fanny Parks also does not write about the observatory in her *Wanderings of a Pilgrim.*[1]

Burrow (Circa 1783)

Burrow also made a reference to the observatory. However, it is doubtful that Burrow visited the observatory site himself. He only writes about one instrument, the Dakṣinottara Bhitti, and does not give any details at that.[2]

Based on the accounts given in literature, we compile the following list of the instruments which had been at the observatory site at one time or another:

1. Large Samrāṭ
2. Small Samrāṭ
3. Digaṁśa
4. Nāḍīvalaya
5. Cakra yantra
6. Dakṣinottara Bhitti I
7. Dakṣinottara Bhitti II
8. Nāḍīvalaya or Incomplete Krāntivṛtta (probably)
9. Palabhā (probably)
10. Agrā (probably)

RESTORATION

The Building

The observatory and the building on top of which it had been built remained in a dilapidated state for a long time, as the accounts of Hooker and Śāstrī point out. Sherring, a resident of Varanasi, also laments that the caretakers of the observatory neglected its upkeep.[3] Eventually, when peace and stability returned to the State of Jaipur, its Maharaja began to take an interest in his property outside the state, such as at Varanasi. Realizing the damage done to the building by years of neglect, he ordered its repairs. The repairs to the building and the bathing *ghāṭa* adjoining it began in 1871 and were completed five years later in 1876.[4]

A new section, facing the river, was added to the building at this time. This new addition can be clearly recognized when viewed from the riverside because

---

[1] Fanny Parks, *Op. Cit.* Ch. 5.

[2] Dharampal, *Indian Science and Technology in the Eighteenth Century*, p. 72, Delhi, 1971.

[3] Sherring gives a detailed account of the observatory. However, there is little new information in his account. His description is based on Hunter and Śāstrī, as he himself acknowledges. Hunter, *Op. Cit.*; and Śāstrī, *Op. Cit.* Concerning the neglect of the observatory, he writes that many Europeans passing through Benares and other sightseers visited the famous observatory, paying a fee to the man in charge. However, the man seemed to be utterly careless concerning the proper maintenance of the instruments. M. A. Sherring, *The Sacred City of the Hindus*, pp. 131-137, London, 1868.

[4] The total cost of the project was Rs.23,421. Public Works Reports for the years 1871-1876, The Museum, Jaipur.

of its relatively modern architecture. After the repairs, the building was converted into a guest house of the Jaipur State.

There are no records to reflect that the observatory instruments were also restored along with the building. At best, the masonry work of the instruments might have been patched up at this time. Ernst B. Havell, writing around 1905, almost 30 years after the renovation of the building, remarks that none of the instruments of the observatory were in working order when he visited the site.[1]

Instruments

After the observatory of Jaipur had been fully restored in 1902, Madho Singh, the Maharaja of Jaipur, ordered a similar restoration of the Delhi and Varanasi observatories and entrusted Gokul Chand Bhavan, who had already restored the observatory of Jaipur, with the responsibility.[2] Bhavan accomplished the restoration work at Varanasi in 1911.[3]

During the restoration, Bhavan drastically altered the scale markings of the observatory. He gave them a Western flavor. He modified them to read according to the Western system of time measurement—in hours, minutes, and seconds. He did not abandon the Hindu system of measurement altogether but gave the system a lesser importance.

Shortly after the independence of the country in 1947, when the princely states merged into the Republic of India, the management of the observatory was transferred to the Archaeological Survey of India, and since then a caretaker staff looks after the building and the observatory instruments.

Present Status

As before, the observatory is located on two different roofs separated by an enclosed courtyard. The roof on which most of the instruments are constructed is about 5 m above ground level and is of irregular plan. It is 31 m long, 9 to 12 m wide, and approached from the northeast corner via a flight of stairs.

In earlier times, as the drawings of Campbell and Hooker show, there were no tall buildings surrounding the observatory that would obscure the sky from the instruments. However, this is no longer the case. The roof, on which most of the instruments are located, is now surrounded on three sides by man-made

---

[1] Ernest Binfield Havell, *Benares, the Sacred City*, 2nd ed., p. 130, London, 1905.

[2] The sanctioned cost of the repairs for the Varanasi observatory was Rs.6,319. R.S.A., File No. 553, Jaipur State.

[3] Bhavan, p. 3. Bhavan was assisted by Candū Lāla, an overseer, and by Bhāgiratha, a *mistrī* or technician to the project.

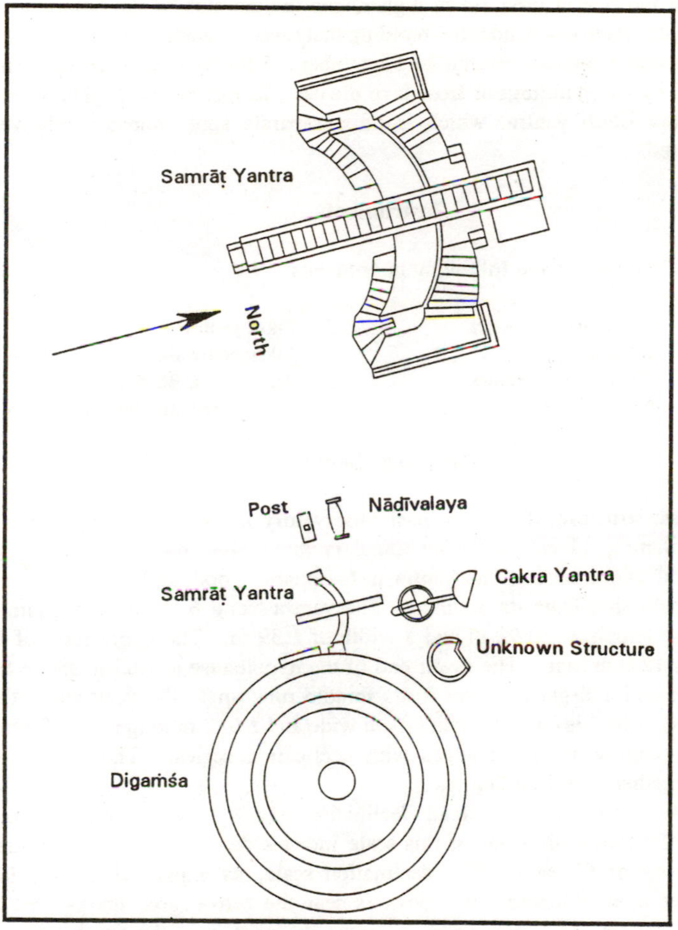

Fig. 8-1 The General Plan of the Varanasi Observatory.[1]

---

[1] based on Kaye, plate XXVI.

structures or tall trees. To the south there is a two-meter high protective wall and a *neem* tree that hinder observations close to the horizon.[1] The building constructed during the renovations of 1871-1876 is also on this side and hinders observing to some extent. To the north, adjacent to the large Samrāṭ, there is a tall tamarind tree at least 2-3 m higher than the gnomon of the Samrāṭ. On the west side, there is a residential building that casts its shadow in the afternoon on the Samrāṭ from late March to September. The only side which is not obstructed by any buildings or trees is to the east, facing the river. The second Dakṣinottara Bhitti yantra, which is on a separate roof, remains relatively unobstructed.

Instruments

The observatory has the following instruments:

1. Large Samrāṭ yantra
2. Digaṁśa
3. Small Samrāṭ yantra
4. Nāḍīvalaya
5. Cakra yantra
6. Dakṣinottara Bhitti I
7. Dakṣinottara Bhitti II
8. Unidentified structure

The Large Samrāṭ

The tallest structure of the Varanasi observatory is its large Samrāṭ yantra. The large Samrāṭ is located near the boundary wall to the west on the main roof. The general appearance of the yantra is reasonably good, except for a single stone slab missing from its gnomon. The yantra has a 6.80 m high gnomon with a base length of 10.92 m and a width of 1.39 m. The hypotenuse of its gnomon is 12.9 m long. The lower end of the hypotenuse is 1.65 m above the roof level, and a flight of stairs in its middle runs up to the very top. The hypotenuse surface is covered with 30 cm wide and 1 to 2 m long slabs of white sandstone, engraved on the edges with declination scales. The stones are secured together with iron cramps.

The scales of the gnomon read declination angles between 69;36 N and 66;24 S. The large divisions of this scale indicate degrees which are divided into 10 parts of 6′ each. The declination scale, as explained earlier, has divisions of uneven lengths. Its divisions near the zero-degree marks are the shortest in length, or 5 mm apart, whereas the ones near the south end are 3.15 cm apart. The divisions are labeled both in the Roman and Devanagari scripts.

---

[1] The wall is of comparatively recent origin.

Fig. 8-2 The Large Samrāṭ of Varanasi.

Fig. 8.3 Digaṁśa Yantra of Varanasi.

The quadrants of the large Samrāṭ are built of interlocking sandstone slabs of varying shades, cemented to a masonry base underneath. The quadrants are 1.755 m wide and have radii of 2.780 m on the west side and 2.785 m on the east side. Steps have been provided on either side of the quadrants for easy access to the scales. Each quadrant has four scales in all—two on the upper surface and two on the sides. The division scheme of the quadrants' scales of the Samrāṭ is quite elaborate. The divisions, engraved on the upper surface of the quadrants, follow the western time-reckoning scheme of hour, minute, and second. Accordingly, the large divisions indicate one-half hours. They are subdivided into minutes which are further divided into four parts of 15 seconds each. The instrument is designed, therefore, to measure time to the nearest 15 seconds. The divisions indicating minutes are 12.5 ± 0.5 mm wide and the divisions indicating seconds are 3 to 3.5 mm wide. The scales on the sides of the quadrants follow the Hindu system of time measurement in *ghaṭikās* and *palas*. Each quadrant is divided into 15 *ghaṭikās*, which are then further divided into 60 *palas*, so that the small divisions read to the nearest *pala*. As a *pala* equals 24 seconds, the divisions for the *palas* are approximately 1½ times as large as those for the seconds.

The large divisions on the surface of the quadrants indicating one-half hour each are labeled by 10 cm long oval-shaped brass plates embedded near the edges. These plates on the north side are inscribed in Devanagari characters, whereas the ones on the south are inscribed in English.

A peculiar feature of the Varanasi Samrāṭ is that the markings on its quadrants do not terminate at 6 o'clock, as they do at Jaipur, but terminate at 6 h, 42 min, 45 sec on the east side, and at 5 h, 13 min on the west. The *ghaṭikā* readings on the sides continue similarly beyond 15-*ghaṭikā* marks and terminate at 16 *ghaṭikā*, 47 *palas*. The reason for this peculiarity is the fact that the instrument is built well above the ground level. Consequently, the horizon sun finds itself well below the horizontal plane passing through the two 6 o'clock marks of the quadrants.

The quadrants of the Samrāṭ have four holes each lined up horizontally with each other. These holes might have been made to facilitate construction of the instrument, for they do not seem to have aided any astronomical observations.[1] Some of these holes have been partially filled in. It should be pointed out that the holes are not seen in Campbell's drawings.[2]

Errors of Construction

Like the Samrāṭs of Jaipur and Delhi, the large Samrāṭ of Varanasi also suffers

---

[1] The Ujjain observatory also has such holes in its Samrāṭ, which, according to G. S. Apte, are for determining the time when a planet or star crosses the prime vertical. See Ch. 9.

[2] See Robert Barker, *Op. Cit.*.

from errors of construction. For instance, a number of divisions on the gnomon of this Samrāṭ have errors of marking up to 2 mm. Besides, its gnomon edges deviate from the linearity requirement. They deviate by as much as ± 2 mm from the straight line joining their two extremes. In addition, at places where stones have moved due to missing cramps, the error can be as large as 10 mm.

According to the theory of the instrument, the horizontal line joining the two 6 o'clock markings on the quadrants on both sides of the gnomon must be tangential to the gnomon surface. However, when this requirement was checked with a string held taut between the two 6 o'clock markings, the gnomon scale to the west was found to be 4 mm below the line.

The two zero markings of the declination scale on the gnomon also do not coincide with the line joining the two 6 o'clock marks. The author found the zero marking on the western edge of the gnomon shifted by $6'$ of arc toward north and that of the east shifted by $3'$, also toward north. It is not certain if this shift was because of improper markings on the gnomon or because of the quadrants not being parallel to the equatorial plane.[1]

Because of the errors just elaborated, the different quadrant scales of the Samrāṭ read time with different degrees of accuracy. The accuracy is best for the scale adjacent to the north edge of the east quadrant. With this scale, on a November afternoon in 1981, the author obtained readings within ± 15 seconds of those displayed by a calibrated digital watch. It may be pointed out that 15 sec. is the least count of the instrument.[2] The west quadrant scales, which are read in the morning hours, are not as accurate. They could be off by as much as one minute.

Another defect of construction is observed around noon. Theoretically, at noon, the shadow of the gnomon should disappear from the west quadrant before it reappears on the east. However, such is not the case at the Samrāṭ. On January 9, 1990, the author noted that around noon, the shadow could be seen simultaneously on the two quadrants. The shadow was first noted at the noon mark of the east quadrant at 12:03:07 I.S.T. At that time the shadow could also be seen on the west quadrant. The shadow disappeared totally from the west quadrant only at 12:05:04.[3,4] Thus the shadow was visible on both quadrants simultaneously for 1 min, 57.6 sec ± 0.5 sec around the noon hour. This observation, for which the reason is unclear, contradicts the observations of the author at the great Samrāṭ of Jaipur. At Jaipur, the shadow of the noon sun

---

[1] The other error in reading time, which an observer faces, is due to the width of the sun's penumbra. On a bright sunny day in winter the penumbra can be as much as 1.5 cm wide in the early afternoon. However, by using a string as described earlier, the uncertainty caused by the penumbra may be eliminated. See Ch. 3.

[2] Apparently, the errors of construction canceled themselves out in the readings.

[3] According to the author's calculations, the noon hour that day arrived at 12:04:50 PM, I.S.T.

[4] For these measurements, the watch was calibrated with a digital clock displayed on Indian television just before the evening-news broadcast.

disappears from the west quadrant before it reappears at the east quadrant.

Digaṁśa

The cylindrical structure of the Digaṁśa is located near the eastern boundary wall of the roof. The instrument, like its sister units elsewhere, consists of two concentric circular walls with entryways and a circular column of masonry at the center. The central column is 1.26 m high and 1.08 m in diameter. At the center of the column is mounted a vertical rod 4.3 cm wide and 1.28 m tall. The inner wall of the Digaṁśa is of the same height as the column. It is about 47 cm thick, and its outer diameter, where the scale is engraved, is 6.4 m. The outer wall of the Digaṁśa, located at a distance of about 1 m from the inner wall, is 2.52 $\pm$0.02 m tall or just about twice as high as the inner wall or the central column. It is approximately 62 cm thick, and its outer diameter is 8.60 m.

The top of the central column of the Digaṁśa has a circular scale around its outer edges. The scale is divided into four quadrants with zero markings along the east-west line and 90-degree markings along the north-south line. The quadrants are divided into degrees which are further subdivided into four parts each, thereby giving the scale a least count of 15′ of arc. The 15′ subdivisions are 2.5 mm in width on the average, but a few subdivisions of width 1.5 to 3.0 mm may also be observed here and there.

The circular scale on the inner wall of the Digaṁśa is divided both according to the Western and Hindu system of measuring the angles of azimuth. For the Western system, the zero mark of the scale is located at the north point and the degree numbers run both clockwise and counter-clockwise down to the 180-degree mark at the south point. The degrees are labeled in English characters. For the Hindu system, with its inscription in Devanagari, the zero markings are at the cardinal points. The circular scale of this wall is divided into degrees or *aṁśas* that are further subdivided into 20 parts. These parts measure 3′ of arc each. Thus the least count of the instrument is 3′ of arc. The small divisions measuring the minutes are 3 $\pm$1 mm wide.

The scale on the outer wall of the Digaṁśa is similarly divided and also has a least count of 3′ of arc. The division length for this scale is 3.5 $\pm$0.5 mm. On the cardinal points of this wall metal pegs are embedded from which one can stretch cross wires.

At the Varanasi Digaṁśa, the vertical plane passing through the object in the sky may be defined in more than one way. For instance, it can be defined by a pair of strings stretched from the lower and upper ends of the rod on the central column. The string tied to the lower end of the rod is stretched over the middle wall, whereas the one tied to the upper end is stretched over the outer wall. For this purpose, the other two ends of the strings are tied to separate pebbles or stones which are suspended on the outer side of the two walls. The observer, with the assistance of a helper, moves the stones around until the strings, the object in the sky, and the top of the rod (as viewed from below the strings) overlap. When this is achieved, the object in the sky lies in the vertical plane defined by the two strings and the rod. The angle of azimuth is then read from either of the scales on the two walls. The scale on the outer-most wall, because of its larger division-length, gives a more precise reading of the azimuth. The scale on the central column, however, gives only an approximate result.

A second method of defining the vertical plane is similar to the one used at Jaipur;[1] and the third, identical with the one used at Ujjain, is described in Chapter 9.

The Small Samrāṭ

To the west of the Digaṁśa, adjacent to its outer wall, is the small Samrāṭ. The east quadrant of the Samrāṭ rests against the outer wall of the Digaṁśa. The gnomon of the small Samrāṭ, with its lower end terminating at a height of 1.7 m above the roof surface, is 3.39 m long, 2.53 m high, and 28 cm wide. Because of the narrow width of the gnomon wall, there are no steps provided for reaching the top end of the gnomon. In this regard, the small Samrāṭ differs from all other Samrāṭs of Jai Singh, which have stairs built into their gnomons. The quadrants of the small Samrāṭ are 51.5 cm wide and have radii of 97 cm.

The small Samrāṭ is similar to the large Samrāṭ in many respects, except that it measures the declination to a maximum of 68^0 N and 57^0 S. Its gnomon scale is divided into degrees, which in turn are subdivided into segments of 10′ each. The least count of the instrument therefore is 10′ of arc. The divisions near the zero marking of the gnomon are about 3 mm wide, whereas near the 54-degree mark they are approximately 9 mm wide.

The quadrants on either side of the gnomon are patterned after those of the large Samrāṭ. As such they are labeled both in the Roman and Devanagari scripts, thereby indicating time in hours and minutes as well as in *ghaṭikās*. The scales on the surface of the quadrants read down to 1 minute. The least count of the *ghaṭikā* scale, which is engraved on the sides of the quadrant, is 1/36 of a *ghaṭikā*, because the *ghaṭikās* are divided into 36 parts.

The small Samrāṭ has two horizontal holes all the way through its quadrants and gnomon, the purpose of which, as explained earlier, appears to facilitate construction.

The small Samrāṭ, in its present condition, suffers from a number of defects similar to its counterpart, the large Samrāṭ. Its gnomon edge deviates from linearity by as much as 3 mm. However, this defect is due to a crack in an aging slab of stone near the top and is not necessarily due to any errors of construction.

---

[1] See Ch. 7.

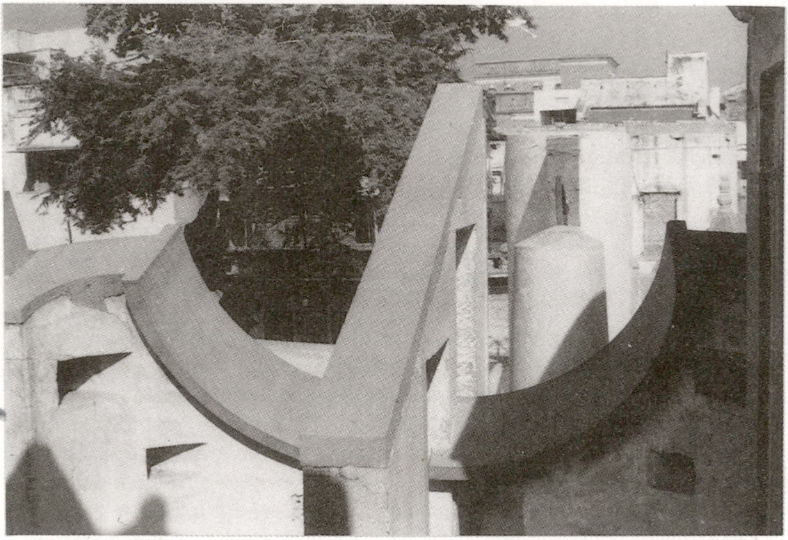

Fig. 8-4 The Small Samrāṭ of Varanasi.

Fig. 8-5 The Nāḍīvalaya of Varanasi.

Further, the surface of the gnomon is 2-3 mm below the line joining the two 6 o'clock markings on the quadrants. The zero markings on the declination scales of the gnomon do not coincide with the line either. The zero markings are about 1½ divisions toward the north from the line. This shift toward the north end should cause an error of 15′ of arc in declination measurements.

A serious shortcoming of the small Samrāṭ is its poor location. The instrument is built too close to the Digaṁśa. Because of the close proximity of the Digaṁśa, the Samrāṭ cannot be used in the morning, just after sunrise, because the shadow of the outer wall of Digaṁśa falls on the gnomon at that time.

Nāḍīvalaya

To the west of the small Samrāṭ, supported on two vertical stone columns, stands the Nāḍīvalaya. The instrument is a single plate or slab of sandstone, a few centimeters thick, circular in shape and inclined parallel to the plane of the equator. The plate has a number of circular scales on its two sides, and iron styli are fixed to the scales' centers. On the north-facing side, the plate has two scales. These scales are concentric and have diameters of 1.38 m and 1.08 m, respectively. The inner scale of these two is divided into *ghaṭikās* and labeled in Devanagari characters. On this scale each *ghaṭikā* is divided into six parts which are further divided into 4 subparts. A subpart therefore reads 1/24 of a *ghaṭikā*, or a minute. The outer scale, on the other hand, is marked in Roman numerals, and its small divisions read one minute each. Apparently, the two side-by-side scales serve the purpose of measuring time simultaneously in both the Hindu and the Western systems.

The south side of the instrument is similar to its north side. It also has two scales, except that the scales are located only on the upper half of the plate. As a result, they are only about one-half as large and have diameters of 54 cm and 66 cm. The least count of these scales is 2½ minutes in modern units and 1/6 *ghaṭikā* in Hindu units.

Behind the Nāḍīvalaya, at a distance of about a meter to the south, is a stone post 1.35 m tall, having a cross-section of 14×13.5 cm^2. About 17 cm below its top end, the post has a north-south hole. This post has been there for quite some time as it appears in Campbell's drawings sketched in 1772. Its purpose is unknown.

Cakra Yantra

Only a few meters to the north of the small Samrāṭ, supported by two oddly shaped columns of masonry and stone, stands the Cakra yantra. The yantra is a circular dial of metal, pivoted along its diameter parallel to the axis of the earth. The instrument is quite rugged in construction. It is fabricated out of a solid beam of iron, 3.5×5 cm^2 in cross section and one meter in diameter. The beam supports a 0.5 cm thick brass strip on one side, on which a measuring scale is engraved. The scale is divided into four quadrants. The 90-degree markings of the quadrants are located near the two pivot points and the zero-degree markings are located halfway in between. The entire scale is divided into degrees and minutes. Its small divisions indicate 15′ of arc. The divisions are fairly uniform and have a width between 2 and 3 mm. They are marked in Devanagari only and are gradually fading away.

Perpendicular to the axis of the Cakra yantra, around its pivot point to the south, is a circular scale engraved on a fixed slab of stone. The diameter of the scale is 29.5 cm. The scale is parallel to the equatorial plane and divided into degrees, which have been left unmarked.

The instrument also has a removable sighting tube mounted at its center with which readings are taken as they are with the Jaipur instrument.[1]

Dakṣinottara Bhitti Yantra I

On the east facing wall of the large Samrāṭ gnomon are the two 90-degree arcs, or quadrants, of a Dakṣinottara Bhitti yantra, or transit instrument. The arcs have a radius of 3.21 m each and intersect one another at a distance of 30 degrees from their bottom end. At the center of the arcs, near the top of the wall, are two 1¼ cm thick iron styli. The scales of the quadrants are engraved on an 8 cm wide sandstone surface and divided into degrees which are further divided into 10 parts of 6′ of arc each. The divisions are labeled in such a manner that the scales indicate both the altitude and zenith distance of an object in the sky. Further, the altitude angles are marked in English characters, whereas the zenith distance is marked in Devanagari. The small divisions of the scale are each 6 ± 1 mm wide. With the Dakṣinottara Bhitti, on November 20, 1981, the author measured the zenith angle of the midday sun as 44;30, which was about 15′ lower than expected. The expected value for the zenith angle for the day was 44;44,46.

With this Dakṣinottara Bhitti, it is difficult to observe a star or planet, particularly if the planet is near the horizon. For planets or stars near the horizon, the upper parts of the two scales would have to be read. However, the upper parts are difficult to access. The upper part of the south scale can only be approached with a ladder, whereas the upper section of the north scale, even with a ladder, is difficult and dangerous to approach. There is hardly any room to set up a ladder on the narrow balcony that overlooks the alley 15 m below.

Dakṣinottara Bhitti Yantra II

The second Dakṣinottara Bhitti Yantra is located on a separate roof, all by itself, some 50 m to the south of the large Samrāṭ. The instrument is in a good

---

[1] See Ch. 7 for the Cakra yantra of Jaipur.

location and can be easily worked with. The yantra is, once again, a double arc on the east-facing side of a meridian wall, 2.76 m long and 3.23 m high with a pin at the center of each arc. The radius of the arcs is about 2.36 m. This yantra is similar to the one just described, except that its least count is only 10′ of arc. The width of its small divisions reading minutes varies between 6.5 mm and 8 mm.

Unidentified Structure

An unidentified structure, or what is left of a former instrument, is located at a distance of 64 cm to the north west of the Digaṁśa. From the structure, the Cakra yantra is about 1.6 m to the west. The instrument is constructed on a platform about 18 cm above the floor of the Digaṁśa and ½ m above that of the Cakra. Of this instrument only a circular channel, 7 to 8 cm wide and about 4 cm deep, remains. The inner diameter of the channel is 1.245 m. Part of the circular structure, to the north, has been chopped off in order to construct a stone railing. An iron pin is embedded in lime plaster at the center of the channel. The function of the structure is unknown.

CHAPTER IX

UJJAIN OBSERVATORY

About 650 kilometers south of Delhi, in the ancient city of Ujjain, Jai Singh built another of his observatories. The observatory is popularly called the Jantar Mantar of Ujjain.

The parameters of the observatory are as follows:[1]

Latitude	23;10,18 N
Longitude	75;46,2 E
Height above sea level	496 m
Local Time	I.S.T. −26 min, 56 sec, or Universal Time +5 hr., 3 min, 4 sec.

History

In Jai Singh's time, Ujjain was the largest city and the capital of the province of Malwa. Jai Singh ruled the province for many years—first as a deputy governor from 1705 to 1706 and then as a governor for almost nine years, between 1712 and 1737. His foremost Hindu astronomer, Jagannātha, calls him *Matsyadeśādhipati* or the Master of *Matsyadeśa* or Malwa. This could be a reason why Jai Singh built an observatory at Ujjain. Another reason for his building an observatory at Ujjain might be that the city had been a center of Hindu astronomy since ancient times and that it is located on the prime-meridian established by Hindu canons of astronomy.

The date of construction of the observatory of Ujjain is uncertain. A plaque erected by Bhavan at the Dakṣinottara Bhitti yantra of the observatory and another plaque near the Samrāṭ state 1719 as the date of construction. The Ujjain observatory, however, could not have been built in 1719, because Jai Singh took his decision to construct observatories only around 1721 and built his first observatory in 1724 at Delhi. His other observatories including that of Ujjain were built afterward.

There are reasons to believe that the Ujjain observatory must have been completed before 1730. Jagannātha mentions the observatory in his *Samrāṭ Siddhānta*, which was written about 1730.[2] After completion, the observatory remained in operation for a decade or so at the most. Its operation definitely

---

[1] Kaye, p. 56.

[2] SSRS, p. 1162, and SSMC, p. 38.

ceased after Jai Singh's death in 1743. The political situation of Malwa, which had been turbulent during Jai Singh's lifetime, deteriorated further after his death, and support for all scientific works evaporated.

Visitors to the Observatory

Hunter was the first European to give a detailed account of the Ujjain observatory. He journeyed to Ujjain in the 1790's and studied its instruments.[1] When Hunter visited Ujjain, there were no signs of any astronomical activity at the observatory site, and signs of neglect were everywhere.

During his visit, Hunter noted seven instruments in all at the observatory; namely: the Samrāṭ, Nāḍīvalaya, Dakṣinottara Bhitti, Digaṁśa, Agrā, and Palabhā. Hunter's account is as follows.

> "The next . . . among those which I have had the opportunity of examining, is the observatory at *Oujein*. It is situated at the southern extremity of the city, in the quarter called *Jeysingpoorah*, where are still the remains of a palace of *Jayasinha*, who was *Soubahdar* of Malwa in the time of *Mohommed Shah*.
>
> "1. *A double mural quadrant* [Dakṣinottara Bhitti], fixed in the plane of the meridian. It is a stone wall, twenty-seven feet high, and twenty-six feet in length. The east side is smooth, and covered with plaster, on which the quadrants are described: on the west side is a stair [way], by which you ascend to the top. At the top, near the two corners, and at the distance of twenty-five feet one inch from one another, were fixed two spikes of iron, perpendicular to the plane of the wall; but these have been pulled out. With these points as centers, and a radius equal to their distance, two arcs of 90 degrees are described intersecting each other. These are divided in the manner represented in the . . . margin [of his article]. One division in the upper circle, is equal to six degrees; in the second, one degree, (the extent contained in the specimens;) in the third, six minutes; and in the fourth, *one minute*. One of these arcs serves to observe the altitude of any body to the north, and the other of any body to the south of the zenith; but the arc which has its center to the south, is continued to the southward beyond the perpendicular, and its center about half a degree, by which the altitude of the sun can at all times be taken on this arc. With this instrument *Jayasinha* determined the latitude to *Oujein* to be 23°, 10′ N."

---

[1] Hunter, pp. 194-200.

Hunter compares Jai Singh's determination of the latitude of Ujjain with his own observations which also came out to be 23;10, without refraction. He continues:

> "This instrument is called *Yām-utter-bhitti-yunter*. With one of the same kind at Dehly, . . . in the year 1729 *Jayasinha* says, he determined the obliquity of the ecliptic to be 23°, 28′. In the following year (1730) it was observed by Godin 23°, 28′, 20″.
>
> "2. On the top of the mural quadrant is a small pillar, the upper circle of which, being two feet in the diameter, is graduated, for observing the amplitude of the heavenly bodies at their rising and setting; it is called *Agrā Yunter*. The circles on it are very much effaced.
>
> "3. About the middle of the wall, the parapet to the eastward is increased in thickness, and on this part is constructed a horizontal dial, called *Puebha Yunter* [Palabhā]. Its length is two feet, four inches, and a half; but the divisions on it are almost totally effaced.
>
> "4. *Dig-ansa Yunter* [Digaṁśa yantra], a circular building, 116 feet in circumference. It is now roofed with tiles, and converted into the abode of a Hindu deity, so that I could not get access to examine its construction; but the following account of it is delivered in the *Sem'rat Siddhanta*, an astronomical work, composed under the inspection of *Jayasinha*."

Next, Hunter describes the principle and operation of the Digaṁśa as learned from the *Samrāṭ Siddhānta*.[1]

> "5. *Nāree-wila-yunter* [Naḍīvalaya], or equinoctial dial is a cylinder placed with its axis horizontally, in the north and south line, and cut obliquely at the two ends, so that these ends are parallel to the equator, (Naree wila). On each of these ends a circle is described, the diameter of which, in this instrument, is 3 feet, 7 inches, and half. These are divided into *ghurries* [*ghaṭikās*] of six degrees, into degrees and subdivisions, which are now effaced. In the center of each circle was an iron pin (now wanting) perpendicular to the plane of the circle, and consequently parallel to the earth's axis. When the sun is in the southern signs, the hours are shown by the shadow of the pin in the south; and when he is in northern signs, by that to the north. On the meridian line on both sides are marked the co-tangent, to the radius equal to the length of the center pin. The shadow of the pin on this line, at noon, points out the sun's declination.

---

[1] See Ch. 4.

> "6. *Semrāt-yunter* [Samrāṭ yantra] also called *Nāree-wila*, another form of equinoctial dial, . . . It consists of a gnomon of stone, containing within it a stair. Its length is 43 feet, 3.3 inches; height from the ground, at the south end, 3 feet 9.7 inches; at the north end, 22 feet, being here broken. On each side is built an arc of a circle, parallel to the equator, of 90 degrees. Its radius is 9 feet, 1 inch; breadth, from north to south, 3 feet, 1 inch. These arcs are divided into *ghurries* [*ghaṭikās*] and subdivisions; and the shadow of the gnomon among them points out the hours. From the north and south extremities of the intersection of these arcs with the gnomon, are drawn lines upon the gnomon, perpendicular to the line of their intersection. These are consequently radii of the arcs; and from the points on the upper edge of the gnomon, where these lines cut it, are constructed two lines of tangents, one to the northward, and another to southward, to a radius equal to that of the arc."

Hunter explains how to measure the declination and the hour angle of a star at night with the Samrāṭ.

Tieffenthaler

Tieffenthaler, who traveled extensively in India for three decades after landing in 1743, also visited the observatory.[1] Tieffenthaler gives only a brief description of the observatory and does not go into details of the instruments. His description of the observatory is more like a travelogue than a scientific account. He says:

> "Not far from there [walled city of Ujjain] is suburb built by Djesing, King of Djepour, a *ci-devant* governor of this province [Malwa]. An astronomical observatory is to be seen there, with instruments, made of cement: namely two equinoctial dials, one large and one small; a gnomon (*axis mundi*) elevated according to the height of the pole at this place, and set in the meridian; and on either side of this is a quadrant of a geometrical circle; also a dial made in lime, and a meridian wall in stone. . . . The geographic latitude of this capital [Ujjain], observed on March 6, 1750, was found to be 23;12."

Fanny Parks

In her book, *Wanderings of a Pilgrim*, written in the early 1800's, Fanny Parks says that "the observatory at Oujein has since been converted into an

---

[1] Tieffenthaler, p. 346.

arsenal and foundry of cannon."[1] Parks does not indicate if she actually visited the observatory. Her statement probably is based on Hunter's remarks: "Urania fled before the brazen fronted Mars; and the observatory was converted into an arsenal and foundry of cannon."[2] Hunter's remarks, however, do not refer to the observatory of Ujjain but to the observatory of Jaipur, which was indeed turned into a cannon factory during Pratap Singh's reign.[3]

G. R. Kaye

Kaye, having visited the observatory in 1915-1916, gave a detailed account of the instruments or what was left of them.[4] Here are some excerpts from his report:[5]

> "The observatory now consists of. . . (a) Samrāṭ yantra (b) Dakṣinottara Bhitti (c) Nari Valaya [Nāḍīvalaya] and (d) Digaṁśa. These are all in a state of ruin. The Samrāṭ Yantra is in a dilapidated state. . . . [Of this instrument] only skeleton remains. . . . The Dakṣinovritti [Dakṣinottara Bhitti] yantra is inclined to the perpendicular at an angle of about 5 degrees. . . . [Of this instrument] the masonry work is fairly intact, but the graduations have disappeared. . . . There are no signs of [Palabhā yantra]. The styles and graduations [of the Naḍīvalaya] have disappeared."

About the Digaṁśa he writes:

> "Digaṁśa Yantra is similar to the one at Benares. . . . Originally there was a pillar at the center, but it has been removed. . . . At the four points of the compass, in the outer and inner walls, were arched openings, but all of those in the outer wall, except that to the west, have been filled up. The outer walls are badly cracked, and a great part of the foundations is now exposed."

Hunter's account, written about 50 years after the death of Jai Singh, and Kaye's account written a century later, make it abundantly clear that there had been no activity at the observatory for a long time, and that it had gradually fallen into ruins.

---

[1] Fanny Parks, *Op. Cit.*, Ch. 5.
[2] Hunter, p. 210.
[3] See Ch. 6.
[4] Kaye, pp. 56-60.
[5] Kaye, pp. 56 ff.

Restoration

Since the early 1900's, Hindu astronomers had been concerned about the plight of Jai Singh's observatories and were anxious to have them restored. They saw in the observatories a means of uplifting their own system of astronomy. The All India Conference of Hindu Astronomers at Bombay thus passed a resolution in 1905 for restoring the observatory of Ujjain.[1] The restoration of the Jaipur observatory, completed only a few years earlier, might have encouraged them to enact the resolution.

Armed with the resolution of Bombay, the *Paṇḍitāśrama Sabhā* (Association of Astronomers) of Ujjain approached Maharaja Madhava Rao Shinde of Gwalior, who ruled the city of Ujjain at the time, and persuaded him to sanction funds for restoring the observatory. The Maharaja granted the request and invited Gokul Chand Bhavan, who had already restored Jai Singh's observatories elsewhere, to supervise the task, which he completed in 1923. Bhavan's brother was the engineer for the project. After the restoration, the facility was renamed the Śrī Jīvājī Observatory.

During the restoration, Bhavan introduced some changes of his own as he had done elsewhere. For instance, he replaced the Hindu system of time measurement on the instrument scales with the Western system of hour, minute, and second.

Two years after the restoration, there was another conference of Hindu astronomers to reform the calendar, held at Pune. The conference asked the Maharaja for additional funds, this time for initiating regular observations at the observatory site. The Maharaja reacted favorably to the request and approved the necessary funds. However, no program of observing was instituted for several years.

The Maharaja, while sanctioning the funds for the restoration, had also mandated that the observatory be used for "correction of the Hindu Almanac by enlightening the *Jośīs* (astronomers), and enabling them to verify the results of their calculations by observations." Further, a *Karaṇa*, i.e., a treatise on practical astronomy, was also to be written in Sanskrit on the lines of old works to enable . . . *Jośīs* to arrive at accurate results."[2]

The *Paṇḍitāśrama Sabhā* of Ujjain was once again expected to carry out these tasks. However, when nothing was done for a long time, the Education Department of the State of Gwalior took over the charge and appointed a permanent staff to run the observatory affairs. Regular observations began to be taken and results published in weekly papers, such as the *Keśarī* of Poona and *Jayājī Pratāpa* of Gwalior. A treatise, with the title *Sarvānanda Karaṇa*,

---

[1] Govind Sadashiv Apte (Govinda Sadāśiva Āpṭe), *A Guide to the Shree Jiwajee Observatory, Ujjain*, pp. 12-13, Ujjain, 1935.

[2] *Ibid*, p. 6.

was published shortly thereafter.[1] Govinda Sadāśiva Āpṭe, the first superintendent of the observatory, declares that the *Karaṇa* proved to be a great help to Hindu astronomers, enabling them to compute results which compared favorably with those in the Nautical Almanac.[2] The observatory also began publishing *Astronomical Ephemeris of Geocentric Places* on a yearly basis for the benefit of Hindu *Pañcāṅga* makers.[3]

The present state of the observatory

At present the observatory is located within a compound on the outskirts of the city in pleasing and relatively quiet surroundings. To the north of the observatory there is a main road and to the south, some 20 to 25 meters below the observatory boundaries, the river bed of Sipra, followed by cultivated fields as far as the eye can see. On the eastern side, near the wall itself, are a dozen or so tall trees, a small drainage ditch, fields and a house. On the west side are also fields and two very small temples. Further to the west a railway bridge can be seen at a distance. Within the boundary wall itself, to the west is a recently constructed small temple, 4 to 5 m tall, dedicated to the Hindu deity, *Hanumāna*. The temple, however, does not obstruct the view of the sky. In contrast to the other observatories of Jai Singh, which have now become surrounded by tall buildings, the Ujjain observatory is still in an isolated place, and no man-made structures seriously obstruct the view of the sky from its instruments.

Currently, the observatory maintains a staff of about 20 people, including 10 professionals such as observers, mathematicians, and a librarian. The rest of the staff looks after the maintenance of the instruments and of the observatory grounds. The instruments are regularly painted with whitewash, and the grounds are well kept with shrubs and flower beds.[4] In 1983, plaques were erected at the instruments explaining their function. The observatory also serves as a weather station, and daily reports of weather are sent to Bhopal, the state capital.

---

[1] Govinda Sadāśiva Āpṭe, *Sarvānanda Karaṇa*, Lahore, 1934; Hindi tr., *Sarvānanda Lāghava*, by Mukunda Vallabha Miśra, Lahore, 1934.

[2] Āpṭe himself was involved in some rudimentary observing. His purpose apparently had been to demonstrate old procedures. He states that, having observed the transit of the *nakṣatras Maghā* (Regulus) and *Revatī* (ζ Piscium), the length of the sidereal year was determined to be 365;15,22,58 days. Similarly, the precession, after five years of observing, was found to be 50.25 sec of arc per year.

[3] The *Ephemeris* was still being published yearly when the author visited the observatory in 1990.

[4] For reasons unknown, the literature contains the erroneous information that the observatory is in ruins. For example, see Blanpied (1974), *Op. Cit.*, p. 97, Ch. 1. Also see Shakti Dhara Sharma, in *History of Astronomy in India*, by S. N. Sen, and K. S. Shukla, *Op. Cit.*, p. 353, Ch. 1.

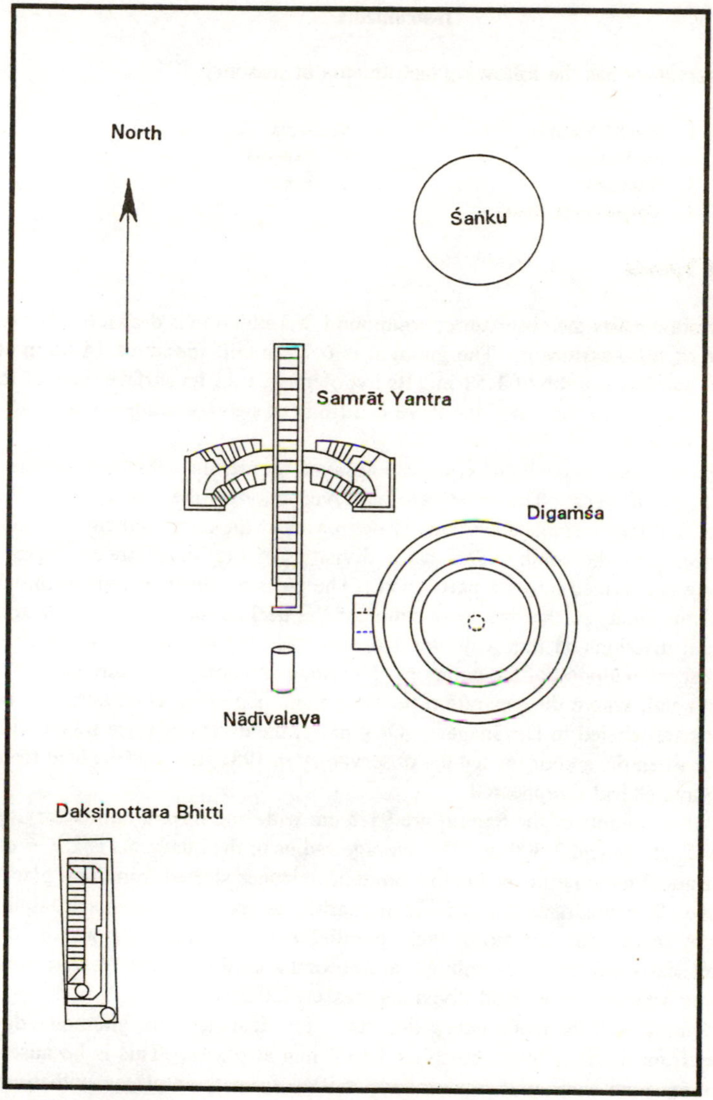

Fig. 9-1 The General Plan of the Ujjain Observatory.[1]

---

[1] based on Kaye, plate XXIV.

Instruments

The observatory has the following instruments of masonry.

1. Samrāṭ Yantra
2. Naḍīvalaya
3. Digaṁśa
4. Dakṣinottara Bhitti
5. Agrā
6. Palabhā
7. Śaṅku

Samrāṭ Yantra

As a visitor enters the observatory compound, his attention is drawn by the tall gnomon of this instrument. The gnomon is 6.75 m tall, measures 14.66 m at its base, and has a width of 1.51 m. Its hypotenuse, i.e., its surface parallel to the earth's axis, is 16 m long, and there is a flight of steps running in the middle up to its very top.

The gnomon has its declination scales engraved on marble slabs on either side of the flight of steps. The scales are engraved only on the upper 4/5 of the gnomon, and they terminate at about 71 degrees near the north end and at about 63 degrees near the south. The major divisions of the scales are in degrees which are subdivided into six parts each. The parts are further divided into 5 subdivisions each, so that the least count of the declination scale is 2′ of arc. The small divisions on this scale are 1.6 mm wide on the average at the zero mark near the midpoint of the gnomon. The same divisions are 7 mm wide near the south end, where the separation between them is greater as expected. The divisions are labeled in Devanagari. Originally, the divisions were inlaid with lead, but when the author visited the observatory in 1981, most of the lead from the engravings had disappeared.

The two quadrants of the Samrāṭ are 91.5 cm wide and have a radius varying between 2.760 m and 2.800 m. The average radius of the quadrants is 2.776 m. The variation in the radius is due to a number of stones shifted from their places with time. The quadrants are surfaced in marble, as are the quadrants at Jaipur, and have scales running along their parallel edges. The divisions on the quadrant scales indicate hour, minute, and second according to the Western time reckoning scheme. The small divisions measure 20 seconds each.

The Samrāṭ suffers from many defects. For instance, its gnomon edge deviates from linearity by as much as 5 to 6 mm at places. This is because a number of stone slabs with scales have shifted from their places with time. Further, the declination scales of the gnomon are non-tangential to the line joining the two 6 o'clock markings on the quadrants flanking the gnomon. The scale on the western edge of the gnomon is 0.5 cm below the line and the one on the eastern edge 1.5 cm below. Further, the zeros on the two declination scales fail to fall on the line joining the two 6 o'clock marks. The zero of the

Fig. 9-2 Ujjain Observatory.

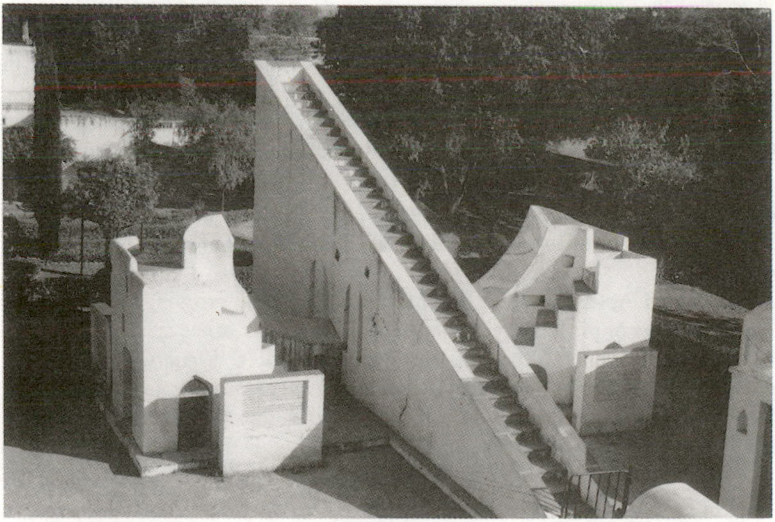

Fig. 9-3 The Samrāṭ Yantra of Ujjain.

west declination scale is displaced by 0.5 cm to the north, whereas the one on the east is displaced by 1.0 cm, also to the north. This zero error could be due either to the quadrants not being parallel to the plane of the equator or due to the declination scales being improperly marked. The zero error alone could cause errors of up to 12′ of arc in the declination readings for an object near the celestial equator. However, for large angles of declination, such as 60 to 70 degrees, where the divisions are farther apart, the effect of the zero error will be minimal.

The workmanship of the Samrāṭ of Ujjain, compared with that of the Samrāṭs elsewhere, is inferior. For example: The arcs on the eastern quadrant's south scale are improperly engraved. Moreover, the small divisions of the quadrant scales are not uniform. Their width varies anywhere from 3 mm to 5 mm.

Although the Ujjain Samrāṭ is designed for measuring time with an accuracy of 20 seconds, it is not as accurate in practice. Due to a combination of factors just elaborated, the readings are usually off by 1½ minutes or more. The author noted that one of his time-readings taken on a November afternoon in 1981, was off by 2½ minutes. This happened when the shadow of a carelessly laid out slab of stone on the gnomon surface fell on one of the east-quadrant scales.

The noon hour at the Ujjain Samrāṭ may also not arrive simultaneously on the two quadrants. Because of the unevenness of the vertical wall of the gnomon, just above the east quadrant, this defect could not be confirmed with certainty. However, it is certain that the noon-hour shadow of the gnomon arrives at the east quadrant no later than 20 to 30 seconds after it leaves the west quadrant. It should be pointed out that 20 sec is the least count or the width of the small divisions of the instrument.

The supporting walls of the west quadrant of the Ujjain Samrāṭ carry the equation of time. The wall supporting the west quadrant has its equation in English and that of the east quadrant in Hindi. In the supporting walls of the quadrants, as well as in the gnomon wall itself, are diamond-shaped holes, 25×25 cm,[2] three each in the quadrants and two in the gnomon. The holes have their diagonals somewhat perpendicular to the declination scales on the gnomon surface. Apparently the holes were there before the restoration. A photograph taken a few years prior to the restoration and reproduced in Kaye's book, clearly shows them.[1]

According to Āpṭe, the holes are for "observing the crossing of the Prime Vertical by stars or planets having northern declinations, since such observances are of importance in . . . the construction of Sacrificial *Kuṇḍas* and altars."[2] Āpṭe does not elaborate how this is to be achieved. In order to ascertain if a planet has crossed the prime vertical one has to define the plane of the prime vertical, say, by stretching horizontal strings through the holes. It will be a

---

[1] Kaye, Fig. 61, plate XXIII.

[2] Āpṭe in *Sarvānanda Karaṇa*, *Op. Cit.*, p. 8.

difficult task for anyone to define such a plane because the holes are not in a vertical plane but are oriented with it at an angle.

At any rate, the interpretation regarding the holes is of recent origin and not described by Jagannātha, the principal astronomer of Jai Singh. In fact, the holes may have no astronomical or religious significance at all. Such holes are found at the two Samrāṭs of the Varanasi observatory also.

Naḍīvalaya

A little south to the Samrāṭ, at a distance of about two meters, is the Naḍīvalaya. The instrument is essentially a masonry cylinder resting on its side along a north-south line on a meter-high masonry stand. The axis of the cylinder points in a north-south direction, and its two flat surfaces lie parallel to the equatorial plane. The surfaces are separated by a distance of 2.16 m, and they have round plates of marble embedded in them, on which circular scales for indicating time are engraved. A style each, at the center of the plates, is fixed for casting its shadow on the scales. Although the short diameter of the flat surfaces is about 1.12 m each, the diameter of the scales is only about one-half as large. Apparently the designers of the instrument did not take full advantage of the surface area by engraving larger scales. The scales indicate time between 5 A.M. and 5 P.M., and their small divisions measure 3 minutes each.

Digaṁśa

The Digaṁśa yantra is located to the east of the Naḍīvalaya. The instrument consists of a central pillar surrounded by a pair of coaxial circular walls of equal height. The outer wall of the pair has a door facing west, through which one enters the instrument. Arched openings provided at the cardinal points of the inner wall permit access to the central pillar. The plan of the present instrument differs from the plan of the original yantra reported by Kaye in 1918. According to Kaye, the two walls surrounding the central pillar did not have the same height. The outer wall was twice as high as the inner.[1] The plan of the original instrument had been more or less similar to that of the Digaṁśa of Varanasi. Apparently Bhavan modified the instrument in 1923. In 1974 the instrument underwent another restoration, but its design was kept intact.

The central pillar of the Digaṁśa is 1.32 m high, whereas the two walls surrounding it are about 2.57 m. The pillar is fairly broad (dia. 1.24 m) and carries a 1.23 m high vertical rod at its center. Because of this length of the rod, the upper end of the rod lies at the same level as the top surface of the walls. The inner diameter of the two walls is 9.86 m and 11.32 m, and these

---

[1] Kaye, p. 57.

Fig. 9-4 The Digaṁśa Yantra of Ujjain.

Fig. 9.5 The Dakṣinottara Bhitti of Ujjain.

both are 0.73 m thick. The pillar has a circular scale along the edges of its top surface, which is graduated in degrees. The inner wall of the instrument has two scales. They are engraved near the wall's top end—one on the inside and the other on the outside. The scales are capable of reading down to 6′ of arc. The outer wall has no scales. Apparently its function is merely to protect the inner structure.

For measuring the azimuth of a celestial object with a Digaṁśa, the observer defines a vertical plane which includes the object as well as the center of the instrument.[1] At the Digaṁśa of Ujjain, such a plane is defined with the aid of the upright rod at the center and a string stretched from its top end to the outer edge of the inner wall. Āpṭe says:

> "The observer stands on the central platform and keeps the vertical rod between his eye and the celestial body whose azimuth, at the moment, is to be determined.[2] . . . An assistant walks on the top of the inner wall, holding one end of the string stretched from the rod. . . . When the observer brings the rod, the object, (and a section of the stretched string) in the same line, the assistant is asked to keep the string on the wall, and the azimuth is read directly on the scale."

The procedure elaborated by Āpṭe is rather imprecise, and it is doubtful that one could fully exploit the precision inherent in the design of the instrument with this procedure. Moreover, the procedure could lead to uncertainties of several degrees because of the finite width (dia. approx. 6 cm.) of the central rod.[3] Besides, looking at the sun directly could have dangerous consequences.[4]

Soon after the restoration of the Digaṁśa in 1923, a locally made portable Turīya yantra, or quadrant, of brass and having a 30-cm radius, was added to the instrument. The quadrant is still extant and is mounted on the central rod occasionally. Altitude and azimuth are determined with it following the conventional procedure. To permit the azimuth to be measured directly with the quadrant, a circular scale in the horizontal plane is permanently attached to the rod's top end. This scale reads to the nearest degree. The Turīya yantra added to the Digaṁśa is a crude device, and its main function appears to be the demonstration of a technique with a portable yantra and not the collection of any serious data.

---

[1] See Ch. 4.

[2] Āpṭe in the *Guide to the . . . Observatory*, *Op. Cit.*, p. 10.

[3] Evidently the procedure is not meant for measuring the azimuth of the sun.

[4] For a celestial object, such as a planet or the moon, a greater degree of accuracy can be achieved by defining the vertical plane with a pair of strings—one tied to the top end and the other to the bottom end of the pole. The loose ends of the strings are tied together and stretched over the inner wall. See Ch. 8.

Dakṣinottara Bhitti Yantra

The Dakṣinottara Bhitti yantra of the observatory is located to the south west of the Samrāṭ, close to the boundary wall. The yantra consists of a pair of 6.13 m radii, 90-degree arcs, with a style each at their center. The arcs are engraved on the east side of a 8×8 m^2 meridian wall of red stone. The wall is square in shape and has a width of 2.45 meter. The arcs on the wall are engraved on white marble and their scales inlaid with lead and divided into degrees. The degrees are partitioned into 10 major divisions which, in turn, are further subdivided into 3 parts each, so that the smallest division of the instrument measures 2′ of arc. Because the zero marks of the scales are at the bottom of the arcs, the instrument measures the zenith angle of a celestial object at the meridian.

Agrā Yantra

On the west-facing wall of the Dakṣinottara Bhitti structure, is a flight of stairs leading to its roof top. On the roof are an Agrā yantra, a Palabhā, and an anemometer belonging to the weather station. Although the Agrā and the Palabhā are described by Hunter and also specified in a report dated October 28, 1913,[1] Āpṭe makes no mention of them in his monograph. He also does not say anything about the anemometer belonging to the weather station. It is possible that the instruments were not restored along with the others in 1923, and that they were added later after Āpṭe had published his monograph in 1935.

The Agrā yantra is built on a 79.5 cm high masonry pillar. It has a circular scale of marble, 45.4 cm in diameter, with a 22.5 cm high and about a centimeter-wide rod at center. The instrument measures the amplitude or the angular distance from the east-west direction of an object at horizon to the nearest degree. Hunter reports the diameter of this instrument to be 2 ft (61 cm) at the time of his visit.[2]

Palabhā

The Palabhā, a sundial, is located on a 80 cm high masonry column close to the Agrā. The column is topped with a marble slab 46 cm in diameter. The instrument has a triangular gnomon similar to the one at Jaipur. The Palabhā scale is divided into hours and minutes such that its least count is 5 min. The scale continues beyond the two 6 o'clock marks on either side of the gnomon by 45 min. The inscription is written in Devanagari.

---

[1] The report of the *Daroghā Imārat Khānā*, R.S.A., File No. 578, Jaipur State.

[2] Hunter, p. 197.

Anemometer

The anemometer is on the same level as the Agrā and Palabhā. It is built on the north-east corner of the roof.

Śaṅku Yantra

The Śaṅku Yantra is located about 10 m to the north-east of the Samrāṭ. The instrument is of recent origin. Both Hunter and Kaye make no mention of it. It also is not included in the list of instruments for repair, prepared by the *Daroghā* of *Imārat Khānā* of Jaipur State in 1913.[1] According to a plaque at the instrument, it was added to the observatory around 1938 while Āpṭe was its superintendent.[2]

The Śaṅku yantra consists of a circular scale with a pole or rod at the center on a 45 cm high masonry platform of 6.8 m diameter. The major divisions of the scale are in degrees and are divided into 6 subdivisions each. Each subdivision is approximately 1 cm wide and measures 10′ of arc. The zero point of the scale is at the north point. The vertical rod at the center of the scale is 1.23 m high and 6 cm in diameter.

The Śaṅku yantra measures the Agrā or amplitude of the sun at sunrise or at sunset, by the shadow of its pole falling on the yantra scale. Using a string, with one end tied to the pole and the other stretched along its shadow, one may determine the azimuth of the sun at any other time as well. By measuring the shadow of the noon hour sun, the declination of the sun can also be determined.

In addition to its circular scale, the instrument has a set of seven curves, laid out from one side of the scale to the other, along which the shadow of the pole moves during certain dates of the year. These dates have been chosen to display the arrival of the equinox on September 23 and March 21, and that of the two solstices on December 22 and June 22. The other dates referring to the curves are July 23, May 22; August 23, April 21; October 22, February 20; and January 21, November 22. The dates represent the approximate time when the sun enters a zodiacal sign. Previously, the curves were labeled according to the sign they represented, but now the names of the signs have been removed.

The Śaṅku yantra of Ujjain has little practical application in the modern age. Nonetheless, it remains an effective tool, in the tradition of Jai Singh, to demonstrate to a lay person certain important aspects of practical astronomy.

---

[1] *Ibid.*

[2] On the edge of the instrument, there is another date, December 22, 1937, which could also be a date associated with the construction of the instrument.

Minor Instruments of Āpṭe

The monograph of Āpṭe lists a Dharātala yantra and a Dik-sādhanā yantra under the title "Minor Instruments." The Dharātala yantra, however, is nothing more than a system of water cross-channels constructed to ensure a level surface for the Samrāṭ. Such channels may also be seen at Jaipur and Varanasi. The Dik-sādhanā according to Āpṭe, was a horizontal platform indicating the cardinal points on a circle. The author did not find any Dik-sādhanā yantra during his visit to the observatory site in 1981. Evidently the structure had been dismantled by then.

CHAPTER X

THE OBSERVATORY OF MATHURA

Jai Singh built his fifth observatory at Mathura, where he had been the *Fauzdāra*, or administrative head of the town, since 1723. The observatory was built within the local fort on the banks of the river Yamuna.

The parameters of the observatory are as follows:

Latitude	27;30 N
Longitude	77;42 E
Height above sea level	approx. 200 m
Local Time	I.S.T. −19 min, 12 sec, or Universal Time +5 hr, 10 min, 48 sec.

No traces of the Mathura observatory remain these days. Even the fort which once housed the observatory has turned into a dirt mound now. The mound is called *Kaṁsa kā Ṭīlā*, or the mound of Kaṁsa, by the local people, who have built houses over it.

The buildings of the Mathura observatory were torn down just before 1857.[1] According to Frederic S. Growse, the observatory buildings had been sold to a government contractor who dismantled them for their material.[2] It is reasonable to assume that by the time the contractor dismantled the structures, the observatory instruments had lost their functional aspect, and their mere skeletons remained.

The French Jesuit Claude Boudier, who was an eye-witness for the astronomical efforts of the Raja, passed through Mathura in 1734 on his way to the Jaipur observatory.[3] He spent a day or two in the town and measured its latitude and longitude; however, he does not say anything about the observatory. Our knowledge of the Mathura observatory is, therefore, largely due to Hunter, who published an account of it in 1799.[4] According to Hunter, the observatory was built on the roof of a building within the fort and had five instruments.

---

[1] The year is well known in Indian history for a popular uprising against the British rule. The uprising is called "*Sepoy* Mutiny" by the British in their history books.

[2] Frederic Salmon Growse, *Mathurā: A District Memoir*, 2nd ed., *India District Gazetteers, North Western Provinces Memoirs, Status Reports etc.*, Northwestern and Oudh Press, Allahabad, 1880.

[3] For Claude Boudier's journey to Jaipur, see Ch. 13.

[4] Hunter, pp. 200-201.

The instruments were:

1. Nāḍīvalaya (1)
2. Agrā yantra (2)
3. Śaṅku or Palabhā (1)
4. Dakṣinottara Bhitti (1)

Hunter writes:

"At Mathura the remains of the observatory are in the fort, which was built by *Jayasinha* on the bank of the Jumna.[1] The instruments are on the roof of one of the apartments. They are all imperfect, and in general of small dimensions.

"1. An Equinoctial Dial [Nāḍīvalaya] being a circle nine feet two inches (2.79 m) in diameter, placed parallel to the plane of the equator, and facing northwards. It is divided into *ghurries* [*ghaṭikās*], of six degrees each: each of these is subdivided into degrees, which are numbered as *palas*, 10,20,30,40,50,60: lastly, each subdivision is further divided into five parts, being 12 minutes, or two *palas*. In the center is the remains of the iron style, or pin, which served to cast the shadow."

The Nāḍīvalaya, according to Hunter, had a single plate only, facing north. Its small divisions must have been almost 5 mm wide and they measured 2 *palas* of time (48 sec) or 12′ of arc each.

"2. On the top of this instrument [Nāḍīvalaya] is a short pillar, on the upper surface of which is an amplitude instrument [Agrā yantra I] (like that described No. 2, Oujein Observatory, called *Agrā-yunter*;)[2] but it is only divided into octants. Its diameter is two feet, five inches (73.7 cm).

"3. On the level of the terrace is another amplitude instrument [Agrā yantra II], divided into sixty equal parts: its diameter is only thirteen inches (33 cm)."

A second instrument described by Hunter on the same terrace appears to have been a Śaṅku, or a Palabhā yantra. Hunter does not give the dimensions of this instrument. It had a circular scale in a horizontal plane and a gnomon similar to that of the horizontal sundial seen atop the Nāḍīvalaya of Jaipur.

---

[1] Hunter mistakenly states that Jai Singh built this fort. It was Māna Singh (1543-1614), an ancestor of Jai Singh, who built the fort.

[2] See Ch. 9.

Fig. 10-1 Mathura Fort. The drawing is based on an engraving by Martin.[1]

[1] R. Montgomery Martin, *Panorama of British India*, reprint, New Delhi, 1984. The original book was published a few years after the uprising of 1857.

From the instrument's scale and from its general appearance, Hunter suspected that the instrument was of recent origin and designed by someone unfamiliar with the principles of horizontal sundials. He writes:

> "4. On the same terrace is a circle, in the plane of the horizon, with a gnomon similar to that of a horizontal dial; but the divisions are equal, and of six degrees, each. It must, therefore, have been intended for some other purpose than the common horizontal dial, unless we may conceive it to have been made by some person who was ignorant of the true principles of the instrument. This could not have been the case with *Jayasinha*, and his astronomers; but the instrument has some appearance of being of later date than most of the others: they are all of stone, or brick, plastered with lime, in which the lines and figures are cut; and the plaster of this instrument, though of the level of the terrace, and consequently more exposed to accidents than the others, is freshest and most entire of all."

The observatory had two Dakṣinottara Bhitti units or meridian dials on either side of a building in the fort. The instruments had scales with arc-lengths larger than semicircles. The arc-lengths in both instruments were perhaps made larger than a semicircle to take advantage of the great height of the observatory location. The instrument could then also measure the meridian angle of objects below the horizontal plane.[1] Hunter says:

> "5. On the east wall, but facing westward, is a segment, exceeding a semi-circle, with the arch downwards. It is divided into two parts, and each of these into fifteen divisions. Its diameter is four feet (1.22 m). On the west wall, facing eastward, is similar segment, with the arch upwards, divided in the same way as the former. Its diameter is seven feet, nine inches (2.36 m)."

Hunter does not describe any styli or pins at the centers of the semicircles, which are necessary for measurement. Perhaps the pins had disappeared by the time he visited the observatory site.

Tieffenthaler

Another visitor to the observatory of Mathura who has left a written account is Tieffenthaler. Tieffenthaler visited the observatory shortly after the death of

---

[1] The arcs of the Varanasi Samrāṭ also have additional arc-lengths for taking advantage of the higher elevation of the instrument.

Jai Singh. Unfortunately, Tieffenthaler's account is sketchy and unclear, but it has some useful information. He writes:[1]

> "Among the buildings of the town, most of which are in ruins, one can make out a fortress, built by an extremely rich Muslim and seated on a hill,[2] from which point one enjoys a view over such a vast plain that one's eyes cannot really grasp its dimensions.
>
> "At the highest point of the fortress, one can see astronomic structures built by the famous Raja Jai Singh, a lover of Astronomy. The main structure is a sundial which represents the *axis mundi*, built of masonry and 12 Parisian feet tall.[3] Additionally, there are equinoctial dials five *palmes* (*spithamarum*) in diameter and other modest structures set up with regard to the latitude of the place.[4] Beyond that, these instruments present various sections of the sphere.
>
> "This observatory is but a poor imitation of the one in Jaipur; but it has the advantage of being located in an elevated spot and overlooking an immense plain: for the observatory of Jaipur, being situated on a plain, does not allow one to see the rising and setting of the stars, except from the top of a masonry dial of prodigious height.
>
> "The fortress we're talking about is very large and solid surrounded by walls with the Djemna [Yamuna] on the east instead of moat. It resembles a mountain skillfully constructed in rock."

The 12-ft gnomon Tieffenthaler mentions may be identified with the Nāḍīvalaya that Hunter wrote about in his paper. Tieffenthaler alludes to many more instruments than were noted by Hunter, two of which he calls equinoctial dials. These dials could have been the Agrā yantras of Hunter.

---

[1] Tieffenthaler, pp. 201-202.

[2] The builder of the fort, Māna Singh, was a Hindu and not a Muslim.

[3] Twelve Parisian feet equal approximately 3.9 m.

[4] A *Palme* was an Italian measure of length varying from 0.274 m to 0.298 m, depending on the region of use. *Nouveau Petit LAROUSSE, Dictionarie encyclopediaque*, p. 734, Paris, 1969.

CHAPTER XI

THE BOOKS AND THE LIBRARY OF SAWAI JAI SINGH

For his multi-faceted program in astronomy, Jai Singh compiled an excellent collection of books on astronomy, astrology, and mathematics for his library and wrote a *Zīj*, or a set of astronomical tables.

ZĪJ-I MUḤAMMAD SHĀHĪ

After conducting observations for many years, Jai Singh prepared a set of astronomical tables, with which astronomers could compute the occurrence of celestial phenomena, such as the new moon, eclipses of the sun and the moon, and the conjunction of heavenly bodies with certainty. He dedicated these tables, or *Zīj*, to the reigning monarch, Muḥammad Shāh. The tables are, therefore, called *Zīj-i Muḥammad Shāhī* or the tables of Muḥammad Shāh. *Zīj-i Muḥammad Shāhī* is in Persian, and over 50 copies of it survive today in various libraries around the globe.[1] According to the Pothikhana records of the Jaipur State, at least one copy of this work was prepared in the Devanagari script.[2]

A cursory examination of the manuscripts of the *Zīj-i Muḥammad Shāhī* suggests that its text underwent at least one revision. The manuscripts of the Raza Library, Rampur, and the Andhra Pradesh State Archives, Hyderabad, belong to the first edition. This edition is smaller in size and has fewer entries in its star catalogs. The well known copy of the *Zīj* in the British Library, on

---

[1] The manuscripts of the *Zīj-i Muḥammad Shāhī* reported at different places are as follows:

1. The British Library, London, Add. ms. 14,373.
2. Cambridge University Manuscript Collection, 742 King 212.
3. Bankipur Library, Patna, No. 11/69.
4. Mulla Firoze Library, Bombay, Ms. RI-53.
5. Majlis Library, Tehran.
6. Arabic and Persian Research Institute, Tonk, Rajasthan.
7. The Raza Library, Rampur, No. 1221.
8. The State Central Library, Andhra Pradesh Archives, Hyderabad, Riyazi No. 300.
9. Sawai Man Singh II Museum, Jaipur, (two copies), Nos. 4 and 5, AG.
10. Maulana Azad Library, Aligarh (two copies), Fasiyah Ulum No. 30 and Sulaiman No. 527/6.
11. Motahari Library, Tehran (three copies), No. 671, 672, 673.
12. Tashkent Oriental Library, (five copies), Nos. 1/230, 517/440, 519/438, 520/411, and 521/439.

[2] *Tozīs*, Pothikhana, Jaipur State, V.S. 1798-1800, R.S.A.

the other hand, is more elaborate and belongs to the second edition. The British Library copy of the *Zīj* is complete, except for a figure or two missing here and there. Its title is *Zīj-i Jadīd-i Muḥammad Shāhī*, or the New Tables of Muḥammad Shāh and is written in the *Nasta'līq* style of the Persian script.[1] We are basing our analysis of the *Zīj* primarily on this copy.[2]

The Date of Completion

Scholars propose various dates for the compilation of the *Zīj-i Muḥammad Shāhī*. Garrett proposes a date of 1723.[3] The catalog of the Raza Library, Rampur, indicates 1140 A.H. or 1727 A.D. *Science and Technology in Medieval India . . .*, published by the Indian National Science Academy, also proposes 1727 A.D.[4] Kaye thinks it was completed in 1728.[5] However, from the evidence within the *Zīj* itself, we can successfully argue that the entire text, or at least certain sections of it, were written in the period between 1731 and 1732.[6] The *Zīj* has a refraction correction table copied from the *Tabulae Astronomicae* of Phillip de La Hire, which was brought to Jaipur from Europe in 1731. Further, the *Zīj* predicts the lunar eclipse of June 8, 1732. It says, "On Sunday, the 15th day of the *Zīl-Ḥijjah* 1144 A.H., a lunar eclipse *will* take place at Jaipur." The text must have been written, therefore, after the acquisition of the *Tabulae Astronomicae* in 1731 but before the lunar eclipse of 1732.

Contents of the Zīj-i Muḥammad Shāhī

The *Zīj-i Muḥammad Shāhī* is divided into three books or *maqālās* of uneven length, and it is patterned after the *Zījes* of the Persian-Arabic school of astronomy, such as the *Zīj-i Sulṭānī* of Ulugh Beg. Its first *maqālā* starts out with the rules and tables for calendar transformations. This section is somewhat short and has only four chapters or *bābs*. In Jai Singh's time, no fewer than

---

[1] The British Library copy of the ZMS is written on fine quality paper. It is bound in a gilt-edged volume 29×20 cm in size. See Mercier, *Op. Cit.*, Ch. 5. On the end fly-leaf, a note says, "Purchased of Major J. B. Jarvis, July 1843, Formerly in Dr. Adam Marshe's Collection, Oriental MSS No. 13." The first few pages of this manuscript have a table of contents in English copied from Hunter.

[2] The other manuscripts of the ZMS investigated for this study exist at: 1. Raza Library, Rampur. 2. State Library of Andhra Pradesh, Hyderabad. 3. Maulana Azad Library, Aligarh. No. 30. 4. Sawai Man Singh II Museum, Jaipur, (two copies).

[3] Garrett, p. 11.

[4] The *Bibliography*, p. 349.

[5] Kaye, p. 89.

[6] A number of writers assert that Khairu'llāh wrote the *Zīj-i Muḥammad Shāhī*. However, there is no direct proof to substantiate this assertion. See C. A. Storey, *Persian Literature*, Vol. 2, part 1, pp. 93-95, London, 1958. We will deal with this subject in detail in Ch. 13.

three separate calendars were in use simultaneously in north India – the Hindu Vikram Saṁvat, the Islamic Hijarī, and the Muḥammad Shāhī era introduced by the reigning monarch, Muḥammad Shāh himself. The Muḥammad Shāhī era began on Monday in the fourth month, or *Rabī al-Sānī*, of the Hijarī calendar in the year 1131 A.H., or on February 20, 1719.[1] The three calendars were mutually incompatible and must have created considerable confusion and uncertainty in the general public. In order to resolve this confusion and uncertainty, the *Zīj* starts out with the rules of calendar transformations. For instance, it states that to find the Muḥammad Shāhī year for the first three months of a year, including the *Rabī al-Awwal* subtract the number 1131 from the current Hijarī Era. For the next nine months, between the *Rabī al-Ākhir* to *Zīl-Ḥijjah*, subtract the number 1130 from the Hijarī era to find the same. For its tables, the *Zīj* uses a modified version of the *abjad* notation of numbers. This system of writing numbers is seen on some astrolabes also engraved in the same general period when the *Zīj* was written. Appendix III illustrates this version of *abjad* and its modern equivalents.

The second book or *maqālā* of the *Zīj-i Muḥammad Shāhī* deals with trigonometry and spherical astronomy. This book has 19 chapters. These chapters are followed by extensive tables such as those of sines of the minutes of arc, equation of daylight for Jaipur and Delhi, and latitude and longitude of a number of cities. In Table 11-1, we reproduce the latitude and longitude of selected cities from the *Zīj*. In this table the longitudes have been measured from the observatory of Delhi, and they are expressed in the units of *ghaṭikā* and *pala*. The Table also gives the modern values of longitudes converted into the *ghaṭikā* and *pala* units for comparison.

In its Chapter 4, the *Zīj* reports the maximum declination of the sun, or the obliquity of the ecliptic,[2] as determined at different places. These values are as follows:

Samarkand	23;30,17
Europe	23;29,6
Jai Singh's observatories	23;28

The third book of the *Zīj*, with its 13 chapters, deals with planets and includes extensive tables and rules for finding the coordinates of the sun, the moon, and the planets. The epoch dates of these tables is February 20, 1719. The planetary tables include the mean elements for the 12 months of an Arabic year

---

[1] The author of the *Zīj* chose February 20, 1719, arbitrarily for the beginning of the Muḥammad Shāhī era. Muḥammad Shāh did not ascend the Mughal throne until September 18, a few months after the chosen date. His predecessor, the emperor Farrukh-Siyar, was assassinated on February 18-19, 1719. The Muḥammad Shāhī era began immediately following the date of the assassination.

[2] ZMS, f. 12.

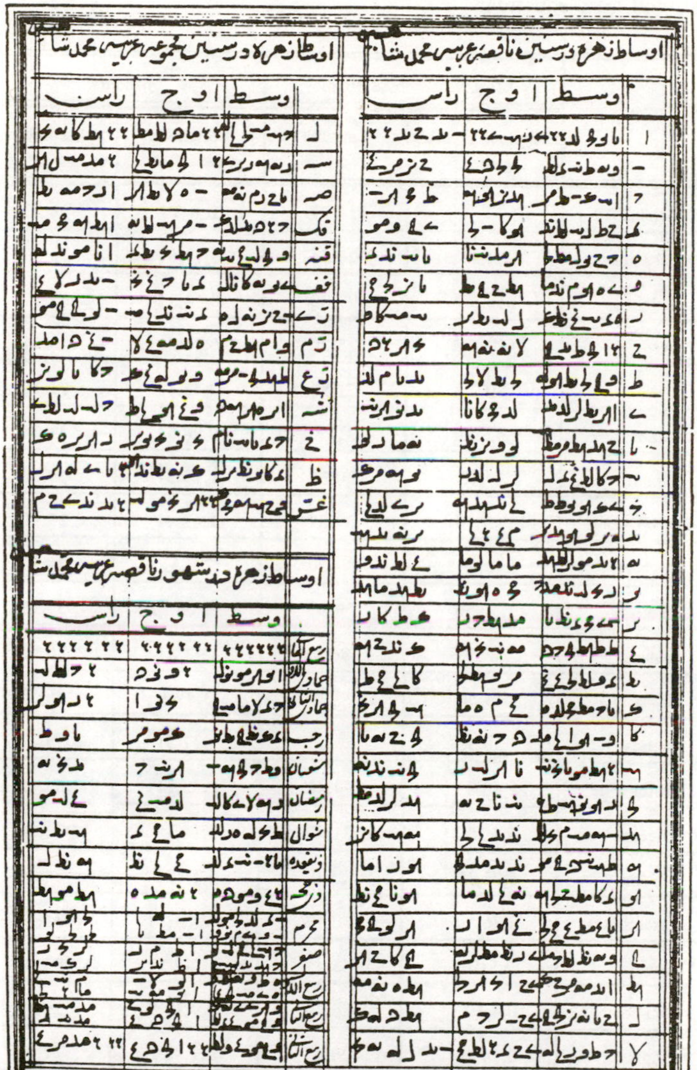

Fig. 11-1 A Page from the *Zīj-i Muḥammad Shāhī*: The Parameters for the planet Venus. (Courtesy, The British Library)

at monthly intervals, for 30 Arabic years at yearly intervals, and for 390 Arabic years at 30-year intervals. The set of tables for the moon has the moon's first, second, and third equations.

Town	Zīj-i Muḥammad Shāhī		Modern	
	Longitude *ghaṭikās* etc.	Latitude	Longitude *ghaṭikās* etc.	Latitude
Shāhjahānābād	0,0	28;37	0,0,0	28;37
Sawai Jaipur	0,12,30	26;54	0,13,51	26;53
Ujjain	0,15,(?)	22;30	0,14,31	23;11
Goa	0,20	15;30	0,12,51	15;31
Surat	0,30,0	21;53	0,43,11	21;10
Lahore	0,42,30	31;50	0,28,31	31;34
Kabul	1,29,10	30;35	1,20,31	34;30
Samarkand	12,23,10	39;37	1,42,41	39;40
Bukhara	2,40,50	39;50	2,7,51	39;47
Tous	11,50,50	33;31	12,58,51	39;10
Basara	4,55,50	30;0	4,53,51	30;30
Maragha	5,15,50	37;20	3,49,1	26;51
Baghdad	5,30,50	33;25	5,27,51	33;20
Moscow	5,40,0	44;48	6,35,11	55;45
Mecca	6,4,10	21;40	6,14,1	21;26
Medina	6,22,30	25;0	6,16,21	24;30
Alexandria	8,36,50	30;18	7,53,1	31;13
Geneva	12,5,50	43;45	11,22,51	44;24
Gwalior	0,4,10	26;29	0,9,19 E	26;12
Agra	0,14,10	26;43	0,7,49 E	27;09
Varanasi	0,37,30	26;55	0,57,49 E	25;20

Table 11-1 Latitude and Longitude of selected towns from the *Zīj-i Muḥammad Shāhī* in units of *ghaṭikā* and *pala*. The modern values in the last columns are for comparison.

For a typical set of planetary elements in the *Zīj-i Muḥammad Shāhī*, see Table 11-2. The table expresses the mean parameters of the sun for 30 Arabic years.

Arabic Years	Mean Tropical Longitude	Longitude of Aphelion or Apogee
Epoch *Zīj* or year 1	10,29;36,1,0,0,0	3,8;26,4,00
2	10,19;30,18,2,11,30	;27,3,48,55
3	10,8;25,26,44,36,9	;28,3,27,42
4	9,27;20,35,27,0,48	;29,3,6,30
5	9,17;14,52,29,12,18	;30,2,55,25
6	9,6;10,1,11,36,54	;31,2,34,31
7	8,26;4,18,13,48,27	;32,2,23,8
8	8,14;59,26,56,13,6	;33,2,1,55
9	8,3;54,35,38,37,45	;34,1,40,53
10	7,23;48,52,40,49,15	;35,1,29,38
11	7,12;44,1,23,13,54	;36,1,8,26
12	7,1;39,10,5,38,34	;37,0,47,14
13	6,21;33,27,7,50,3	;38,0,36,9
14	6,10;28,35,50,14,42	;39,0,14,57
15	5,29;23,54,32,39,21	;39,59,53,44
16	5,19;18,1,34,50,51	;40,39,42,17
17	5,8;13,10,17,15,30	;41,59,39,27
18	4,28;7,27,19,27,0	;42,59,10,22
19	4,17;2,36,1,51,49	;43,58,49,10
20	4,50;57,44,16,28	;44,58,27,58

Table 11-2 The longitude of the sun and the longitude of its aphelion at one year intervals in the *Zīj* beginning 1131 A.H. The epoch date of the *Zīj* is Feb. 20, 1719.

Arabic Years	Mean Tropical Longitude	Longitude of Aphelion or Apogee
21	3,25;52,1,46,27,58	;45,58,6,45
22	3,14;47,10,28,52,37	;46,57,55,40
23	3,3;42,19,11,17,16	;47,57,34,28
24	2,23;36,36,13,28,46	;48,57,13,16
25	2,12;31,44,55,53,25	;49,57,2,10
26	2,2;26,1,58,4,55	;50,56,40,59
27	1,21;21,10,40,19,34	;51,56,29,54
28	1,10;16,19,22,54,13	;52,56,8,42
29	1,0;10,36,25,5,43	;53,54,47,30
30	0,19;5,45,7,30,22	;54,55,36,24
31	0,8;0,53,49,55,1	3,8;55,55,15,12

Table 11-2 cont. The longitude of the sun and the longitude of its aphelion at one year intervals in the *Zīj* beginning 1131 A.H.

For a detailed listing of the tables of the *Zīj*, the reader should refer to Appendix 4.[1] In the conclusion or *Khātmā* section of book 3, the *Zīj* gives detailed procedures, with examples, for calculating the occurrence of eclipses of the moon and the sun.[2] In the following, we summarize the calculated results for the eclipses of the moon and the sun, given as an example in the *Zīj*.

Lunar Eclipse

Place of Observation: Jaipur

Date: Sunday, the 14th day of the *Zīl-Ḥijjah* month of the 14th year of the Muḥammad Shāhī era, i.e., 1144 A.H. Or the *Pūrṇimā* of the month of *Jyaiṣṭha*, 1789 V.S., or 1654 S.E. The date translates into June 8, 1732. (Gregorian)

Time: The end of the umbra phase at 18 *ghaṭikās*, 31 *palas* (7 hr, 24 min, 24 sec) after midday or 5 *ghaṭikās*, 30 *palas* (2 hr, 12 min) after sunset.

---

[1] The Appendix is based on Mercier, *Op. Cit.*, Ch. 5.

[2] ZMS, f. 175 ff.

We checked these timings and found them to be accurate within 3 minutes of the values calculated with a computer.

Solar Eclipse[1]

Place of Observation: Jaipur
Date: Monday, the 30th day of the eleventh month of the 16th year of the Muḥammad Shāhī era, i.e., 1146 A.H. Or the *Amāvasyā* of the month *Vaiśākha*, 1791 V.S., or 1656 S.E. The date translates into April 2, 1734 (Gregorian)
Time: 7 *ghaṭikās*, 47 *palas*, and 30 *vipalas* (3 hr, 7 min) after midday.

Finally, the *Zīj* has tables of the coordinates of 1018 stars arranged according to the order of the constellations found in the *Zīj-i Ulugh Beg*. The epoch date of these tables is 1138 A.H. or 1725-26 A.D. The manuscripts of the Raza Library, Rampur; Andhra Pradesh Archives, Hyderabad; and Maulana Azad Library, Aligarh; have a fewer number of stars as stated earlier. According to the text of the *Zīj*, its author contemplated originally a star table of some 60 stars only. It is possible that these stars were those whose coordinates Jai Singh's astronomers had actually measured.[2] The astronomers must have taken their observations around 1138 A.H. (1725-26 A.D.) as it is the epoch date of the catalog. Subsequently, the astronomers decided to extend the tables further by including the rest of the stars from Ulugh Begh's *Zīj*. This they did by adding 4;8 to the star longitudes.[3] It should be pointed out that it was customary with the compilers of Islamic *Zījes* to update their star charts by simply applying a precession correction to their longitudes. In Table 11-3, we reproduce the coordinates of the stars in the constellation of Ursa Minor from *Zīj-i Muḥammad Shāhī*.[4]

Jai Singh wrote his *Zīj-i Muḥammad Shāhī* keeping the *Zīj-i Sulṭānī* of Ulugh Beg as a model. We found that a large number of Jai Singh's tables run parallel to those of the *Zīj-i Sulṭānī*.[5] The tables in the second book of the *Zīj-i Muḥammad Shāhī*, concerning the spherical trigonometry are identical with those

---

[1] ZMS, f. 182 ff.

[2] Kaye examined a Devanagari version of the star catalog of the *Zīj-i Muḥammad Shāhī* and noted that the longitudes of 68 stars in it could not be obtained by adding a correction factor to Ulugh Beg's star catalog. It is possible that these 68 stars included those whose coordinates were measured by Jai Singh's astronomers. See Kaye, p. 115. We were not able to locate the catalog Kaye refers to at the Sawai Man Singh II Museum, Jaipur.

[3] A spot check applied to the star tables of the *Zīj-i Muḥammad Shāhī* revealed this indeed is the case. Kaye's photograph of the first-page of the *Jaipur Catalog*, also has a statement to this effect.

[4] ZMS, f. 194.

[5] We compared the British Museum copy of the ZMS with the *Zīj-i Ulugh Beg*, No. 252, at Osmania University, Hyderabad.

in the Ulugh Beg's *Zīj*.[1] The *Zīj* of Ulugh Beg contains a number of astrological tables, but Jai Singh excluded these from his *Zīj*.[2] In the star tables of the *Zīj*, Jai Singh reproduced the stellar magnitudes from Ptolemy and from al-Ṣūfī both.[3] The star tables also have a column on the "temperament" of the stars that Ulugh Beg does not have.

Constellation – Ursa Minor						
	Description	Longitude	Latitude	Magnitude (al-Ṣūfī)	Magnitude (Ptolemy)	Mezāj (Temperament)
1.	Star on the edge of the Bear's tail	2;24,23	66;27 N	3	3	30,5
2.	Second star on the tail, next to it	2;26,33	70;0	4	4	30,5
3.	Third star on the tail, next to it	3;5,3	73;45	4	4	30,5
4.	Star on the left hind foot: 2 stars on the fore-leg, one to the right of the other	3;21,21	74;36	4	4	30,5
5.	Star on the right hind foot, one to the north	3;28,23	78;0	5	4	30,5
6.	Star on the left fore-leg, to the south	4;9,33	73;0	2	2	30,5
7.	Star on the right paw to the north	4;18,3 (?)	75;9 N	3	2	30,5

Table 11-3 The stars of the constellation of Ursa Minor in the *Zīj-i Muḥammad Shāhī* and *Zīj-i Sulṭānī* of Ulugh Beg

[1] Mercier, *Op. Cit.*, Ch. 5.

[2] David Pingree, "Indian and Islamic Astronomy at Jayasimhā's Court," Defferant to Equant, *Annals of New York Academy of Science*, pp. 313-328, New York, 1987.

[3] Knobel believes that "Ṣūfī's catalog is simply that of Ptolemy, in which the longitudes are brought up to the epoch by the addition of 12;42." Edward B. Knobel, *Ulugh Beg's Catalog of Stars*, p. 14, Washington, D.C., 1917.

The Telescope

The *Zīj-i Muḥammad Shāhī* lacks data taken with a telescope. Its author, however, is aware of the instrument as he discusses its capabilities. In the introduction to the section on the visibility of the moon, he says:

"The rules [regarding the visibility of the moon] are based on naked-eye observations only, although the telescope is now constructed in our kingdom. The telescope enables one to see bright stars in broad daylight, say, around the noon hour. It also enables one to see the moon when there is hardly any light in it, or when its face is totally dark and invisible to the unaided eye. And the same is true about the common stars [planets], i.e., regarding their visibility and invisibility....

"With the telescope we have noted certain facts in contradiction to the well known texts. For instance, we have seen with our own eyes that Mercury and Venus get their light from the sun the way the moon does. When these planets are close to the sun, they appear bright, and when they are away, they appear faint, indicating that they do indeed receive light from the sun.

"Further, the planet Saturn appears oval in shape; an oval whose lower half is larger than the rest. Around Jupiter are seen four bright stars circling the planet. On the face of the sun there are spots. And, as the sun rotates, they rotate with it, completing a turn in one year."

He goes on to add:

"Since the telescope is not readily available to an average man, we are going to base our rules of computation for naked eye observations only, which, in turn, are based on earlier texts."[1]

Does the Zīj have European Astronomical Findings?

In the third book or *maqālā* of the *Zīj-i Muḥammad Shāhī*, its author points out that the astronomers of Europe were attempting to reduce difference between observation and theory regarding the motion of the sun. In the same context he remarks that the orbit of the sun is oval in shape.[2] From these remarks, some scholars have deduced that the writer of the *Zīj* was aware of the laws of Kepler—particularly his first law. In his first law, Kepler describes the orbits of planets to be elliptical. Kepler's laws are based on a heliocentric view of the

[1] ZMS, f. 189.
[2] ZMS, f. 132 ff.

solar system, but the writer of the *Zīj*, is still struggling with the geocentric picture. Moreover, the author of the *Zīj* does not extend his remarks about the orbital shape of the sun to other planets. Hence the deduction that the author of the *Zīj* was aware of the true nature of the planetary orbits is not convincing.

Although the *Zīj* has one or more tables borrowed from the Tables of French astronomer de La Hire, Jai Singh does not seem to be familiar with Western techniques of observing. For example, he does not describe the use of the satellites of Jupiter for determining the longitude of a location. He still elaborates upon the age-old method of observing a lunar eclipse from two different locations for the exercise of determining the longitude difference between two locations on the globe.[1]

The Originality of the Zīj-i Muḥammad Shāhī

Since the publication of the *Zīj-i Muḥammad Shāhī*, doubts have been cast about the originality of its data and suggestions made that the *Zīj* is based on the *Tabulae Astronomicae* of Phillip de La Hire. Pons, a Jesuit priest, who visited the Raja in 1734, wrote in 1740 to a friend in Europe, that the Raja would publicize or popularize the *Tabulae Astronomicae* of de La Hire under his own name or patronage.[2] Hunter, the first European to study the *Zīj-i Muḥammad Shāhī* around 1785, also suspected the originality of the *Zīj's* contents.[3] Additionally, in a recent article, it has been declared that all the planetary tables of the *Zīj* are mere adaptation of de La Hire's Tabulae Astronomicae and that they "in no way depend on observations made in India." [4,5]

These doubts contrast with Jai Singh's claims about his own work. In the preface to the *Zīj*, he writes:

> "... in this place [Delhi], ... astronomical instruments were constructed with all the exactness that the heart can desire, and the motions of the stars constantly observed with them for a long period of time.... (And then) mean motions and equations were established which were consistent with observation. He [Jai Singh] found the

---

[1] ZMS, ff. 20v-21r.

[2] Lettre du P. Pons au P. du Halde, Careical', sur la còte de Tanjaour, aux Indes orientales, 23, Nov. 1740., *Lettres*, p. 645.

[3] Hunter, p. 205.

[4] Mercier contends, "... all the tables of the *Zīj* concerning the sun, moon, and planets are taken from La Hire's work, ... They in no way depend on observations made in India, except in so far as the meridian of reference has been shifted from Paris to Delhi." Mercier, *Op. Cit.*, Ch. 5.

[5] Mercier's assertion that the tables of *Zīj-i Muḥammad Shāhī* are based on *Tabulae Astronomicae* rests primarily on his "longitude" argument. He states that, for the *Zīj*, the rates of motion for the mean parameters have been calculated first for the Paris meridian for the epoch date of February 20, 1719, and then, applying a longitude correction, changed for the meridian of Shāhjahānābād or Delhi. *Ibid*.

> calculations [with these] agreed perfectly with the observation.... A table under the name of His Majesty [Muḥammad Shāh], ... comprehending the most accurate rules and most perfect methods of computation was constructed...."

In order to verify the validity of Jai Singh's claim, as well as to confirm the doubts cast on it by others, we analyzed the planetary tables of the *Zīj*. On the basis of our analysis, we conclude that the planetary tables of the *Zīj* are indeed genuine and independent of the *Tabulae Astronomicae* of de La Hire.[1]

Tabulae Astronomicae of de La Hire

De La Hire (1640-1718), a highly competent observer and a member of the Academie de Sciences at Paris, published his *Tabulae Astronomicae* in 1687. The second edition of the text came out in 1702 and a reprint 25 years later in 1727.[2] The second edition of the *Tabulae* is in two books. Its first book describes how to apply the tables given in book II for solving astronomical problems, such as the calculations of solar and lunar eclipses. The second book has a wide variety of charts and tables, beginning with the table of conversion of arc to time, equation of time, and geographical coordinates of 126 cities and towns around the globe. These are followed by extensive tables for the mean longitude, aphelion and the ascending node of the moon and of the planets.

Sawai Jai Singh had acquired both editions of the *Tabulae Astronomicae* of de La Hire. His assistants brought the first edition in 1731 from Portugal,[3] and then du Bois copied the second edition a year later in 1732 at Jaipur.[4] The manuscript of du Bois may still be seen at the Sawai Man Singh II Museum of Jaipur. Du Bois writes that Jai Singh ordered the translation of the *Tabulae* as soon as it reached Jaipur and that Jai Singh entrusted him to complete the task. The Sawai Man Singh II Museum preserves an incomplete translation of the *Tabulae Astronomicae* in Sanskrit under the title *Firaṅgī Candravedhopayogī Sāraṇī*.[5]

---

[1] For a detailed discussion of this subject see Virendra Nath Sharma, "The *Zīj-i Muḥammad Shāhī* and the Tables of de La Hire," *Indian J. Hist. Sci.*, Vol. 25 (1-4), pp. 34-44, (1990).

[2] *Tabulae Astronomicae Ludovici Magni* ... , of Philippus de La Hire, Paris, 1702; second edition Paris, 1727.

[3] For Sawai Jai Singh's delegation to Europe see Ch. 13, also Virendra Nath Shrama, "Jai Singh, His European Astronomers and the Copernican Revolution," *Indian J. Hist. Sci.*, Vol. **17** (2), 345-352, (1982). Also George M. Moraes, "Astronomical Missions to the Court of Jaipur," *J. Bombay Roy. Asiatic Soc.*, **27**, 61, (1951).

[4] Du Bois, *Op. Cit.*, Ch. 3.

[5] Formerly mislabeled as *Firaṅgī Candravedhopayogī Sāraṇī*, Catalog No. 5609, of the Sawai Man Singh II Museum. See Bahura II, p. 63.

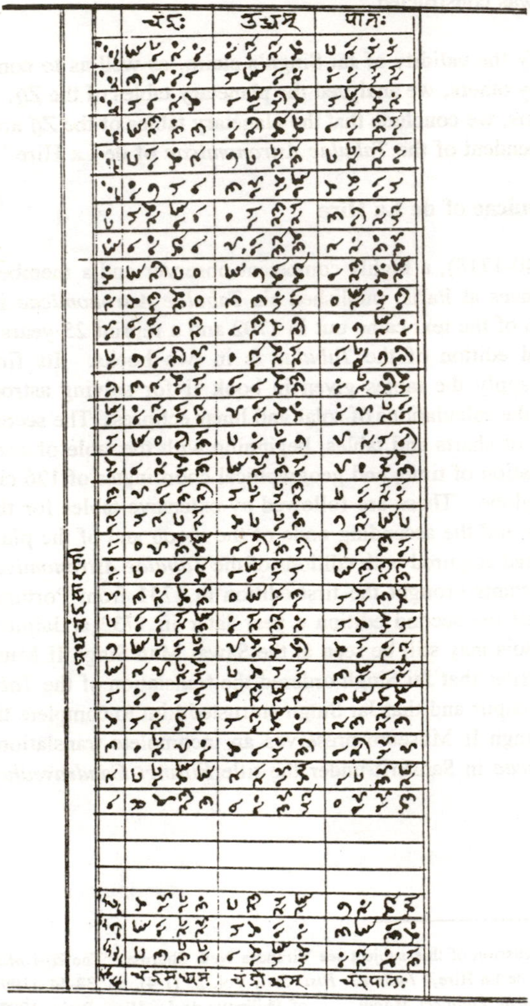

Fig. 11-2 A Page from the *Dṛkpakṣa Sāraṇī*: the Parameters for the Moon.

The Oriental Research Institute of Baroda also preserves a copy of the translation listed under the title *Dṛkpakṣasāraṇī*.[1] This manuscript is also incomplete and includes tables and explanatory text only. In Figure ... we reproduce a table for the parameters of the moon. A section of the translation called *Dṛkpakṣasāraṇyam Sūryagrahaṇam* is located at the Bhandarkar Oriental Research Institute of Pune.[2] The Pune copy, as its name accurately describes it, concerns with only the calculations of solar eclipses and has no tables.

The Tabulae Astronomicae and Zīj-i Muḥammad Shāhī

In order to verify the assertion that Jai Singh based his planetary tables of the *Zīj-i Muḥammad Shāhī* on de La Hire's *Tabulae Astronomicae*, we calculated the mean parameters of the planets from the *Tabulae Astronomicae*. We did this for a number of Arabic years following the date of the *Zīj*'s epoch, February 20, 1719.

We carried out these calculations in more than one way. The *Tabulae Astronomicae* has tables with mean values of the parameters for 1-20 years calculated for every planet. We reasoned that if Jai Singh and his astronomers had, in fact, decided to transform the tables of the *Tabulae Astronomicae* for their own *Zīj*, the simplest approach for them, for at least some of their data, would have been to use de La Hire's 1-20 year tables directly. By applying a simple equation they could have obtained their desired results. For example, the mean longitude of the sun for the *n*th year after the epoch year should be obtained as follows.

Mean L for the nth year = (mean L in the *Tabulae* for n years)$\times$(354;22)$\div$(365;15) + 329;36,1

In the above equation, 354;22 is the number of days in an Arabic year; 365;15 is the length of a Julian year, used by de La Hire; and 329;36,1 is the mean L at the epoch date in the *Zīj-i Muḥammad Shāhī*. For calculating the mean L, the number of complete revolutions calculated from the sidereal period of the planet should also be taken into account.

Following this procedure, we calculated yearly mean parameters for the sun, the moon, and the planets and compared them with those in the *Zīj*. We give a typical set of calculations for the planet Venus in Table 11.4. In literally scores of such calculations that we carried out for our investigation, we did not find a single entry in the *Zīj* that agreed with our calculated results.

---

[1] *Dṛkapakṣa Sāraṇī*, Catalog. No. 3162, 29 ff, Oriental Research Institute, Baroda.

[2] *Dṛkapakṣasāraṇyam Sūryagrahaṇam*, No. 926 of 1886-92, Bhandarkar Oriental Research Institute, Pune. The translator of the works is Kevalarāma, a prominent assistant of the Raja.

Julian Years	Mean Motion in *Tabulae* (*L*)	Arabic years following epoch ZMS	Mean Motion Computed from *Tabulae* (*L*)	Mean Motion in ZMS (*L*)
		0	------	11,16;23,34
1	7,14;47,36	1	6,13;45,40	6,15;9,52
2	2,29;35,13	2	1,11;7,47	1,12;20,20
4	6,0;46,33	4	3,7;25,14	3,8;16,31
5	1,15;34,9	5	10,4;47,20	10,5;26,41
7	4,15;9,21	7	11,29;31,32	0,1;23,9
10	3,1,;8,18	10	8,23;11,6	8,24;27,48
15	4,16;42.27	15	7,11;34,52	7,13;32,55
17	7,17;53,46	17	9,7;52,19	9,9;29,23
19	10,17;28,58	19	10,21;52,54	11,3;49,44
20	6,3;52,43	20	6,1;31,54	6,2;26,2

Table 11-4 The mean longitude, *L*, for the planet Venus for the Arabic years following the epoch date of the *Zīj-i Muḥammad Shāhī* (ZMS) calculated from *Tabulae Astronomicae* of de La Hire. The very first number in the columns represents the number of the signs elapsed. The numbers after the semicolon follow the sexagesimal scheme of expressing angles.

Next we calculated the same parameters for a number of years following the epoch date from de La Hire's 1600-year motions. We reasoned that perhaps Jai Singh ignored the yearly data from the *Tabulae Astronomicae* and instead based his calculations on the data of a larger number of years, such as 1600 Julian years.[1] However, even with the 1600-year data, we failed to find any agreement. We repeated the procedure with a much smaller number of years, for example, 20. Once again the results were negative. We also converted the yearly difference for the parameters of de La Hire into those for the Arabic years of the *Zīj* but could find no agreement there either.

---

[1] There is some difference between the parameters computed on the basis of the yearly tables and those computed on the basis of a large number of years such as 1600 Julian years. However, this difference cannot account for the difference between the tables of de La Hire and those of Jai Singh.

	Element	de La Hire	ZMS	Modern
Sun	*L*	349;16,49,51	349;16,49,46	349;16,50,1
	Γ	0;0,59,40	0;0,59,38	0;1,0,3
Moon	*L*	12ʳ349;16,32,35	12ʳ349;16,30,56	12ʳ349;16,32,35
	Γ	39;28,47,32	39;28,47,34	39;28,40,25
	Ω	−18;45,55,53	−18;30,7,18	−18;45,54,23
Mercury	*L*	4ʳ10;12,18,27	4ʳ10;12,18,28	4ʳ10;12,7,13
	Γ	0;1,35,51	0;1,35,55	0;0,54,20
	Ω	0;1,22,42	0;1,22,45	0;0,41,24
Venus	*L*	1ʳ207;45,24,59	1ʳ207;50,51,26	1ʳ207;40,18,39
	Γ	0;1,23,38	0;1,23,41	0;0,49,11
	Ω	0;0,44,42	0;0,42,31	0;0,32,0
Mars	*L*	185;42,47,4	185;42,47,15	185;42,48,4
	Γ	0;1,4,29	0;0,52,44	0;0,60,48
	Ω	0;0,35,47	0;0,35,55	0;0,26,56
Jupiter	*L*	29;27,29,59	29;27,30,8	29;27,29,52
	Γ	0;1,31,36	0;0,55,35	0;0,56,14
	Ω	0;0,13,42	0;0,13,42	0;0,35,18
Saturn	*L*	11;52,7,24	11,52,7,14	11,13,54,33
	Γ	0;1,19,15	0;1,7,37	0;1,8,24
	Ω	0;1,9,20	0;1,9,25	0;0,30,30

Table 11-5 The mean planetary elements per Arabic year calculated from the *Tabulae Astronomicae* and the *Zīj-i Muḥammad Shāhī* (ZMS). The entries in the columns are according to the sexagesimal scheme.

Our investigations thus left us little choice but to suspect that the parameters in the *Zīj* and in the *Tabulae Astronomicae* have been obtained from two different sets of elements. Our suspicions were confirmed when we calculated the mean yearly motions of the elements from the two texts. We give the results of these calculations in Table 11-4. We display the fractions beyond the seconds of arc merely to point out the difference between the mean elements of the two astronomical texts. They have no real significance. The instruments of Jai

Singh and those of de La Hire as well did not justify even the retention of seconds. The last column in the table shows the modern values provided for comparison purposes.

Tables in the Zīj-i Muḥammad Shāhī from Tabulae Astronomicae

Although the planetary tables, as shown in Table 11-5, are independent of the *Tabulae Astronomicae*, it would be erroneous to think that none of the information in the *Zīj* is based on de La Hire's *Tabulae*. Islamic *Zījes*, by and large, lack refraction-correction tables, but the *Zīj-i Muḥammad Shāhī* has such a table. Jai Singh borrowed this table directly from the *Tabulae Astronomicae*.[1] There is no evidence that the Raja himself determined any refraction correction data for his *Zīj*. Another table that he could have adapted from the *Tabulae Astronomicae* is the equation of time. In the *Zīj*, the equation of time is in *ghaṭikā* and *pala* units, which, after transformation into minutes and seconds, becomes identical with the table of the *Tabulae Astronomicae*. We should point out, however, that tables for equation of time are not unique to European works; they are also found in just about every *Zīj* written in the Islamic world during the middle ages.

Commentaries on the Zīj-i Muḥammad Shāhī

The *Zīj-i Muḥammad Shāhī* is written in a difficult to understand style and, consequently, there were three commentaries written on it.

Tas'hīl-i Zīj-i Muḥammad Shāhī

Azīm al-Dīn Muḥammad Khān Abdu'llāh, alias Mahārat Khān, wrote *Tas'hīl-i Zīj-i Muḥammad Shāhī*.[2,3] We do not know the date when Mahārat Khān wrote his commentary. Rahman calls it an 18th century composition.[4] However, any direct evidence in support of Rahman's date is lacking. A copy of the commentary of Mahārat Khān, preserved at Hyderabad, is in 184 folios. It is written in *Nastaᶜlīq* script and is divided into 3 main books or *maqālās*, a preface, and a conclusion.[5] In the preface, the author explains first the

---

[1] ZMS, f. 146.

[2] *Tas'hīl-i Zīj-i Muḥammad Shāhī* by Azīmu'ddīn Muḥammad Khān ᶜAbd Allah known as Mahārat Khān, No. 297, Riyāzī, Andhra Pradesh State Library, Hyderabad.

[3] In literature, Mahārat Khān is erroneously spelled as Mahābat Khān. For instance, see Blanpied (1974), p. 107, *Op. Cit.*, Ch. 1.

[4] The *Bibliography*, p. 275.

[5] The *Bibliography*, p. 276, lists two other places where the manuscripts are located: Khudabaksha Library, Bankipur, Patna, No. 1057, H.L. No. 1050. 113 ff; and Edinburgh (Univ. of ?) PMC-375; No. 417, 217 ff. A copy of the *Tas'hīl* is probably preserved at the Mothari

terminology related to an observatory. Then he goes on to list 5 instruments generally found at an observatory. In these instruments he does not include the instruments invented by Jai Singh such as the Samrāṭ yantra. He also does not include the telescope. His instruments are: 1. *Libnā* (Meridian mural arc), 2. *Suds-i Fakhrī* (Sixty degree meridian arc), 3. *Dhāt al-Thuqbatayn,*[1] 4. *Dhāt al-Shuʿbatayn,*[2] and 5. *Dhāt al-Ḥalaq.*[3]

Next, Mahārat Khān names three famous observatories of the past: the observatory of Hipparchus, that of Ptolemy at Alexandria, and of Ulugh Begh at Samarkand. He mentions four famous Zījes. According to him the famous *Zījes* are, *Zīj-i Jāmiʿ-i Kūshyār*, *Zīj-i Gumdālī Shāh*, *Zīj-i Kasharafī-i Muzahārī* and *Zīj-i Shāhjahānī* of Mullā Farīd al-Dīn. Mahārat Khān does not include the *Zīj-i Ulugh Beg* in his list of the famous *Zījes*, but has in its place the *Zīj-i Shāhjahānī*, which is based on Ulugh Beg's *Zīj.*[4]

In the first *maqālā* of his commentary, which has 22 sections, Mahārat Khān includes the topics related to spherical astronomy. He also discusses a procedure for finding the direction of the *Qiblā*, or the shrine at Mecca in the Arabian peninsula. The latter topic is not included in the *Zīj-i Muḥammad Shāhī*. The subchapters of Mahārat Khān's text run parallel to that of the *Zīj*. In the third *maqālā*, which has 13 sections, Mahārat Khān deals with subjects such as eclipses, visibility of the moon and the rising and setting of the fixed stars.

Khulāsah-i Zīj-i Muḥammad Shāhī

Muḥammad bin Ashūshtarī (al-Shūshtarī ?) wrote a *Khulāsah* or gloss on *Zīj-i Muḥammad Shāhī* a little before 1280 A.H. or in 1863 A.D. The *Khulāsah* has 215 pages.[5] The Raza Library, Rampur, has a copy of it, and it is in Ashūshtarī's own pen. The *Khulāsah* of Ashūshtarī has a table of ascendants calculated for his ancestral town, Shūshtar(?) in Iran. In the *Khulāsah* the author does not follow the subject order given in the *Zīj*, however.

Library of Tehran also.

[1] See Ch. 2.

[2] *Ibid.*

[3] *Ibid.*

[4] Mahārat Khān remarks that since the word *Zīca*, with its roots in the word *Zīj*, has become quite common, its use is acceptable to denote an astronomical text or *Zīj*.

[5] *Khulāsah-i Zīj-i Muḥammad Shāhī* by Muḥammad bin Ḥasan Ashūshtarī, No. 1222, Raza Library, Rampur.

Sharḥ-i Zīj-i Muḥammad Shāhī

A third commentary, or Sharḥ, on the *Zīj* is said to have been written by Abu'l Khairu'llāh, a contemporary of Jai Singh.[1] we were unable to locate this work.

The Zīj-i Muḥammad Shāhī and the Traditional Scholars of India

Jai Singh published his *Zīj-i Muḥammad Shāhī* after a great deal of labor. However, it turned out to be of little value to the world of astronomy at large. Its information was based on naked-eye observations, and it made no use of the neo-astronomy of Europe. At that time, far better tables with their data collected with telescopic-sight-fitted sextants and quadrants had already been published in Europe. However, to a large number of traditional scholars of India, to whom western science was out of reach, it proved to be a great boon. Indian scholars trained in the Islamic or Arabic tradition of the subject readily adopted it. They made copies of it which found their way to faraway places such as the libraries of Iran and Central Asia. In India it remained in use for almost 150 years after its first publication in 1731-32. Major Jarvis observed in 1845:

> "By these tables eclipses are calculated and almanachs composed at Dehli and in the Northern Provinces of India at this day."[2]

Further, in 1885, by when the country was in complete sway of the British rule, J. P. Stratton, the Resident of Jaipur, noted:

> "... *Zīj-i Muḥammad Shāhī*, the tables of Muḥammad Shāh, are still used by Indians in their computations, and in preparing the elaborate calendars needed for the many religious festivals of this country."[3]

Although there is little doubt that the Islamic world adopted the *Zīj* enthusiastically, it is uncertain if Hindu astronomers showed the same degree of enthusiasm in accepting it and its new parameters. For instance, the weekly motions of the sun and the moon given by Kevalarāma in his *Pañcāṅga Sāraṇī* are not the same as calculated on the basis of the *Zīj*. It remains to be seen if Kevalarāma was an exception or the representative of a trend. The answer to this question can only come from the analysis of a number of *Sāraṇīs* prepared at Jaipur and elsewhere in the 18th and the 19th centuries following the

---

[1] The *Bibliography*, p. 285, quotes *Jāmi'-i Bahādur Khānī*. The reference is unclear however. Khān Ghorī, on the other hand, gives the following reference: *Jāmi'-i Bahādur Khānī* of Ghulām Husain Jaunpurī, p. 579, Calcutta, 1835. See S.A. Khān Ghorī, "Impact of Modern European Astronomy on Jai Singh," *Indian J. Hist. Science*, Vol. **15** (1), pp. 50-57, (1980).

[2] ZMS, flyleaves.

[3] Purohita Hari Nārāyana Collection, Oriental Research Institute, Jaipur, No. 309/11,

publication of the *Zīj*.

Yantrarājaracanā

Another text attributed to Sawai Jai Singh is the *Yantrarājaracanā*, or a treatise on the astrolabe. The treatise has now been edited by Kedāranātha and published by the Oriental Research Institute of Jodhpur.[1] The *Yantrarājaracanā* has two chapters—one on the construction of the instrument and the other on taking observations with it. The text published by the Rajasthan Oriental Research Institute also has a short tract by Śrīnātha Chāgāṇī, in which Chāgāṇī renders the first chapter of the *Yantrarājaracanā* into 29 verses. It is not certain if Chāgāṇī was a contemporary of Jai Singh. The Columbia Library manuscript does not carry the verses written by Chāgāṇī.[2] The text published by the Rajasthan Oriental Research Institute also includes an explanatory essay in Sanskrit at the end by its editor, Kedāranātha.[3]

Sūryasiddhānta Vyākhyā

Jai Singh wrote *Sūryasiddhānta Vyākhyā*, or commentary on *Sūryasiddhānta*, the canon of Hindu astronomy. The Oriental Research institute, Jodhpur, preserves a copy of it.[4] Another copy, entitled *Sūryasiddhāntavicāra* and in the pen of three different scribes, is at the Sawai Man Singh II Museum.[5,6]

The royal library

Sawai Jai Singh, in his efforts to revive astronomy in the country, not only built observatories but also compiled an excellent library on astronomy-astrology and mathematical subjects. When he became king, Jai Singh inherited only 32 books on astronomy, astrology, and mathematics in his ancestral library. However, during his lifetime, he acquired an excellent and unique collection of

---

[1] *Yantrarājaracanā* of Jai Singh, Rajasthan Puratan Granthmala, Vol. 5, ed., Kedāranātha, Jaipur, 1953.

[2] 1. *Yantrarāja-racanā-prakāra*, of Savāī Jayasiṁha, 3 ff., David Eugene Smith Collection, Indic ms. No. 168. The manuscript is apparently incomplete. 2. *Yantrarāja*, David Eugene Smith Collection, Indic ms. No. 73. A note on the manuscript states, "Treatise on the astrolabe by the Emperor, Jey Sing, copied by a priest of the court of the present Maharaja of Jeypore. Purchased by Professor Smith from Pundit Joshi, Jeypore, Christmas, 1907." Both manuscripts are preserved at the Columbia University Libraries, New York.

[3] Kedāranātha was the supervisor of the Jaipur Jantar Mantar until 1953.

[4] *Sūryasiddhānta Vyākhyā* of Maharaja Jai Singh, No. 29498, 29 ff, Rajasthan Oriental Research Institute, Jodhpur.

[5] *Sūryasiddhāntavicāra*, Bahura II, p. 126. Bahura writes that it was written for Jai Singh. It has a date of 1763 V.S. or 1708 A.D.

[6] *Sūryasiddhāntavicārapustakam*, No. 4955, The Museum.

almost 200 books on these subjects. Jai Singh enriched his library, the Pothikhana (Pothīkhānā), with books brought from all over the country and from faraway places such as Europe. His library had books in Sanskrit, Hindi, Bangala, Persian, Arabic, Latin, French, German, and English. He acquired some of these books, particularly the ones from abroad, at enormous cost. For example, he purchased the Grosser Atlas of the world, printed in Germany in 1725, at a cost of Rs. 520.[1] Some of the books in his library, such as the *Zīj-i Shāhjahānī Nityānandī* belonged to the Imperial Library at Delhi at one time as it bears a seal to this effect.

Jai Singh developed an interest in collecting astronomical books at an early age. When he was only 13, he had *Bhāsvatī*, an astronomical text in Sanskrit, copied for him as pointed out earlier. Later when he decided to revise the planetary constants, he ordered well-known texts and commentaries copied, translations done, and a number of original works written by his pundits.[2]

Inventory of the Royal Library

The inventory records of the Raja's personal library give a good glimpse of Jai Singh's collections. There were two inventories done of the library stock. The first began in 1741 during Jai Singh's lifetime and was completed in 1743, the year of his death; and the second was undertaken four years later during the rule of Sawai Īśvarī Singh, Jai Singh's successor. The second inventory, which was less elaborately done than the first, has a date of 1804-1805 V.S. or 1747-48 A.D. It has a total of 3,191 volumes listed under 1,293 different titles.[3] The inventory lists 206 books on astronomy-astrology and mathematics.[4] We are giving the list these books in Appendix 5. It should be emphasized that the inventory does not include all the books collected by the Raja because some of them were kept elsewhere. For example, neither of the inventories includes the books translated by Kevalarāma or by Nayanasukha Upādhyāya. Also a number of the Persian books, which were definitely bought by Jai Singh, are missing from the two inventories. Apparently the royal library housed books for the Raja's personal use. The astronomers had separate libraries for their own use at or near the site of the observatories.

---

[1] *Tozīs*, Pothikhana, Jaipur State, *Op. Cit.*

[2] See Ch. 12.

[3] At another place, the inventory lists the total no. of the volumes as 3,668 and the number of titles 1,038. Pothikhana records, Jaipur State, R.S.A.

[4] The inventory gives, along with the title of a book, its numbers of copies, folios, sentences or *granthas* (a set of 32 letters); whether bound or unbound, the type of binding, and finally the cost. For example: The inventory lists *Ratnamālā* text as having 7 copies; one of its copies had 41 folios, 750 sentences (*granthas*), and its cost was 12 *annas*. The inventory does not give the name of the author(s) unless the name happens to be a part of the title. The cost of the books, according to the inventory, varied anywhere from 2 *paisas* to Rs. 520.

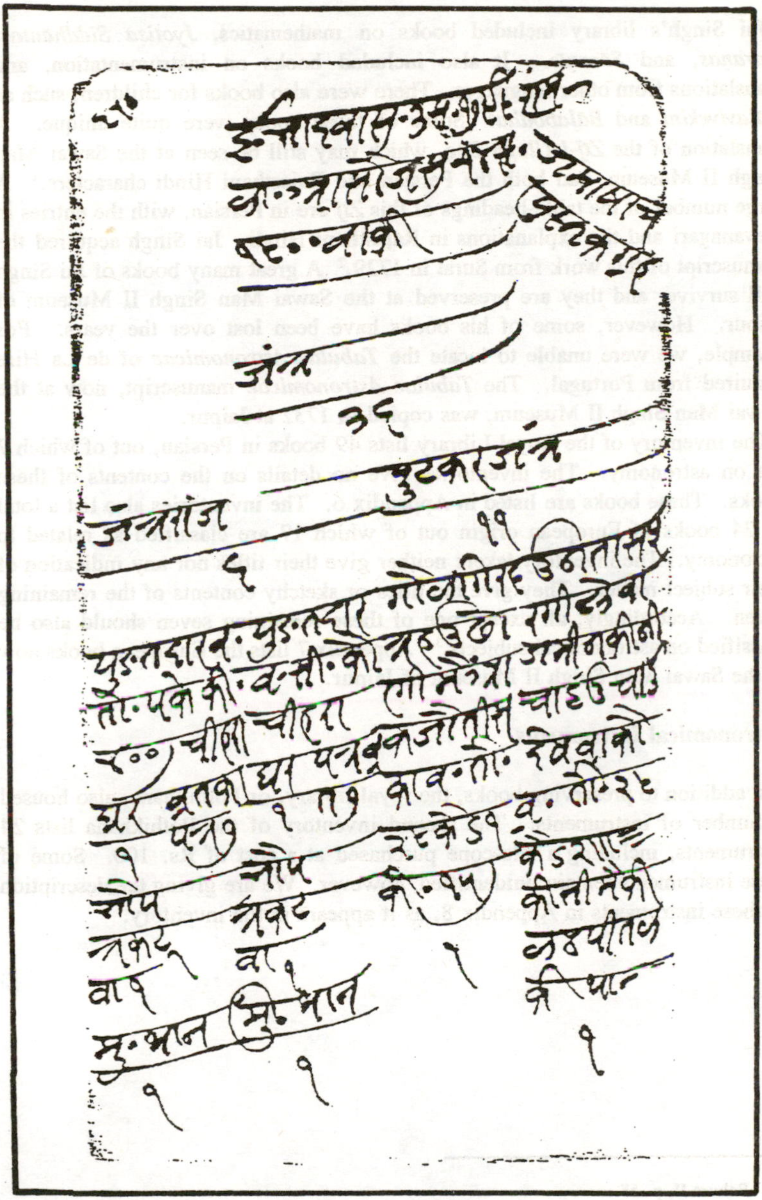

Fig. 11-3 A Page from the Inventory of Jai Singh's Library.

Jai Singh's library included books on mathematics, *Jyotiṣa Siddhāntas*, *Karaṇas*, and *Sāraṇīs*. It also included books on instrumentation, and translations from other languages. There were also books for children, such as *Bālavivekinī* and *Bālabodha*. Some of these books were quite unique. A translation of the *Zīj-i Ulugh Beg*, which may still be seen at the Sawai Man Singh II Museum, had both the Persian and Rajasthani Hindi characters.[1] A large number of the table-headings of this *Zīj* are in Persian, with the entries in Devanagari and the explanations in Rajasthani Hindi. Jai Singh acquired the manuscript of this work from Surat in 1729.[2] A great many books of Jai Singh still survive, and they are preserved at the Sawai Man Singh II Museum of Jaipur. However, some of his books have been lost over the years. For example, we were unable to locate the *Tabulae Astronomicae* of de La Hire acquired from Portugal. The *Tabulae Astronomicae* manuscript, now at the Sawai Man Singh II Museum, was copied in 1732 at Jaipur.

The inventory of the Royal Library lists 49 books in Persian, out of which 7 are on astronomy. The inventories give no details on the contents of these books. These books are listed in Appendix 6. The inventories also list a total of 24 books of European origin out of which 17 are classified as related to astronomy. The inventory takers neither give their titles nor any indication of their subject matter. They give the titles or sketchy contents of the remaining seven. Accordingly, all except one of these remaining seven should also be classified on astronomical subjects.[3] Appendix 7 lists the European books now in the Sawai Man Singh II Museum of Jaipur.

Astronomical Instruments

In addition to preserving books, the royal library, or Pothikhana, also housed a number of instruments. The second inventory of the Pothikhana lists 24 instruments, including a telescope purchased at a cost of Rs. 100. Some of these instruments remain unidentified, however. We are giving the description of these instruments in Appendix 8, as it appears in the inventory.

---

[1] Bahura II, p. 58.

[2] The Rajasthan State Archives in its file No. 424/1 preserves a deed of the purchase. The deed papers reveal that Nandarāma brought the manuscript of 100 folios from Surat. The cost was Rs. 20.

[3] This book does not fit into the category of astronomy because it is on the art of war or battles.

CHAPTER XII

SAWAI JAI SINGH'S HINDU ASTRONOMERS

For his ambitious program in astronomy, Jai Singh assembled a large group of scholars with backgrounds in different traditions of astronomy. At his observatories, Hindu pundits, Muslim *munajjimūn* (astronomers), and European Jesuits worked side by side. Jai Singh's early education, similar to other Rajput princes' education, had been solely under Hindu pundits, and it is from them he first learned his astronomy. Later, the pundits became the mainstay of his program, translating and copying astronomical texts, erecting observatories, and carrying out the day-to-day operations of his observatories.

JAGANNĀTHA SAMRĀṬ

Jai Singh's most favored Hindu astronomer was his religious guru Jagannātha Samrāṭ. Jagannātha's father's name was Gaṇeśa and his great-grandfather's was Viṭhṭhala.[1,2] Jagannātha came in contact with the Raja at an early date, long before his observatories were planned, and remained with him until the very end. Tradition has it that he was born in a Brahmin family of a village in Maharashtra and was discovered by Sawai Jai Singh as he was returning from a campaign against the Marathas sometime during 1702-1703 A.D.[3,4] Recognizing the talents of young Jagannātha, the Raja persuaded him to move up north and study Persian and Arabic, the two languages prominent at the imperial court but neglected by the tradition-bound Brahmin scholars of Sanskrit. Jagannātha soon became well-versed in the languages, and later on when the Raja initiated a vigorous program in astronomy he put his knowledge to good use by translating astronomical and mathematical works from Arabic into

---

[1] B. N. Temani, "An Account of Maharaja Sawai Jai Singh's Works in Astronomy," R.S.A., file No. 1425, dated Oct. 30, 1939. Temani does not give the source of his information. As superintendent of the office of *Dīvān-i Huzūrī* (State Archives) of the Jaipur State, he had access to the archival records of the State.

[2] Bahura II, p. 54. Bahura's information is based on an article of Temani. G. Bahura, private communication. Temani, Ref. 1.

[3] Sudhākara Dvivedī confuses two Jai-Singhs, that is, between the Mirzā Raja and his great grandson, the builder of the observatories. Further, Dvivedī speculates 1652 A.D. as the date of birth of Jagannātha. He does not give any evidence in support of the date, however. See Sudhākara Dvivedī, *Gaṇaka Taraṅgiṇī*, p. 109, reprint, Varanasi, 1933.

[4] Dvivedī's account of Jagannātha's birth date is disputed by Bhārgava who concludes it to be 1680. However, Bhārgava's own date is also not based on any solid archival evidence. See Puruṣottamalāla Bhārgava, *Contribution of Jaipur to Sanskrit Literature*, Ph. D. thesis, (unpublished), Rajasthan University, pp. 237-242.

Sanskrit. Jagannātha outlived his patron by about a year and died in 1744 A.D.[1]

Jagannātha was an eyewitness as well as an active participant in Jai Singh's astronomical adventures. Writing in his *Samrāṭ Siddhānta*, he points out how Jai Singh's early experiments with metallic instruments, constructed according to the Islamic school of astronomy,[2] failed and how Jai Singh opted for instruments of masonry and stone as we have elaborated in Chapter 2. From Jagannātha we learn that the Raja sent several astronomers overseas to collect data.[3]Jaganna ¯tha might have had a role in designing Jai Singh's observatories. Decades later, the local Brahmins of Varanasi told John Lloyd Williams that it was Jagannātha who had designed the observatory of Varanasi.[4]

Jagannātha's Works

Samrāṭ Siddhānta

Jagannātha's major work is the *Samrāṭ Siddhānta* or *Siddhānta Sārakaustubha*. Twenty-five to thirty copies of this work survive in the libraries and archives of the country.[5] Because the *Samrāṭ Siddhānta* mentions Jai Singh sending observers to distant islands, the text could have been completed only after 1730 A.D.[6] The *Samrāṭ Siddhānta* is based on Naṣīr al-Dīn al-Ṭūsī's version of the *Matematikē Suntaxis* (*Mathematical Composition*) or the *Almagest* of Ptolemy, and its first thirteen chapters run parallel to the thirteen books of the *Almagest*.[7] In its introduction Jagannātha explains that the text is a Sanskrit rendition of the Arabic work, *al-Majesṭī*, and that he has written it in a style so that even a "novice can comprehend its contents (easily)."[8] Jagannātha goes on to add that his text has 13 chapters, 141 sections, and 196 illustrations.[9]

---

[1] R.S.A., Dastūr Kaumvār Records of Jaipur Rajya, as quoted by Bhārgava, p. 242, *Op. Cit.*

[2] SSRS, pp. 1162-1163, and SSMC, pp. 38-39.

[3] SSRS, p. 1064, and SSMC, p. 41. We will deal with this topic in Ch. 13.

[4] John Lloyd Williams was a resident of Varanasi in the 1790's. For his account see Ch. 8.

[5] CESS, Vol. A3, p. 57; also Sen (1966), *Op. Cit.*, Ch. 3.

[6] SSRS, p. 1164, and SSMC, p. 41. Contrary to this internal evidence, Bahura writes that the text was completed in 1728 A.D. See Bahura II, p. 58. The author was unable to consult the Museum manuscript on which Bahura's assertion is based.

[7] *Ptolemy's Almagest*

[8] SSRS, p. 5.

[9] For a brief summary of the *Samrāṭ Siddhānta*, see M. L. Sharma, "Jagannātha Samrāṭ's Outstanding Contribution to Indian Astronomy in Eighteenth Century A.D.," *Indian J. History of Science*, 17, pp. 244-251, (1982).

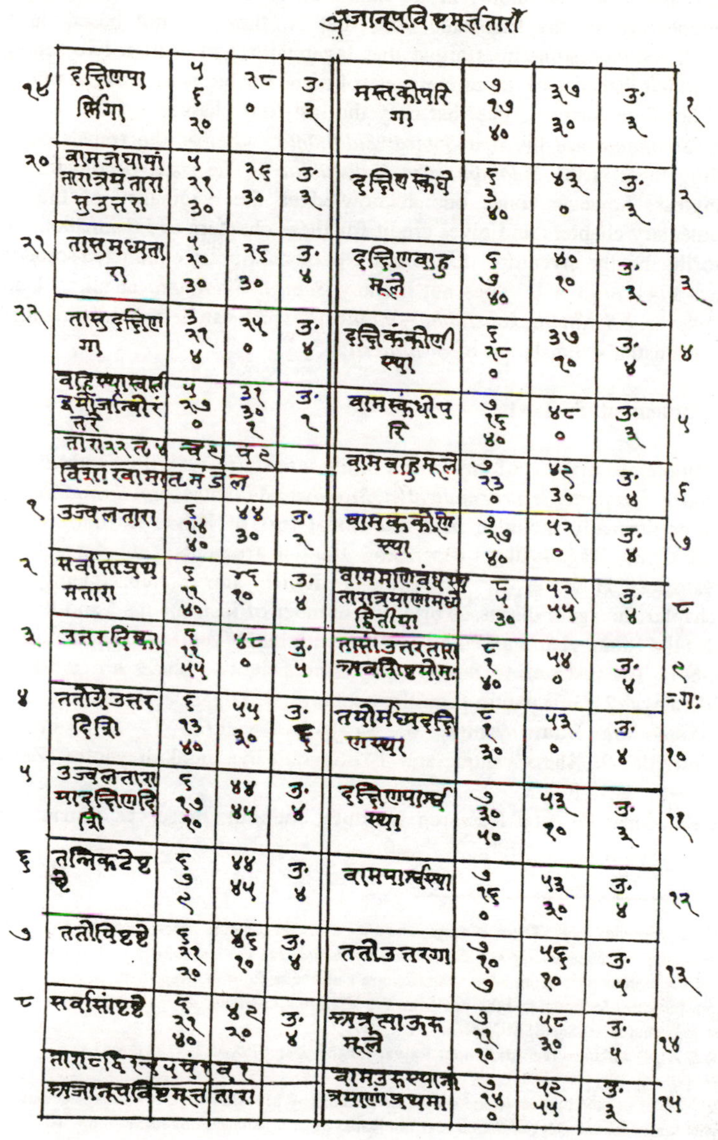

अजानूपविष्टमत्ततारा

२४	दक्षिणपार्श्विगा	५ ६ २०	२८ ०	उ. ३	मस्तकोपरिगा	७ २७ ४०	३७ ३०	उ. ३	१
२०	वामजंघायां तारात्रयं तासु उत्तरा	५ २१ ३०	२६ ३०	उ. ३	दक्षिणस्कंधे	६ ३ ४०	४३ ०	उ. ३	२
२१	तासु मध्यतारा	५ २० ३०	२६ ३०	उ. ४	दक्षिणबाहुमूले	६ ७ ४०	४० १०	उ. ३	३
२२	तासु दक्षिणगा	३ २२ ४	२५ ०	उ. ४	दक्षिणकंकोणीस्था	६ २८ ०	३७ १०	उ. ४	४
	वह्निस्यास्वार्द्धमीजन्वीरंतरे	५ २७ ०	३१ ३० १	उ. १	वामस्कंधोपरि	७ १६ ४०	४८ ०	उ. ३	५
	ताराः २२ तत्र ५	ध्रुव १	पं २		वामबाहुमूले	७ २३ ०	४२ ३०	उ. ४	६
	विशाखामत मंडल								
१	उज्वलतारा	६ २४ ४०	४४ ३०	उ. २	वामकक्षकोणिस्था	७ २७ ४०	५२ ०	उ. ४	७
२	सर्वासां प्रथमतारा	६ २२ ४०	४६ १०	उ. ४	वाममणिबंधस्थ तारात्रयाणां मध्ये द्वितीया	८ ५ ३०	५२ ५५	उ. ४	८
३	उत्तरदिक्का	६ २२ ५५	४८ ०	उ. ५	तासां उत्तरतारा त्ववशिष्टयोः	८ १ ४०	५४ ०	उ. ४	९ =गाः
४	ततोऽपि उत्तर दिक्स्था	६ १३ ४०	५५ ३०	उ. ६	तयोर्मध्ये दक्षिणस्था	८ १ ३०	५३ ०	उ. ४	१०
५	उज्वलताराया दक्षिणदिक्स्था	६ १७ १०	४४ ४५	उ. ४	दक्षिणपार्श्वस्था	७ ३० ५०	५३ १०	उ. ३	११
६	तन्निकटस्थे	६ ७ ९	४४ ४५	उ. ४	वामपार्श्वस्था	७ १६ ०	५३ ३०	उ. ४	१२
७	ततोऽपि दृष्टे	६ २१ २०	४६ १०	उ. ४	ततो उत्तरगा	७ १० ७	५६ १०	उ. ५	१३
८	सर्वासां पृष्ठे	६ २१ ४०	४२ २०	उ. ४	[illegible] ऊरुमूले	७ २२ २०	५८ ३०	उ. ५	१४
	ताराः ८ दि १	५ पं ४	व १		वामऊरुस्थानां त्रयाणां प्रथमा	७ २४ ०	५२ ५५	उ. ३	१५
	अजानूपविष्टमत्त	तारा							

Fig. 12-1 A Page from the *Samrāṭ Siddhānta*: Star table.

As a supplement to the *Samrāṭ Siddhānta*, Jagannātha has added four chapters of traditional Hindu astronomy in the *siddhānta* style. Because these chapters are independent of the translated text, that is, they are not based on the *Almagest*, scholars sometimes argue that Jagannātha had written two separate books, which, because of the carelessness of some scribe, were merged into one book. Scholars further argue that only the last four chapters of the so-called *Samrāṭ Siddhānta* are the real *Samrāṭ Siddhānta*, whereas the translated text preceding them is the *Siddhāntasāra Kaustubha*.[1] The issue becomes rather meaningless, however, once one acknowledges the originality of the four supplementary chapters and gives credit for these chapters to Jagannātha. It is noteworthy that the inventory of Jai Singh's personal library conducted sometime between 1741-1743 A.D. does not list any *Samrāṭ Siddhānta* at all.[2] It does list, however, a *Siddhāntakaustubha—majasatī*, which can be certainly identified with Jagannātha's translation of *al-Majesṭī*.[3]

The Supplementary Text

The supplementary text of the *Samrāṭ Siddhānta* starts out with *Yantrādhyāya*, followed by *Jyotpatti*, *Tripraśnādhyāya*, *Madhyamādhikāra*, and *Spaṣṭādhikāra*, the topics normally included in a traditional text of Hindu astronomy.[4] In *Yantrādhyāya*, Jagannātha describes the instruments at Jai Singh's observatories.[5] However, he leaves out the Rāma yantra. Subsequently, in a later chapter, he again discusses briefly a number of instruments,[6] and there he includes the Rāma yantra and the Sarvadeśīya Jarakālī yantra (Zarqālī universal astrolabe). The instruments discussed by him at the two places are as follows: 1. Nāḍīvalaya, 2. Gola yantra (armillary sphere), 3. Digaṁśa, 4. Samrāṭ yantra, 5. Dakṣinottara Bhitti yantra, 6. Ṣaṣṭhāṁśa yantra, 7. Jaya Prakāśa, 8. Krāntivṛtta, 9. Rāma yantra, and 10. Sarvadeśīya Jarakālī yantra (Zarqālī astrolabe).

The Rajasthan Oriental Research Institute, Jodhpur, preserves a manuscript

---

[1] M. L. Sharma, *Ibid*. There is some substance to the arguments raised by M.L. Sharma, because a number of manuscripts of the *Samrāṭ Siddhānta* lack the last four chapters. For instance, the manuscripts obtained by Muralīdhara Caturveda from the Rajasthan Oriental Research Institute, Jodhpur, and the one from the Vikram University of Ujjain, did not include the last four chapters. See SSMC, p. 1.

[2] R.S.A., Pothikhana records, Jaipur Rajya, for the year 1798 V.S. (1741 A.D.).

[3] *Ibid*.

[4] The order of the chapters in the supplement edited by Muralīdhara Caturveda is somewhat different than in the *Samrāṭ Siddhānta* text published by R.S. Sharma. See SSRS, pp. 1031 ff., and SSMC.

[5] SSRS, pp. 1031-1048, also SSMC, pp. 2-14. There are two copies of the *Yantrādhyāya*, Nos. 24(i), 7460-17213; and 24(i) 2905-36223, preserved at the RORI.

[6] For Rāma yantra, see SSRS, p. 1163, and SSMC, p. 39; and for Jarakālī yantra see SSRS, pp. 1252-1260, also SSMC, pp. 96-105.

entitled *Yantraśāstra* with Jagannātha quoted as its author. However, the work is identical with the *Yantrādhyāya* and the *Jyotpatti* chapters of the *Samrāṭ Siddhānta*, and, therefore, is not an independent composition.[1]

It is noteworthy that while discussing the instruments for an observatory, Jagannātha makes no mention of the telescope, although the instrument had been available to the astronomers of Jai Singh.[2] Perhaps to Jagannātha, the *yantras* of an observatory meant the instruments that directly led to some numerical data. He also leaves out the astrolabe, even though the instrument had been popular with Indian astronomers, and Jai Singh had a large specimen of it made.[3] Jagannātha praises Jaya Prakāśa and the Rāma yantra, and leads us to believe that the two were the most accurate instruments at Jai Singh's observatories. However, a close examination of Jai Singh's instruments reveals that this is not the case.[4] Besides, the two are cumbersome to use at night. The most precise instruments of Jai Singh's observatories were the large Samrāṭs and the Ṣaṣṭhāṁśas constructed at Delhi and Jaipur.[5]

Rekhāgaṇita

Jagannātha's other work, also done for his patron, is *Rekhāgaṇita*—a text on geometry—translated from Naṣīr al-Dīn al-Ṭūsī's *Taḥrīr al Ukaledas*, the Arabic version of Euclid's *Stoicheia* or *Elements*.[6] Jagannātha must have completed his *Rekhāgaṇita* by 1727 A.D. or earlier because the earliest manuscript of this work has a copying date of 1727 A.D. (1784 V.S.).[7] The copy was prepared

---

[1] *Yantraśāstra* of Jagannātha, RORI, No. 35964, 45ff. The *Jyotpatti* section is incomplete in this ms.

[2] There is evidence to believe that Jai Singh's astronomers were familiar with the telescope. In the *Zīj-i Muḥammad Shāhī*, Jai Singh discusses the telescope and what can be observed with it. ZMS, f. 18. See Ch. 11.

[3] Virendra Nath Sharma, "The Great Astrolabe of Jaipur and Its Sister Unit," *Archaeoastronomy*, No.7, *J. Hist. Ast.*, Vol. XV, pp. 126-128, (1984). Also see Ch. 7.

[4] Jaya Prakāśa has a varying degree of accuracy. The Jaipur instrument, for instance, measures time with an uncertainty of $\pm\frac{1}{2}$ to $\pm 1\frac{1}{2}$ min, and the zenith distance and declination both with an uncertainty of $\pm 3'$ of arc. The uncertainty in the measurement of azimuth and right ascension could be anywhere from $\pm 3'$ to ± 1 deg. With the Rāma yantra, a precision of $\pm 6'$ of arc is the best one can expect for angles of 40° to 45°. See Ch. 4.

[5] The Samrāṭ of Jaipur, if properly constructed can measure time with an accuracy of ± 3 second or better, and the right ascension and declination both with accuracy of $\pm 1'$ of arc. See Ch. 6.

[6] *The Rekhāgaṇita or Geometry in Sanskrit, Composed by Samrāḍ Jagannātha*, Vol. I, Books I–VI. Undertaken for Publication by the Late Harilāl Harshādarāi Dhruva, ed. by Kamalāśankara Prāṇaśankara Trivedī, Nirṇaya-Sāgara Press, Bombay, 1901. For the manuscript listings of *Rekhāgaṇita*, see CESS, Vol. A3, pp. 56-57, and Sen (1966), pp. 89-90, *Op. Cit.*, Ch. 3.

[7] Dhruva and Trivedī, *Op. Cit.*, p. 7. According to Trivedī, the library of Sanskrit College, Varanasi, has a copy of it.

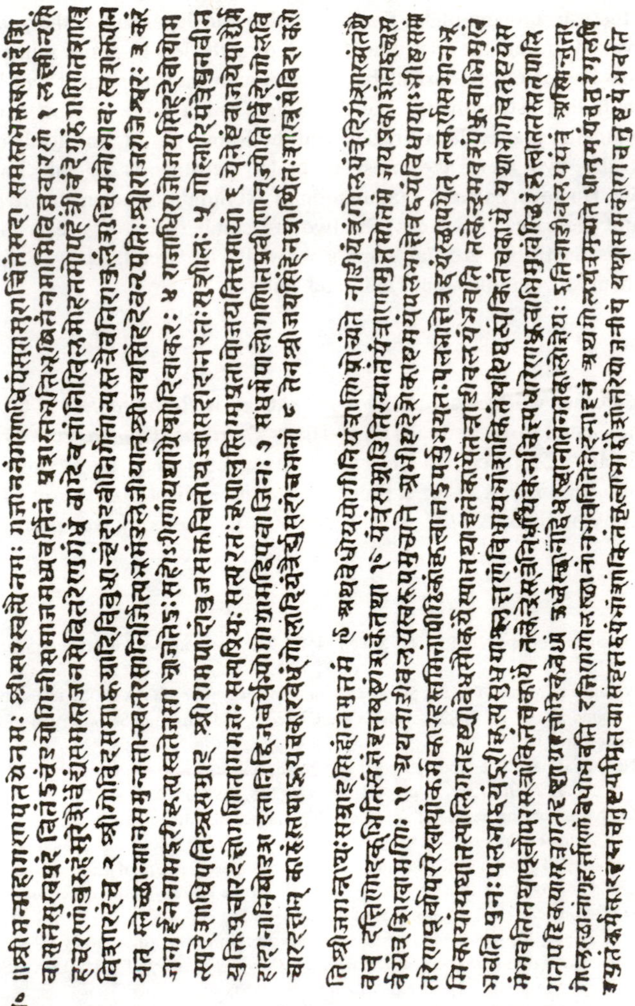

Fig. 12-2 Two pages from the Supplementary text of *Samrāṭ Siddhānta.*

for Jai Singh himself and is preserved at Varanasi.[1] At the Sawai Man Singh II Museum of Jaipur, where most of Jai Singh's library is preserved, there are two copies of *Rekhāgaṇita.*[2] The *Rekhāgaṇita* has 15 chapters or "books" dealing with subjects related to plane geometry, theory of numbers, and solid geometry. Because Jagannātha had very few equivalent terms available to him in the existing literature for his translation, he found it necessary to prepare a glossary of more than 100 terms for his task.[3] These terms were later adopted by virtually every other author in India writing on mathematical topics.

Yantraprakāra

Yantraprakāra is a text on instrumentation. Only two copies of this work are in existence. One is preserved at the Sawai Man Singh II Museum, Jaipur, and the other is at the Rajasthan Oriental Research Institute, Udaipur branch.[4] Evidently, Jagannātha or one of the students under his guidance wrote the *Yantraprakāra*, as it is identical in parts with the *Yantrādhyāya* of the *Samrāṭ Siddhānta*. Its title says, "The instruments constructed by Śrī Mahārājādhirāja are being written." A close scrutiny of *Yantraprakāra* reveals that some of its sections could only have been written after 1729 A.D., because its author quotes a set of observations conducted on the night of April 18, 1729 A.D.[5] The *Yantraprakāra* and the supplementary sections of the *Samrāṭ Siddhānta* are of the same general period. On the very first page, the *Yantraprakāra* gives a list of instruments constructed by Jai Singh at his observatories. The list is reproduced in Table 12-1.

It is interesting to note that the list does not include the Samrāṭ yantra by name. A reason for this omission could be that the Samrāṭ was also known as Naḍīvalaya in those days, and most likely it is included in that category. But more puzzling than the omission is that the instrument is not discussed in the text.[6] The Śaṅku, although included in the list, is not described in the text either. Instead, two other instruments, namely, the Sarvadeśīyakapāla yantra

---

[1] *Ibid.*

[2] Bahura II, p. 432, mss. No. 5372 and 5373. One of these manuscripts was completed in S.E. 1650, on *Vaiśākha Śukla Pūrṇimā*, or on April 23, 1728 A.D.

[3] For the compilation of the terms coined by Jagannātha, see B. L., Upadhyaya, *Prācīna Bhāratīya Gaṇita*, (in Hindi), pp. 371-374, New Delhi, 1971. For a comparison of the terms coined by Jagannātha with their Greek equivalents, see J. Ludo Rocher, "Euclid's Stoicheia and Jagannātha's Rekhāgaṇita; A Study of Mathematical Terminology," *J. Oriental Inst.*, Vol. 3, pp. 236-256, Baroda, (1953-1954).

[4] Ms. No. 3156, Rajasthan Oriental Research Institute, Udaipur Branch, City Palace, Udaipur. For a detailed analysis of the *Yantraprakāra*, see *Yantraprakāra* (S).

[5] *Yantraprakāra* (J), pp. 23-24; and *Yantraprakāra* (S), p. 85. In the *Yantraprakāra* (S), the date has been translated as May 7, 1729, however.

[6] A possibility exists that the instrument section of the *Yantraprakāra* was written before any of the Samrāṭ yantras were erected at Jai Singh's observatories.

and Cūḍā yantra have been described in the text that follows the list. The Pratirāśinām Krāntivṛttāni, as described in the text is a portable instrument and not built out of masonry and stone.

1.	Jaya Prakāśa	4
2.	Nāḍīvalaya	7
3.	Krāntivṛatta	1
4.	Palabhā yantra	1
5.	Digaṁśa yantra	1
6.	Śara yantra (Celestial-latitude dial)	1
7.	Agrā yantra (Amplitude dial)	1
8.	Yāmyottarabhitti (Dakṣinottara Bhitti)	2
9.	Jātulhalaka (Dhāt al-Ḥalaq or Armillary sphere)	1
10.	Yantrarāja (Astrolabe, for time reckoning)	2
11.	Jātuḥśukavataina (Dhāt al-Thuqbatayn or Dioptra)	1
12.	Jātuśuvataina (Dhāt al-shu^{c}batayn or Triquetrum)	1
13.	Sudasphakarī Ṣaṣṭhāṁśa	1
14.	Śaṅku yantra (Upright rod)	(unknown)
15.	Pratirāśinām Krāntivṛttāni (Rāśivalayas)	12

Table 12-1 Instruments of Jai Singh according to the *Yantraprakāra*.

A comparison with the *Yantrādhyāya* of the *Samrāṭ Siddhānta*, the *Yantraprakāra* is rather elaborate and includes computations, some incidental data, a number of tables, and descriptions of instruments such as *Dhāt al-Shu^{c}batayn* of the Islamic school. The instruments of the Islamic school might have been based on some text of Arabic or Persian which had been translated for the Raja when the observatories were being planned.[1] But this conjecture needs to be further investigated. The *Yantraprakāra* is important for the reason that it is the only text in which Jai Singh's early instruments built according to the Islamic school of astronomy are discussed in any detail.

Jagannātha as an Observer

In addition to translating from Arabic texts, Jagannātha was probably involved in collecting data at Jai Singh's observatories. In the *spaṣṭādhikāra* chapter of *Samrāṭ Siddhānta*, while explaining the procedure of finding solar parameters, Jagannātha selects a set of readings taken with a Ṣaṣṭhāṁśa at the Delhi observatory.[2]

---

[1] S. M. R. Sarma has identified five instruments, namely, Yāmyottara yantra, Yāmyottarabhitti yantra, Dhāt al-Ḥalaq, Dhāt al-Shu^{c}batayn, and Dhāt al-Thuqbatayn as taken directly from Naṣīr al-Dīn al-Ṭūsī's version of the *Almagest*. See *Yantraprakāra* (S), p. 4.

[2] See SSMC, pp. 81-83, also SSRS, p. 1216-1221, p. 1240 ff.

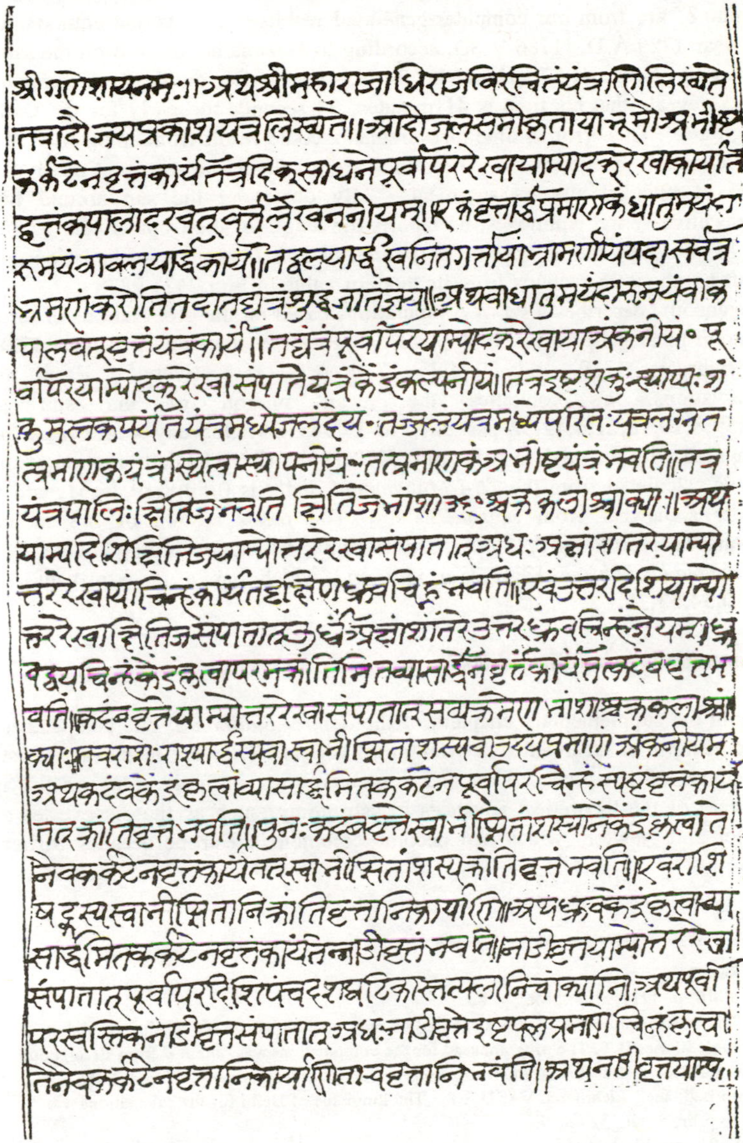

श्रीगणेशाय नमः ॥ अथ श्रीमहाराजाधिराजविरचितयंत्राणि लिख्यते
तत्रादौ जयप्रकाशयंत्रं लिख्यते ॥ आदौ जलसमीकृतायां भूमौ अभीष्ट
कर्कटेन वृत्तं कार्यं तत्र दिक्साधनेन पूर्वापररेखा याम्योदग्रेखा कार्या त
द्वृत्तं कपालोदरवत् वृत्तं खननीयम् ॥ एकं हस्तार्द्धप्रमाणं धातुमयं वा
द्रुमयं वा कलशार्द्धं कार्यं ॥ तत्कल्याद्धं खनितगर्त्तायां आमरणीयं यदा सर्वत्र
आमरणं करोति तदा तद्यंत्रं शुद्धं जातं ज्ञेयं ॥ अथवा धातुमयं दारुमयं वा क
पालवत् वृत्तं यंत्रं कार्यं ॥ तद्यंत्रं पूर्वापरयाम्योदग्रेखायां अंकनीयं ॥ पू
र्वापरयाम्योदग्रेखासंपाते यंत्रे केन्द्रं कल्पनीयं ॥ तत्र इष्टशंकुः स्थाप्यः ॥ शं
कुमस्तकपर्यंतं यंत्रमध्ये जलं देयं ॥ तज्जलं यंत्रमध्ये परितः यत्र लग्नं त
त्प्रमाणकं यंत्रं स्थिरं स्थापनीयं ॥ तत्प्रमाणकं अभीष्टयंत्रं भवति ॥ तत्र
यंत्रपालिः क्षितिजं भवति क्षितिजं अंशा ३६० चक्रकलाअंक्या ॥ अथ
याम्यदिशि क्षितिजयाम्योत्तररेखासंपातात् अधः अक्षांशांतरे याम्यो
त्तररेखायां चिन्हं कार्यं तद्दक्षिणध्रुवचिन्हं भवति ॥ एवं उत्तरदिशि याम्यो
त्तररेखाक्षितिजसंपातात् उर्ध्वं अक्षांशांतरे उत्तरध्रुवचिन्हं ज्ञेयम् ॥ ध्रु
वद्वयचिन्हं केन्द्रं कृत्वा परमक्रांतिमितव्यासार्द्धेन वृत्तं कार्यं तत्कदंबवृत्तं भ
वति ॥ कदंबवृत्तयाम्योत्तररेखासंपातात् सव्यक्रमेण अंशाश्चक्रकलाअंका
क्याः ॥ अत्र राशेः राश्यार्द्धस्य वा स्वाभीप्सितांशस्य वा उदयप्रमाणं अंकनीयम् ॥
अथ कदंबकेन्द्रं कृत्वा व्यासार्द्धमितकर्कटेन पूर्वापरचिन्हस्पृष्टं वृत्तं कार्यं
तत् क्रांतिवृत्तं भवति ॥ पुनः कदंबवृत्ते स्वाभीप्सितांशस्थाने केन्द्रं कृत्वा त
नैव कर्कटेन वृत्तं कार्यं तत् स्वाभीप्सितांशस्य क्रांतिवृत्तं भवति ॥ एवं राशि
षट्कस्य स्वाभीप्सितानि क्रांतिवृत्तानि कार्याणि ॥ अथ ध्रुवकेन्द्रं कृत्वा व्या
सार्द्धमितकर्कटेन वृत्तं कार्यं तन्नाडीवृत्तं भवति ॥ नाडीवृत्तयाम्योत्तररेखा
संपातात् पूर्वापरदिशि पंचदश घटिकाः तत्फलानि चांक्यानि ॥ अथ पूर्वा
परस्वस्तिक नाडीवृत्तसंपातात् अधः नाडीवृत्ते इष्टपलप्रमाणं चिन्हं कृत्वा
तेनैव कर्कटेन वृत्तानि कार्याणि ॥ ताः अहोरात्रवृत्तानि भवंति ॥ अथ नतवृत्तयाम्ये

Fig. 12-3 A page from the *Yantraprakāra*.

We checked Jagannātha's readings and found them to differ by less than a minute of arc from our computer-generated results.[1] The vernal equinox for the year 1729 A.D. (1786 V.S.), according to Jagannātha, arrived on the same day at 18 *ghaṭīs* and 57 *palas* (7 hr, 34 m, 48 sec) after midday.[2] A computer check reveals that his time is 41 minutes, 34 seconds too early for the Delhi longitude.[3] This error in time measurements could result from an error of 40″ in declination measurements. The results are excellent and represent nearly the limit of unaided eye measurements. By observing the sun around two consecutive vernal equinoxes, he reports the length of the tropical year as 365 d., 14 *ghaṭikās*, and 31 *vighaṭikās* (*palas*), or 365.24194 d., a calculation that is off by about 25 seconds from its modern value of 365.2421 9878 d.

Jagannātha describes a set of computations based on the observations of three separate lunar eclipses that took place one after the other on 19 August 1728, 13 February 1729, and 29 July 1730 A.D. respectively.[4] From the observations, he computes the mean motion of the sun as 0:0;59,8,19,18,21,4,42,32 per day, or 0°.985 644 935/day which may be compared with the modern value of 0;59,8,19,49.47.[5] The daily motion of the sun as calculated from the *Zīj-i Muḥammad Shāhī* is 0;59,8,19,46,51, and it differs somewhat from Jagannātha's.[6] The mean motion of the moon, Jagannātha reports as 0,13;10,35,2,9,51,0,38, which, according to the *Zīj-i Muḥammad Shāhī* is 13;10,35,1,38.4 and which has a modern value of 13;10,34,53,26.

Jagannātha's Scientific Beliefs

From the supplementary chapters of the *Samrāṭ Siddhānta*, one gets a fairly good glimpse of Jagannātha's astronomical beliefs. They are medieval at best, like those of his predecessors and contemporaries in India. Jagannātha is unaware of the theoretical advances in astronomy, such as the discoveries of Kepler and Newton, which had become common in Europe decades earlier.

---

[1] See Ch. 5.

[2] SSRS, p. 1217, also SSMC, p. 80.

[3] The vernal equinox of 1729 A.D., based on our computer generated results, arrived at 15 hr, 7 min, 30 sec, U.T. The program used for the computations was obtained from Bretagnon, Pierre, and Simon, Jean-Louis, *Planetary Programs and Tables from −4000 to +2800*, publ., William-Bell, Inc., Richmond, VA, U.S.A. The longitude of Delhi for the calculations was taken as 5 hr, 8 min, 52 sec, E.

[4] SSRS, pp. 1240−1246.

[5] SSRS, p. 1246. The modern value is for a tropical year.

[6] The reason for the difference might be that Jagannātha is merely illustrating a procedure, whereas the parameters in the *Zīj-i Muḥammad Shāhī* are based on multiple observations. See Virendra Nath Sharma, "*Zīj-i Muḥammad Shāhī* and the Tables of de La Hire," *Indian J. Hist. Sci.*, Vol. 25 (1-4), pp. 34-44, (1990). Also see Ch. 11.

Instead, he believes in the astrological effects of planets.[1] He seems to admire Ulugh Beg and the astronomical and mathematical advances of the Islamic world. Jagannātha displays a strong belief in the importance of observing. In fact, to him "observation" is the *pramāṇa* or deciding factor when doubts (discrepancies) arise between theory and observation.[2]

Strangely, he appears to suggest that there could not be a theory which would reconcile itself totally with observation, and he displays an undue faith in the canons of Hindu astronomy. He comments, "The observed motion of the planets in the sky is different than obtained with the canons (siddhāntas). (What is more), even from the texts written by the *ṛṣis* (sages), such as the *Brahma Siddhānta* and the *Sūrya Siddhānta*, the results for the planets differ from one another." Searching for a reason for this disagreement, he goes on to propose, "In the sky there is no uniformity (consistency). That is, with time and place, because of the inconsistency, disagreements develop between theory and observation. If there were consistency in the sky (as regards to planetary motion), then (predictions of) the *Siddhāntas* would always agree."[3] Jagannātha's statement indicates that he does not suspect any shortcomings in the planetary theories of the *siddhāntas*, and, consequently, he makes no suggestions to improve them. Jagannātha has an unjustified faith in the knowledge of ancient authors, whom he calls "the divinely inspired" or *ṛṣis*![4]

Kevalarāma

Kevalarāma was a native of Modesa, a village in Gujarat, and had already become famous as an astronomer in his native state when Jai Singh brought him to Amber in 1725 A.D.[5,6] His father, Baija Nātha, was an astronomer-astrologer,[7] and it was from him Kevalarāma learned astronomy. At Amber, Kevalarāma soon earned the favor of his patron, Jai Singh, and received the title of "Jyotiṣarāya," or "the astronomer royal." Kevalarāma's year of birth

---

[1] "A planet which has set is weak in its (astrological) influence," Jagannātha says. SSRS, p. 1064, and SSMC, p. 40.

[2] SSRS, p. 1165, and SSMC, p. 41.

[3] *Ibid.*

[4] *Ibid.*

[5] Temani, *Op. Cit.*

[6] Kevalarāma should be distinguished from Kevalarāma Pañcānana (*fl.* 1728-1762 A.D.), his contemporary from Bengal. Pañcānana resided first at the court of Kṛṣṇacandra of Navadvīpa (1728-1780 A.D.) and then at the court of Sawai Madho Singh (1750-1767 A.D.), the third son of Jai Singh. CESS Vol. A2, p. 63. Kevalarāma who was associated with Jai Singh often carries the title of *Jyotiṣarāya* (astronomer royal) with his name in many of the works attributed to him. Kevalarama Pañcānana wrote a number of books. See CESS, Vol. A2, p. 63. It is easy to confuse the works of the two authors, particularly when a manuscript does not have its author indicated.

[7] Bhārgava, p. 258, *Op. Cit.*

is not known, but from the *Dastūr Kaumvār* records of the Rajasthan State Archives, it is certain that he died in 1782 A.D. (1839 V.S.).[1] Kevalarāma was a prolific writer and is credited to have written the following books:

(1) *Jayavinodī Pañcāṅga Sāraṇī*, (2) *Pañcāṅga Sāraṇī*, (3) *Tithi Sāraṇī*, (4) *Jīvāchāyā Sāraṇī*, (5) *Vibhāga Sāraṇī*, (6) *Dṛkpakṣa Sāraṇī*, (7) *Brahmapakṣanirāsa*, (8) *Rekhāgaṇita*, (9) *Dṛkpakṣaspaṣṭagrahānayana*, (10) *Bhāgavata-jyotiḥ-śāstrayorbhūgolakhagola Parihāra* (11) *Abhilāṣāśataka*, and (12) *Gaṅgāstuti*.

Jayavinodī Pañcāṅga Sāraṇī[2]

Kevalarāma apparently completed *Jayavinodī Pañcāṅga Sāraṇī* around 1657 S.E. (1735 A.D.), as the earliest entry in the *Sāraṇī* begins with that year. The Smith Indic manuscript of this *Sāraṇī* at the Harvard University contains tables of *yogas* for 1 to 30 years. The manuscript also has yearly motions for *yogadhruvas*, *yogavāras*, *yogeravikendras* and *yogecandrakendras*. The Rajasthan Oriental Research Institute manuscript No. 28484 of this *Sāraṇī*, on the other hand, is more elaborate and has, in addition, tables for *tithiśeṣāṅkas*, *tithikṣayas*, and *nakṣatras*.[3]

Pañcāṅga Sāraṇī[4]

In the *Pañcāṅga Sāraṇī*, Kevalarāma includes the tables for the longitude of the apparent sun, the daily motion of the sun, and the moon's apparent motion and its equation of center. In the book's introduction, Kevalarāma emphasizes that his *Sāraṇī* is not against the *Sūrya Siddhānta*, probably to ward off criticism that he was going against established tradition. In the *Sāraṇī*, he quotes the daily motion of the sun as 0,0;59,8,10/day.[5] The daily motion according to the

---

[1] R.S.A., DK, Vol. 16, p. 595.

[2] RORI, Nos. 11839, 36ff, and 28484, 39ff. The second manuscript has a copying date of V.S. 1953 or 1896 A.D. The other copies according to Pingree are: Smith Indic Collection, No. 61, 23ff, Harvard Univ., and Calcutta Sanskrit College, 17, 19ff. See David Pingree, "Sanskrit Astronomical Tables in the United States," *Transactions of the American Philosophical Society*, New Series—Vol. 58, Part 3, pp. 66-67, (1968).

[3] Mehra, having examined the tables of the Harvard copy, concludes that the parameters used in there are from the *Sūrya Siddhānta*. See Anjani Kumar Mehra, *Indian J. Hist. Sci.*, Vol. 17, No. 2, pp. 257-259, (1982). According to Pingree, on the other hand, the parameters are from *Pañcāṅgavidyādharī* composed by Vidyādhara at Jīrṇagaḍha in Saurāṣṭra in 1643 A.D. See Pingree (1987), *Op. Cit.*, Ch. 11. For the constants of *Pañcāṅgavidyādharī*, see David Pingree, (1968), "Sanskrit Astronomical Tables in the United States," pp. 60-61, *Op. Cit.*, Ch. 11.

[4] *Pañcāṅga Sāraṇī*, RORI, No. 12615, 31ff.

[5] The modern value for the daily motion of the sun is 0;59,8,19,49.47. The daily motion of the sun calculated from the 30 Arabic year motion given in the *Zīj-i Muḥammad Shāhī* also differs considerably with Kevalarāma's values. The daily motion of the sun from the *Zīj* is 0,59,8,19,32 per day. See ZMS, f. 130. Also see V. N. Sharma, (1990), Ref. 40.

Zīj-i Muḥammad Shāhī is 0;59,8,19,46,51 per day. The weekly mean motions of the sun and the moon according to Kevalarama are as follows: The sun: 0,6;53,57,11,10 or 6°.899 218 364 (sidereal), the moon: 3,2;14,4,4,26 or 93°.234 464 97 (sidereal),[1] and the *Candrakendra*: 3,1;37,17,13,37 or 91°.621 451 93.

The yearly motion of the ascending node of the moon, the *Rāhu*, is given as −0,19;21,13,30. The modern value of this constant for the sidereal year is −0,19;20,30,19.5. The length of the sidereal year calculated on the basis of Kevalarāma's constants, turns out to be 365 d, 6 hr, 12 min, 37.5 sec, which is almost identical with the value given in the *Sūrya Siddhānta*.[2] This value is 3m 28 sec longer than modern measurements.

Because Kevalarāma based his computations on the tradition of the *Sūrya Siddhānta*, it is easy to understand why his constants differ from those of Jagannātha's in the *Samrāṭ siddhānta* and also from the *Zīj-i Muḥammad Shāhī*.[3] Considering the fact that both astronomers worked at the same location, and that they must have had access to the data collected there, one would expect them to have used the data in their work, but they did not. Kevalarāma chose not to use the data because he was following a tradition and was not necessarily concerned with the new parameters worked out at the observatories.

Tithi Sāraṇī[4]

Kevalarāma's *Tithi Sāraṇī,* preserved at the Rajasthan Oriental Research Institute, with three folios to it, is probably an incomplete copy of his *Pañcāṅga Sāraṇī*.[5]

Jīvāchāyā Sāraṇī[6]

His *Jīvāchāyā Sāraṇī* consists of tables of the logarithms of sines (*Jīvā*) and tangents (*Chāyā*) of angles in 8 digits, in increments of a minute of arc.[7] The tables were copied from a European work brought by the Jesuits. A fairly good

---

[1] The accepted weekly sidereal values of these parameters are: sun = 6°.899 263 797 per 7 days, moon = 93°.234 508 17 per 7 days.

[2] D. A. Somayaji, *A Critical Study of the Ancient Hindu Astronomy*, p. 72, Karnatak Univ., Dharwar, 1971.

[3] SSRS, p. 1246.

[4] RORI, No. 3125, 3ff.

[5] According to Bhārgava, the *Sāraṇī* is based on the *Sūrya Siddhānta*. See Bhārgava, thesis, *Op. Cit.*, p. 271.

[6] RORI, No. 23958, 91ff. The manuscript was copied in 1887. There is no colophon to confirm that the work is indeed by Kevalarāma.

[7] The *Jīvā* of angle $\theta = 10 + \log \sin \theta$. For example, the *Jīvā* of 31;51 = 9.7223848. It is listed as 97223848, the decimal after 9 is understood. Similarly the *Chāyā* of 31;51 = 10 + log tan 31;51, and it is listed as 97932560.

copy of this work is preserved at the Rajasthan Oriental Research Institute Jodhpur.

Dṛkpakṣa Sāraṇī[1]

As we have already pointed out, Kevalarāma translated the *Tabulae Astronomicae* of de La Hire under the name *Dṛkpakṣa Sāraṇī*. Fragments of this work are preserved at various places. One fragment of the translation is preserved at the Sawai Man Singh II Museum under the title *Firaṅgī Grahavedhopayogī Sāraṇī*.[2] The Oriental Research Institute of Baroda also preserves a copy of the translation listed under the title *Dṛkpakṣa Sāraṇī*.[3] This manuscript includes tables and explanatory text. A table for the parameters of the moon from this manuscript is reproduced in Figure 12-1. Another section of the translation, in 103 verses, entitled *Dṛkpakṣasāraṇyāṁ Sūryagrahaṇam*, is stored at the Bhandarkar Oriental Research Institute of Pune.[4] The Pune manuscript is a short tract on solar eclipses, and it is based on the *Firaṅgī* (European) treatment of the subject, according to Kevalarāma himself. It has no diagrams, tables or *Sāraṇīs* and also no date of compilation. However, it is reasonable to assume that Kevalarāma completed the work after 1731 A.D., the year Jai Singh's delegation returned from Europe, and that it was done with the assistance of some Jesuit such as du Bois in the service of the Raja.[5]

Dṛkpakṣaspaṣṭagrahānayana[6]

Kevalarāma's *Dṛkpakṣaspaṣṭagrahānayana* is preserved at the Rajasthan Oriental Research Institute, but it is not certain if the manuscript is another copy of his *Dṛkpakṣa Sāraṇī*. The manuscript has only four folios to it, and it has a date of 1824 V.S. (1767 A.D.).

---

[1] *Dṛkpakṣa Sāraṇī*, BORI, 926 of 1886/92, 10ff. Pingree in his CESS, Vol. A2, mistakenly identifies the author of this work as Kevalarāma Pañcānana. The error has been rectified, by him in one of his papers. See Pingree (1987), *Op. Cit*, Ch. 11.

[2] Formerly mislabeled as *Firaṅgī Candravedhopayogī Sāraṇī*. See the Museum, No. 5609. Also Bahura II, p. 63.

[3] *Dṛkpakṣa Sāraṇī*, No. 3162, 29ff, Oriental Research Institute, Baroda.

[4] *Dṛkpakṣasāraṇyām Sūryagrahaṇaṁ*, Bhandarkar Oriental Research Institute, Pune, No. 926 of 1886-92.

[5] Du Bois, *Op. Cit.*, Ch. 3.

[6] RORI, No. 11259/1, 4ff.

Fig. 12-4 A page from *Dṛkpakṣasāraṇyām-sūryagrahaṇam*.

Brahmapakṣanirāsa

Kevalarāma's *Brahmapakṣanirāsa* is a short essay on the *Brahmapakṣa* of astronomy. The copy of this essay at the Rajasthan Oriental Research Institute has only six folios to it and is in the pen of the scribe Govinda Rāma, son of Gaṇeśa.[1] Although the manuscript is undated, it is certain from the script that the copy belongs to the post-Jai Singh era.

Jayavinodī Pañcāṅga

Kevalarāma undertook the responsibility of publishing the yearly calendar of Jaipur, the *Jayavinodī Pañcāṅga*, named after his patron Jai Singh. According to Bahura, these *Pañcāṅgas* received wide acceptance in India,[2] and a good many of them are still preserved at the Sawai Man Singh II Museum of Jaipur.[3] After the death of Kevalarāma, his descendants continued publishing the *Pañcāṅga* under the original name—*Jayavinodī Pañcāṅga.*[4]

Vibhāga Sāraṇī

Kevalarāma is said to have prepared *Vibhāga Sāraṇī* consisting of logarithm tables.[5] However, no copy of this *Sāraṇī* has been located thus far.[6] Hunter saw a copy of this work at Ujjain around 1786 A.D. in the possession of Kevalarāma's grandson. Hunter identified the *Sāraṇī* as the "Annex to Cunn's or Commadine's Edition" of some mathematical text.[7] He also noted that, in the *Sāraṇī* the "inventor" of the logarithm was called Don Juan Napier, which led him to conclude correctly that the translation must have been done with the assistance of some Portuguese astronomer.[8] The Portuguese source on which the Sanskrit translation is based was most likely brought by Fr. Figueredo from Portugal.[9]

---

[1] RORI, No. 28628, 6ff.

[2] Bahura II, p. 59.

[3] Bahura II, pp. 54-55.

[4] *Jayavinodī Pañcāṅga* is currently published by Manihāron kā Rāstā, Jaipur. The publishers claim to be the direct descendants of Kevalarāma.

[5] Soonawala, p. 10, *Op. Cit.*, Ch. 6.

[6] An elaborate set of mathematical tables in Devanagari for astronomical computations is preserved at the Jantar Mantar of Jaipur. According to a note on the tables, they were transcribed by Govinda Nārāyaṇa Chaurāsiā, from some tables in English, and these tables should not be identified with the *Vibhāga Sāraṇī* of Kevalarāma. Chaurāsiā completed the work of rendering the tables in Devanagari on June 26, 1926.

[7] Hunter, p. 209.

[8] *Ibid*. The Portuguese astronomer-physician at the court of the Raja was Pedro de Silva. See Ch. 13.

[9] For Jai Singh's delegation to Europe, see Ch. 13.

Bhāgavata-Jyotiḥ-śāstrayoḥ-bhūgolakhagola Parihāra

Kevalarāma's *Bhūgolakhagola Parihāra* is an essay, in which he explains the discrepancy between the geography or astronomy given in the *Bhāgavata Purāṇa* and the well-known astronomical facts of his times.[1] He concludes his essay by saying that he wrote the work at the order of Sawai Jai Singh.

Tārā Sāraṇī

A number of authors have indicated that Kevalarāma translated *Zīj-i Ulugh Begī* under the title *Tārā Sāraṇī*.[2] However, there is no such *Sāraṇī* listed in the published catalogs of the Rajasthan Oriental Research Institute, or, for that matter, in the collection of the Sawai Man Singh II Museum of Jaipur. Possibly the Sanskrit version of *Zīj-i Ulugh Begī*, brought by Nanda Rāma from Surat has been mistakenly identified with *Tārā Sāraṇī*.[3] The Sawai Man Singh II Museum, in its Puṇḍarīka collection, does have a star table—*Krāntivṛtta dhruvāṅka*—with coordinates for 256 selected stars. The stellar coordinates in this manuscript refer to the year 1726 A.D. (1783 V.S. or 1648 S.E.) and are identical with those given in the *Zīj-i Muḥammad Shāhī*.[4] This work, therefore, may not be identified with Kevalarāma's *Tārā Sāraṇī*.[5]

Finally, Bühler reports Kevalarāma having written *Rekhāpradīpa*.[6] The manuscript of this text has yet to be traced. Similarly, the *Rāmavinoda Sāraṇī* supposedly written by him has yet to be located.[7] At any rate, the Sawai Man Singh II Museum, where most of Jai Singh's library is preserved, does not have any copy of this *Sāraṇī*.

---

[1] *Bhāgavata-Jyotiḥ-śāstrayoḥ-bhūgolakhagola Parihāra*. 1. Oriental Research Institute, Baroda, No. 11049. 2. The Bhandarkar Oriental Research Institute, Pune, No. 956 of 1886-92, 13ff. This manuscript does not have the name of the author. 3. PUL II 3731, 20ff, CESS, Vol. A2, p. 63.

[2] Soonawala, p. 10, *Op. Cit.*, Ch. 6; Bahura II, p. 59; Bhārgava, p. 269, *Op. Cit.*; and Bhatnagar, p. 328, all state that Kevalarāma wrote a *Tārā Sāraṇī*. The source of their information appears to be the monograph of Soonawala, published in 1952, *Op. Cit.*, Ch. 6.

[3] *Zīj-i Ulugh Begī*, the Museum, No. 45; Bahura I, pp. 58-59.

[4] The Museum, No. 21 of its Puṇḍarīka collection, 4ff.

[5] The manuscript that Kaye reports having seen, and which is reproduced by him in his book, is a Devanagari rendition of the star catalog of the *Zīj-i Muḥammad Shāhī*, and is definitely not a creation of Kevalarāma. See Kaye, pp. 98-118.

[6] G. Bühler, "A Catalog of Sanskrit Manuscripts in the Private Libraries of Gujarat, Kathiawad, Kachchh, Sindh and Khandes," Bombay, 1871-1873, as quoted by Sen, p. 110, *Op. Cit.*, Ch. 3. Also CESS, Vol. A2, p. 63. The ms. has 4 folios to it.

[7] Bahura II, p. 59.

Abhilāṣāśataka and Gaṅgāstuti[1]

In addition, Kevalarāma is said to have written two non-astronomical religious works—namely, *Abhilāṣāśataka* and *Gaṅgāstuti*.[2]

Nayanasukha Upādhyāya

Nayanasukha Upādhyāya, or Nayanasukhopādhyāya, hailed from the town of Mālāvatī, where his father Narahari was an astronomer-astrologer. Nayanasukha could very well have been one of the scholars attracted to Jai Singh's court because of the generosity Jai Singh extended towards men of letters. Nayanasukha's brother, Hīrānanda, also an astronomer, in his *Ṭhākuradāsavilāsa* completed in 1783 A.D. (1705 S.E.), tells us that Nayanasukha was awarded the title of *Paṇḍitarāja* by the emperor at Delhi.[3] According to Hīrānanda, Nayanasukha was involved in observing the planets under the patronage of Jai Singh.[4]

Nayanasukha Upādhyāya made a valuable contribution to the program of the Raja by translating a number of texts of the Persian-Arabic school of astronomy. In this task he was assisted by one or more astronomers of the Islamic school. The Muslim scholars read and explained a text to Nayanasukha, which he then rendered into Sanskrit.[5]

With the assistance of Muḥammad Ābid, a Muslim scholar, Nayanasukha Upādhyāya translated *Tadhkira* of Naṣīr al-Dīn al-Ṭūsī with al-Birjandī's *Sharḥ* in 1729 A.D. (1786 V.S.).[6] The translation involves the eleventh chapter of the second book of *Tadhkira* only.[7] In this chapter the lunar and the planetary models of the "School of Maragha" have been described. In the translation, Nayanasukha elaborated the difficult passages of the book. A copy of the

---

[1] *Abhilāṣāśataka*, RORI, No. 11204, and *Gaṅgāstuti*, RORI, No. 3300, 7ff.

[2] Bhārgava, based on an article by Kedar Nath Sharma, states that Kevalarāma wrote *Jaya Siṁha Kalpalatā*, an incomplete work on planetary computations, and that the work is preserved at the Sawai Man Singh II Museum. However, no such work is listed in Bahura I or Bahura II. See Bhārgava, *Op. Cit.*, p. 268. The article quoted by Bhārgava is: Kedāra Nātha Śarmā, "Āmera ke Mahārājā Sawā'ī Jaya Siṁha ke Grantha aur Unakī Vedhaśālāyen," (in Hindi), *Nāgarī Pracāraṇī Patrikā*, Navīna Saṁsakaraṇa, Vol. 3, p. 403.

[3] *Ṭhākuradāsavilāsa*, mss. Nos. 5523/3009, *Rajasthan Puratana Granthamala* No. 151, general ed. P. D. Pathak, *Catalog of Sanskrit and Prakrit Manuscripts*, (Alwar collection), Part XXI, ed. O.L. Menaria, et al, RORI, Jodhpur, 1985. Most probably it was Muḥammad Shāh, the Mogul emperor between 1719-1748 A.D., who awarded the title of *Paṇḍitarāja* to Nayanasukha.

[4] *Ibid.*

[5] The contribution of Muslim astronomers is discussed in Ch. 13.

[6] *Tadhkira of Naṣīr al-Dīn al-Ṭūsī in commentary of ʿAlī al-Birjandī* by Nayanasukhopādhyāya, ms. No. 46 AG, 56ff, the Museum. Also Bahura I, pp. 62-63, 101-102; and Bahura II, p. 58.

[7] Pingree (1987), *Op. Cit.*, Ch. 11.

translation was admitted to the royal library in 1730 A.D. Nayanasukha definitely translated two other works, namely, *Ūkaragrantha* and *Yantrarāja risālā bīsa bāba.*[1] *Ūkaragrantha*, which deals with spherical geometry, is based on an Arabic copy of *Spherics* of Theodosius and was completed, once again, with the assistance of Ābid. *Ūkaragrantha* has been published along with the *Samrāṭ Siddhānta* of Jagannātha.[2] The manuscript of the Ūkaragrantha at the Sawai Man Singh II Museum, copied in 1730 A.D. (1787 V.S.), is profusely illustrated with appropriate geometrical figures.[3] *Yantrarāja risālā bīsa bāba* is a translation of Naṣīr al-Dīn al-Ṭūsī's *Risālah bīst bāb*. It has been published under the title—*Yantrarājavicāraviṁśādhyāyī.*[4] The *Risālā* consists of 20 chapters and, despite being a text on the astrolabe, has no illustrations at all. The copy of the *Risālā* at the Sawai Man Singh II Museum is in the pen of Kṛpārāma and the date of its copying is unknown. A fourth book, *Jarakālīyantra*, in 13 folios, may also have been translated by Nayanasukha, although definite proof of this is lacking.[5,6] *Jarakālīyantra* is a text dealing with the astrolabe *Saphaea Arzachelis*. The Jantar Mantar observatory of Jaipur preserves a fine specimen of this instrument, which was fabricated in 1680 A.D.[7]

In his *Ṭhākuradāsavilāsa*, mentioned earlier, Hīrānanda asserts that his brother Nayanasukha revised the *Siddhānta*, called *Jayasiṁhakaustubha*. However, *Jayasiṁhakaustubha* has yet to be located.[8] The inventory list of Jai Singh's personal library does not include any such text.

---

[1] *Ūkaragrantha*, (copied 1729, acquired 1730 A.D.), 46ff. No. 44 AG, Bahura I, pp. 58-59. *Yantrarāja risālā bīsa bāba*, No. 42 AG, 28ff., Bahura I, pp. 60-61.

[2] SSRS, pp. 1260-1328.

[3] Bahura I, No. 245E, 44ff, the Museum. The other two copies of this work are preserved at 1. Baroda, No. 8926, entitled *Kaṭara* from Arabic *quṭr*. 2. Calcutta Sanskrit College, *Jyotiṣa*, No. 118. The Calcutta copy was made in 1730 A.D. and most likely at Jai Singh's own court. For these copies and for others, see CESS, Vol. A3, p. 132 and CESS, Vol. A5.

[4] *Yantrarājavicāraviṁśādhyāyī* by Nayanasukha Upādhyāya, ed., Vibhūtibhūṣaṇa Bhaṭṭācārya, Sampurnanand Sanskrit University, Varanasi, 1979. Although the author of the work is identified as Nayanasukha, the inscription to this effect on the manuscript is in a different hand and might have been added at a later date. See photograph No. 2, and p. 33. The Jaipur Museum copy of this text does not have the name of the author.

[5] Bahura II, p. 35, No. 5483, 13ff, the Museum. Also the Museum's Puṇḍarīka Collection, No. 28, 8ff.

[6] Pingree, having compared a manuscript at Trinity College, Cambridge, (Cat. No. R.15.139), with a manuscript of Nayanasukha's *Ūkara*, suggests that Nayanasukha wrote the *Jarakālīyantram*. See Pingree (1987), *Op. Cit.*, Ch. 11.

[7] For a description of the astrolabe, see Kaye, pp. 27-30.

[8] Pingree suggests that this work is identical with Jagannātha's *Samrāṭ Siddhānta*. Pingree (1987), *Op. Cit.*, Ch. 11.

Kṛpārāma

Jai Singh's astronomers usually wrote their texts in Sanskrit, the language of the Brahmin scholars of medieval India. However, Kṛpārāma, a Nāgara Brahmin, wrote his *Samayabodha* in Hindi in 1715 A.D.[1,2]

Anonymous Translators

Nothing is known about the translator of a European monograph on perspective drawing. This translation, entitled *Pratibimba Siddhānta*, is a small booklet in the *Khaḍī bolī* dialect of Hindi spoken in the cultured circles of Delhi and Agra at the time. Jai Singh had the text translated for builders, engineers, technicians, artists, and draftsmen. Because this group of professionals did not necessarily know Sanskrit, Jai Singh had the translation done in Hindi. A unique copy of *Pratibimba Siddhānta* is preserved at the Ḳhasmohar collection of Sawai Man Singh II Museum of Jaipur.[3] The text of *Pratibiṁba Siddhānta* deals with the subject of perspectives or the theory representing solid objects on a flat surface in such a way as to convey the impression of depth and distance. Its subject matter is similar to chapters on perspectives in engineering graphics courses taught to first-year students in engineering schools in the United States. The author of the text starts out by defining technical terms such as the horizon (*Kṣitija rekhā*) and the ground line (*jamīna kī rekhā*). Next, he discusses one-point perspectives, two-point perspectives, and arcs-in-perspective with appropriate diagrams. The author also displays shading techniques to indicate depth. Finally, he shows how to draw perspectives of complicated figures such as a table or a chair.

The penmanship of the booklet is excellent, and it has 56 illustrations related to the subject. In the manuscript, the translator does not identify himself. But a remark in Rajasthani on the very first page translates: "The book belonging to Peḍro Jī (which has been) translated (in here)." The Peḍro Jī mentioned by the writer of the note could have been Pedro de Silva, the physician-astronomer who came from Portugal in 1730 A.D. and settled in Jaipur.[4] He could also have been the Peḍro Jī who went to Portugal as a member of the scientific fact-finding mission of the Raja. Since both of these Pedros were Portuguese, the

---

[1] CESS, Vol. A3, p. 22.

[2] Pingree suggests that this Kṛpārāma is someone other than the favorite scribe of Jai Singh with the same name. Pingree (1987), *Op. Cit.*, Ch. 11.

[3] *Pratibimba Siddhānta*, the Museum, No. 2016. The title *Pratibimba Siddhānta* is given to the booklet by the Museum. Although the book does not have any date, its script suggests it to be of the Jai Singh period. For details of the *Pratibimba Siddhānta*, See Virendra Nath Sharma, "*Pratibimba Siddhānta* of Jai Singh's Library," to appear in Indian J. Hist. Sci.

[4] According to a genealogy given to the author by the de Silva family of Jaipur, there was only one Pedro among their ancestors—the one who came originally from Portugal in 1731 A.D. The genealogy had been obtained from the R.S.A. for a court case.

original text must have been either in Portuguese or in Latin. The author believes that *Pratibimba Siddhānta* is the very first book on any technical subject translated from a European work into the *Khaḍī-Bolī* dialect of Hindi.

Laiyara Vedha Patrāṇi are daily tables for some unidentified parameters of the sun and the moon, and, according to the title, they are based on the *Tabulae Astronomicae* of de La Hire. The exact nature of the tables in *Laiyara Vedha Patrāṇi* is not clear.[1] The tables are for the years 1727-1738 A.D., and there are two copies of them at the Sawai Man Singh II Museum; the second copy being entitled *Navīna Vedha Patrāṇi*.[2]

Anonymous Authors

A Devanagari version of the *Zīj-i Muḥammad Shāhī*, mentioned in the Pothikhana records of the Rajasthan State Archives, must have also been prepared by one or more Hindu scholars, whose names are not known.[3]

Procurement of Books

Jai Singh had inherited only a few books on astronomical subjects when he succeeded to the throne of Amber. As a matter of fact, as late as 1715 there were only 32 books on *Jyotiṣa*, or, astronomy-astrology, in the royal library. However, Jai Singh left behind an excellent collection at his death. An inventory of his *Pothīkhānā*, or the library, which began in 1741 and was completed after his death in 1743, lists 188 books on different aspects of *Jyotiṣa*. Jai Singh built his excellent library by collecting books from far and wide. He received his Sanskrit books via Brahmins. For example, in 1714, he received from Nīlāmbara Bhaṭa a book on *Jyotiṣa* in 145 folios.[4] He procured an incomplete Sanskrit-version of Ulugh Beg's famous *Zīj* via Nandarāma.[5]

Astrologers

In Jai Singh's India there was little difference between an astrologer and an astronomer. The Sanskrit term *Jyotiṣī* applied to both astronomers and astrologers. In fact, the Hindu pundits who wrote on their favorite topics in astronomy were mostly astrologers by trade.

---

[1] *Laiyara Vedha Patrāṇi*, the Museum, No. 5183, 12ff.

[2] *Navīna Sāraṇī Vedha*, the Museum, No. KM 43, 28ff.

[3] R.S.A., Pothikhana records of Jaipur Rajya.

[4] R.S.A., file No. 424/1, Pothikhana-Suratkhana, Jaipur State.

[5] The copy was purchased for Rs. 20. The R.S.A. preserves the deed papers of this purchase. *Ibid*.

Hari Lāla Miśra

Hari Lāla Miśra was more an astrologer than an astronomer.[1] He hailed from a family of *Jyotiṣīs* who were natives of Rajasthan.[2] Hari Lāla himself had apparently lived for a while in the Shekhavati region of the Jaipur state. His astrologer father, Vaṁśīdhara, however, had moved to Vrindavan, and it is from him that Hari Lāla learned his trade. Apparently, Hari Lāla came in contact with Jai Singh while Jai Singh was the administrator of Mathura and the governor of the province of Agra (1723 - ?).[3] Evidently, Hari Lāla must have also studied under Jagannātha Samrāṭ because he calls him his *guru* or teacher.

Hari Lāla composed *Muhūrtaśiromaṇi*,[4] a work on astrology related to astrologically auspicious moments for various observances at the time the city of Jaipur was founded.[5] The work was completed on *Phālguna Śukla* 3, 1793 V.S., or Monday, March 4, 1737 A.D. "This was the day when the new city of Jaipur attained its full bloom after its completion," he writes.[6] Hari Lāla's other composition, similar in content to the *Muhūrtaśiromaṇi*, according to Bahura, is *Muhūrtakalpadruma*.[7]

The other work of Hari Lāla, according to Führer, had been *Tithyuktiratnāvalī*. It concerns religious observances for various *tithis*. The manuscript of this work was seen by Führer in 1885 with the descendants of Madhusūdana, a Brahmin, much honored at the court of the Sikh ruler Ranjit Singh. However, the manuscript was poorly taken care by its owner at the time, and its whereabouts are now unknown.[8] Perhaps the same Hari Lāla also copied the first eight chapters of *Mantrabhāṣyam* of *Vājsaneyīsaṁhitā*, a religious text by Uvvaṭa, the son of Vajraṭa.[9]

Yaśasāgara

Another astrological work written during Jai Singh's rule is by Yaśasāgara.

---

[1] *Muhūrtaśiromaṇi* of Hari Lāla Miśra, the Museum, No. 5017. See Bahura II, pp. 406-408.

[2] In *Muhūrtaśiromaṇi*, Hari Lāla recounts six generations of his ancestors. In ascending order they are: Bhānu Paṇḍita, Kṛṣṇa Śarmā, Harivaṁśa Tripāṭhī, Dayālu, Sukhadeva, and Vaṁśīdhara. Bhānu Paṇḍita was Hari Lāla's grandfather.

[3] See Ātmā Rāma's *Savāī Jaya Siṁha Carita*, p. 129, Jaipur, 1979. Also Bhatnagar, pp. 162-163.

[4] The Museum, No. 5017, 49ff and No. 5502.

[5] This work is listed in the inventory of Jai Singh's library. The manuscript had 49 folios to it. R.S.A., Pothikhana records, Jaipur State.

[6] Bahura II, pp. 406 and 408.

[7] *Muhūrtakalpadruma*, Bahura II, No. 5502, p. 83. The library of Jai Singh also had a copy of it in 35 folios. R.S.A., Pothikhana records, Jaipur State.

[8] A. Führer, "Ueber indisches Bibliothekswesen II," *Centralblatt für Bibliothekswesen* 2, pp. 41-58, (1885).

[9] *Mantrabhāṣyam*, the Museum, No. 4523, and Bahura II, pp. 223, 398.

Yaśasāgara wrote *Jātakasārapaddhati*, a text on nativity-horoscope in 1705 A.D. (1762 V.S.).[1] Further, some anonymous author wrote *Svarasiddhahaṁsa*, apparently a work on astrology. The date of composition of this text is 1709 A.D. (1766 V.S.).[2]

Astronomers on Daily Wages at the Jaipur Observatory

Jai Singh employed a large number of astronomers at his observatories. In 1735 A.D., the Jaipur observatory alone had 22 Hindu astronomers employed at daily wages.[3] The names of the astronomers, as they appear in the Rajasthani dialect of Hindi, are: Udayanī, Gaṇapatī, Rāma Kisana, Govyanda Bhaṭa, Muljī Bhaṭa, Gaṅgāvīsana, Govyanda Bhaṭa (II), Āṇada Rāma, Nanda Rāma, Sundarjī, Devakīsana, Devasura, Jetakāra (?), Ratana Sīṅgha, Fate Canda, Jivana Jo. (Jyotiṣī), Āsādhara, Gaja Sīṅgh (?), Māyā Rāma, Maujī Rāma, Tulā Rāma, and Harī Rāma. These daily-wage earners helped erect masonry instruments and took observations of the sun, moon, and the planets. They were paid up to Rs. 31 per month, depending on the amount of time they put in.[4] Claude Boudier, an eyewitness in 1734 A.D. to the observatory operations, writes that Brahmins were busy day and night observing at Jaipur.[5] The observatory's financial accounts for the year 1734-35 A.D. confirm Boudier's remarks. At Delhi, Mathura, Ujjain and Varanasi, where Jai Singh had his other observatories, there must also have been similar teams of astronomers employed. However, the names of these daily-wage earners are not known.

Scribes

Jai Singh employed a number of competent scribes for copying books for his library because there were no printing presses in India then. These scribes, because of the technical nature of the subject matter they copied, had to have some training in astronomy and often called themselves *jyotiṣīs*. They copied the manuscripts in Sanskrit using the Devanagari script. They frequently initialled the manuscripts and wrote down the date of completion of their work. Accordingly, the names of a number of such scribes participating in the astronomical program of the Raja have survived. Kṛpārāma copied *Sharḥ Tadhkira* of al-Birjandī. Tulārāma made copies of *Makaranda Jyotiṣa-tippaṇam*

---

[1] *Jātakasārapaddhati* of Yaśasāgara, the Museum, No. 5402, and Bahura II, p. 255.

[2] The Museum, No. 5575, and Bahura II, p. 128.

[3] R.S.A., Imaratkhana Records for the year 1734 A.D. (1791 V.S.). In addition to the 22 Hindu names in the records there are two Muslim names also, namely, Kisandī Khān and Ghulām Ḥusain.

[4] *Ibid*.

[5] *Lettres*, p. 778.

in 1706 A.D. (1763 V.S.),[1] *Bhāsvatī* of Śatānanda in 1701 A.D. (1758 V.S.),[2] and *Nalikābandhakramapaddhati* of Rāmakṛṣṇa in 1706 A.D. (1763 V.S.). He also copied *Kāla Jñānam*.[3] Another scribe, Lakṣmīdhara, was responsible for *Ūkaragrantha* which he copied in 1729 A.D. (1786 V.S.). He also copied *Vakramārgavicāra* sometime before 1787 V.S.[4] Ṭīkārāma copied *Hayatagrantha*, a Sanskrit translation of a Persian work.[5] The manuscript of *Hayatagrantha* was added to Jai Singh's library in 1730 A.D. (1787 V.S.).[6,7] Lokamaṇi was responsible for making a copy of Jagannātha's *Rekhāgaṇita* in 1728 A.D.[8] Of the three copies of *Zīj-i Nityānandī-i Shāhjahānī* at the Sawai Man Singh II Museum, at least one was definitely copied by Gaṅgārāma, a native of Kashmir.[9,10] The work was completed on Thursday, the *Vaiśākha, Kṛṣṇa* 11, 1784 V.S. (1727 A.D.), at noon, as he notes.

Assessment

The astronomical activities of Jai Singh's Hindu astronomers may be divided into four categories: (1) erecting instruments and taking data at the observatories, (2) writing texts and commentaries, (3) translating from other languages, and (4) collecting or copying books for the royal library.

The Hindu pundits helped Jai Singh erect stone and masonry instruments. Their efforts in this regard are praiseworthy. The Great Samrāṭs and the Ṣaṣṭhāṁśa yantras of Delhi and Jaipur, that they helped erect, had a very high degree of accuracy. The readings taken with the Delhi Ṣaṣṭhāṁśa deviated less than $\pm 1'$ from the true values.[11] One minute of arc is considered to be the limit for non-telescopic observations. Jagannātha, the principal astronomer of Jai Singh, displayed a strong belief in observing, and it is quite likely that his belief was shared by others as well.

The Hindu pundits, with the encouragement and active support of their patron Jai Singh, wrote a large number of books and commentaries. However, these texts and commentaries were based mostly on the works of their predecessors.

---

[1] Bahura II, p. 75.

[2] Bahura II, p. 54.

[3] Bahura II, p. 20.

[4] Bahura I, p. 63.

[5] Bahura I, pp. 62-63, and Bahura II, p. 58.

[6] Bahura II, p. 58.

[7] Pingree has identified *Hayatagrantha* with the *Risālah dar hay'at* (Monograph on astronomy) of al-Qushjī. Further, he believes that the translation into Sanskrit was done in the 17th century. See Pingree (1978), Ch. 1.

[8] The Museum, No. 5372 and 5373; Bahura II, p. 432.

[9] Bahura I, p. 99.

[10] Bahura II, p. 241. One of the copies at the Museum bears the seal of the imperial library, the *Kutubkhānā* of Shahjahan.

[11] See Ch. 5, and 6.

In other words, there is little original material in their works. A cause of this lack of originality may lie in the scholars' undue faith in the *Siddhāntas*, or canons of Hindu astronomy.[1] It appears as if Jai Singh's pundits did not wish to break away from a tradition which for all practical purposes had become stagnant. They also believed in the astrological effect of planets as did the astronomers of Europe in medieval times.

The Hindu scholars translated a great number of texts. However, the credit for these translations, in part at least, should go to Jai Singh. He realized more than any one else that the astronomy of the country needed an infusion of fresh ideas, and that there was much to be learned from other traditions, such as from the Islamic and the European. Having reached this conclusion, he made arrangements for translating astronomical and mathematical works into Sanskrit and divided the task between his principal assistants: Jagannātha, Kevalarāma, Nayanasukha Upādhyāya, and others. However, these translations turned out to be of those works which, in retrospect, had become outdated in the astronomical circles of Europe.

The pundits came in contact with Europeans—the Jesuit priests mostly—and thus became acquainted with better methods of computations. But this association produced only fragmentary infiltration of European thought. It did not introduce them to the epoch-making theories of Kepler and Newton and, consequently, did not initiate the modern age of astronomy in the country.[2]

---

[1] A similar situation once prevailed in medieval Europe, when the works of Ptolemy and Aristotle were considered infallible and the final authority.

[2] This subject is treated in detail in Ch. 13.

CHAPTER XIII

THE MUSLIMS AND THE EUROPEANS

MUSLIM ASTRONOMERS

The Muslim astronomers of India rendered useful service to the cause of the Raja. Information about Jai Singh's Muslim astronomers comes primarily from the *Dastūr Kaumvār*, a 32-volume set of books in the Rajasthan State Archives, which records the favors, honors, and gifts that the rulers of Amber or Jaipur granted over a period of several generations. The *Dastūr Kaumvār* records go back to the time of Māna Singh (1550-1617); however, they are more detailed for the Jai Singh period. The records list the Muslim astronomers under the category *Kaum Musalamān* (Muslim caste) and identify them by the title *nujūmī* (astronomers or astrologers).[1] Jai Singh gave his *nujūmīs* a variety of gifts, such as a *siropā*, or ceremonial dress; a horse; or amounts of cash ranging from Rs.1 to Rs.1,000. The *Dastūr Kaumvār* records occasionally elaborate on the reasons for the gift, while at other times they simply list the amount spent on the gift and give no details. We will assume that the gifts, awards, and honors, which the *nujūmīs* received, were primarily for their services related to astronomy. However, the assumption is open to question until substantiated by other sources. Some information regarding the Muslim astronomers also appears in Jagannātha Samrāṭ's *Samrāṭ Siddhānta* and some incidental remarks appear in literature such as the *Safīnā-i Khushgo* of Brindāvana, a contemporary of Jai Singh.[2]

Dayānat Khān and other Nujūmīs

Jai Singh's most favored and decorated Muslim astronomer, according to *Dastūr Kaumvār*, was Dayānat Khān.[3,4] He met the Raja before any of the observatories were completed and remained associated with him for more than two decades. In 1718 A.D. (1775 V.S.) he received from the Raja his first gift, a very generous sum of Rs.300, according to the records of *Dastūr Kaumvār*.

---

[1] The Hindustani word *nujūmī* has its origin in the Persian word *nujūm*, meaning astrology.

[2] *Safīnā-i Khushgo* of Brindāvana, ms., Khudabaksha Library, Bankipur, Patna. Published, Patna, 1959.

[3] DK, Vol. 19, p. 563.

[4] Dayānat Khān is mentioned in a manuscript copied for him and which is now preserved at the Museum, Jaipur. The manuscript is a commentary on the *Kitāb al-Manāẓir* (optics) of Ibn al-Haytham. See Appendix VII. It was copied for al-Shāh Qiyād ibn ʿAbd al-Jalīl al Ḥārithī al-Badkhshī known as Dayānat Khān. See King (1980), *Op. Cit.*, Ch. 3.

Then in 1724, he was honored again with a *siropā*, or a ceremonial robe, and some other items as a farewell gift at Mathura. During the period of the more than twenty years that he remained associated with the Raja, Dayānat Khān was decorated at least six different times. The very last gift he received, according to the records in *Dastūr Kaumvār*, was in 1739.[1] One may reasonably assume that Dayānat Khān played a major role in the astronomical program of the Raja.

Little is known about the contributions of the other *nujūmīs* who received gifts and honors from the Raja. Nizām Khān *nujūmī* received some gift in 1774 V.S. (1717 A.D.).[2] Mirzā ʿAbdu'r-raḥmān was honored with a horse on *Māgha Śukla* 11, 1777 V.S. (1721 A.D.).[3] Sheikh Asadu'llah was decorated with a *siropā* in 1777 V.S. (1720 A.D.) and with other gifts, in 1775 and 1783 V.S. (1718, 1726 A.D.).[4] The other *nujūmīs* listed in the records are Sheikh Asatu'llah (1719, 1720 A.D.),[5] Muḥammad Ābid Mullā (1725 A.D.),[6] Sheikh Aḥmad (1725 A.D.),[7] Sayyid Muḥammad (1725-1726 A.D.),[8] Sheikh Muḥammad Shafī (1725, 1729 A.D.),[9] and Va'iz Muḥammad Mehdī (1731 A.D.).[10] Mullā Imāmu'ddīn *nujūmī*, according to *Dastūr Kaumvār*, received a sum of Rs.100 in 1725, apparently for his expertise in astronomy.[11] Then again, in 1726, he received a sum of Rs.200.[12] His name is found on the *Sharh Zīj-i Ulugh Begī*.[13] He probably wrote the *Sharḥ*. Imāmu'ddīn, according to *Safīnā-i Khushgo* of Brindāvana, was a highly respected scholar of mathematics. He resided at Delhi and died in 1733.[14]

Contribution of Muslim Astronomers

The Muslim astronomers constructed the early instruments of the Raja, which, according to the *Zīj-i Muḥammad Shāhī*, were based on the Islamic books.[15] The astronomers probably were also involved to some extent in erecting the masonry instruments of the observatories of Delhi and Jaipur. According to the

---

[1] Ref. 3.

[2] DK, Vol. 19, p. 563.

[3] DK, Vol. 18, p. 557.

[4] DK, Vol. 18, p. 540.

[5] DK, Vol. 18, p. 554. It is quite possible that Asatu'llah and Asadu'llah are one and the same person. The scribes of Jaipur State were not always careful with their spellings.

[6] DK, Vol. 18, pp. 590-591. Muḥammad Ābid Mullā might have been the person who collaborated with Nayanasukha Upādhyāya in translating Arabic works in Sanskrit. See Ch. 12.

[7] DK, Vol. 18, p. 502.

[8] DK, Vol. 20, pp. 193-194.

[9] DK, Vol. 20, p. 604.

[10] DK, Vol. 20, p. 192.

[11] DK, Vol. 18, p. 745.

[12] *Ibid*.

[13] Bahura I, pp. 74-75.

[14] *Safīnā-i Khushgo* of Brindāvana, f. 123, *Op. Cit*.

[15] ZMS, f. 1.

Imaratkhana (*Imāratakhānā*) records, Ghulām Ḥusain and Kisandī Khān (?), along with other observatory employees, were receiving daily wages in 1733-34 at Jaipur when observatory instruments there were receiving finishing touches there.[1]

Jai Singh's primary aim in erecting the observatories had been to prepare a set of astronomical tables, i.e., a *Zīj*. It is very likely that some of the astronomers mentioned above were on the team that compiled the tables for his *Zīj-i Muḥammad Shāhī*. However, the *Zīj* does not mention any astronomers or their contributions to the text.

During the time of Jai Singh, or somewhat before him, there were some excellent astrolabe makers in India. Jai Singh himself was a collector of fine astrolabes. His Persian astrolabes were presumably engraved or procured by his Muslim assistants.[2] The author has been unable to trace any names in the Rajasthan State Archives that would definitely establish the makers of the astrolabes in the private collection of the Raja.

Astronomical Texts

Jai Singh's early training had been solely under Hindu pundits from whom he learned the Hindu school of astronomy. He soon developed an interest in the Persian-Arabic school of the subject and began acquiring books and patronizing its scholars. In 1716 he received the first Persian book for his library, *Turīya Jantra*, from Sheikh Abdu'llah.[3] At the same time, the Sheikh also brought another book, *Turīya Jantra Pīlkī*, probably also in Persian. It is noteworthy that the books were on instrumentation, indicating the Raja's early interest in observational astronomy. Before 1716, according to the inventory of his library of 1715, he had only Sanskrit texts.[4] Gradually, Jai Singh acquired quite a few books in Persian and Arabic on both astronomy and astrology. He acquired in 1725 *Jīz-i Shāhjahānī* and *Al-Tafhīm li-awā'il ṣinaᶜat al-tanjīm* through Hakīm Mīsī Jamān Khān who was a resident of Jūsī.[5] Many of these books have survived, and one may still see them in the collection of the Sawai Man Singh II Museum of Jaipur.[6] A list of these books is given in Appendix VII. The Persian-Arabic books for the royal library were either purchased from their

---

[1] *Tozīs* Imaratkhana, V.S. 1791, Jaipur Rajya, R.S.A.

[2] For Jai Singh's astrolabes, see Ch. 7.

[3] File No. 424/1, Jaipur Rajya, R.S.A. The records in the file do not give the Persian titles of the books. It was common with Jai Singh's librarians to list a book by its content, particularly if it was in a language other than Sanskrit or Hindi. On the other hand, there is a remote possibility that the two books brought by Abdu'llah were Persian renditions of some Sanskrit texts. The word *jantra* is a colloquial derivation of the Sanskrit term *yantra* or instrument.

[4] *Ibid*. The inventory lists a total of 32 books on astronomy in Sanskrit.

[5] Bahura I, p. 75.

[6] Bahura I, pp. 72-76. King has given a brief description of the manuscripts. King (1980), *Op. Cit.*, Ch. 3.

owners or copied from the existing texts. ʿAbdu'llah provided a copy of *Jāmiʿ-i Shāhī*, a work on astrology.[1]

Translations into Sanskrit

The Hindu astronomers carried out their work in Sanskrit. Primarily for the benefit of these scholars, Jai Singh had a number of works translated into Sanskrit. As we have discussed earlier in Ch. 12, Nayanasukhopādhyāya translated a number of books with the assistance of Muḥammad Ābid. The Muslim astronomers must have also contributed to the preparation of the Devanagari version of the *Zīj-i Muḥammad Shāhī*.[2,3]

Judging by the gifts in the *Dastūr Kaumvār*, it appears that the involvement of the Muslim astronomers in Jai Singh's program began sometime around 1715, reached a peak a decade later, and then tapered off sometime before his death in 1743.[4]

Abu'l Khair alias Khairu'llah

Khairu'llah has been called the chief assistant of the Raja and the author of the *Zīj-i Muḥammad Shāhī*.[5,6] However, these titles appear doubtful when one does not find any mention of gifts or honors bestowed upon him in the *Dastūr*

---

[1] *Ibid*. Bahura I, pp. 72-73.

[2] At least one copy of the *Zīj-i Muḥammad Shāhī* was prepared in Devanagari. See Ch. 11.

[3] The other Sanskrit translations acquired by Jai Singh were: (1). *Zīj-i Nityānandī Shāhjahānī*, probably based on the *Zīj-i Shāhjahānī* of Ibrāhīm al-Dihlawī, (acquired 1727), No. 23 AG. (2). *Zīj-i Ulugh Begī*, (tables only; acquired in 1729 from Surat), No. 45 AG. See Ch. 11.

[4] In the *Dastūr Kaumvār* there are no gifts listed for the years 1732-1734, 1736-37, and 1739-1743 A.D.

[5] For instance, see S. A. Khan Ghori, "The Impact of Modern European Astronomy on Raja Jai Singh," *Indian J. Hist. Sci.*, Vol. 15 (1), p. 53, (1980).

[6] Storey (1958), p. 95, *Op. Cit.*, Ch. 11. Also see Nadavi (1946), p. 281, *Op. Cit.*, Ch. 1. Khairu'llah allegedly wrote a commentary on *Zīj-i Muḥammad Shāhī*. See the *Bibliography*, p. 285. Also see Ch. 11. In the *Bibliography*, we have the following account about Khairu'llah:

> "He was the son of Lutfu'llah Muhandis, and a grandson of Ustād Aḥmad. He resided at Bangor Mau in the Unnao district. He received his education from his father and from his elder brother. He was appointed director of the observatory at Delhi in 1718, by the Emperor Muḥammad Shāh, where he wrote his astronomical observations and compiled his tables. Among his works are: *Sharḥ-i Zīj-i Muḥammad Shāhī, Taqrību'l-Taḥrīr and Qānūnu'l-Wafq*. In the *Sharḥ* he has brought out some arguments contradicting the earlier theory regarding the orbits of the sun and its satellites which he believed to be elliptical and not circular."

The author has found little evidence, however, that Khairu'llah was the director of the Delhi observatory of Jai Singh. Besides, the observatory work did not start until after 1721-22. Moreover, his *Sharḥ* on the *Zīj-i Muḥammad Shāhī* has yet to be located. The Raza Library, Rampur, has two manuscripts on mathematics and astronomy by Khairu'llah but a *Sharḥ* on the *Zīj-i Muḥammad Shāhī* is not one of them.

Kaumvār. It is inconceivable that the Raja, generous as he was in bestowing honors upon his scholars, would totally ignore his "chief assistant" and the "author" of an important work such as the *Zīj-i Muḥammad Shāhī*. The only reference that this author has been able to trace about Khairu'llah in the contemporary literature is one by Brindāvana. Writing about Khairu'llah and Jai Singh in his *Safīnā-i Khushgo*, Brindāvana remarks that Jai Singh spent two million rupees in the course of 20 years on his astronomical pursuits and that he expended these sums on the advice of Khairu'llah.[1] Scholars perhaps have interpreted the statement of Brindāvana more liberally than it deserves. The author believes that Khairu'llah played some role in the beginning, perhaps by urging the Raja to undertake the ambitious task of revising the astronomical tables. He might also have acted as an occasional advisor to the Raja. It is doubtful, however, that he was ever involved in the program of Jai Singh to the same extent as Dayānat Khān might have been who was profusely decorated from time to time by the Raja.

Delegation to Europe and Observers to Distant Islands

In 1727, Jai Singh dispatched a fact-finding scientific mission to Europe, of which Sheikh Asadu'llah *nujūmī* was a member. The delegation returned to Jaipur in 1731. We will be discussing the delegation to Europe shortly. Jai Singh believed that better astronomical results depended on observations taken from different locations around the globe. According to Jagannātha Samrāṭ, his command was: "In every country, in the east, the south, the west and the north, everywhere observations are to be made."[2] Accordingly, Muḥammad Sharīf was sent by the Raja to the *Firaṅga* country (overseas).[3] After having stayed there, he went to the island of "Mahailā" and determined its latitude to be 4;12 South.[4] Jagannātha writes:

> "In this part of the globe where the south celestial pole is above the horizon, he (Sharīf) observed the shapes of the constellations, drew them on paper and brought back the depictions. He also observed the longitude, latitude, and colatitude of the places he visited."

The *Dastūr Kaumvār* records do not mention any gift to a "Sharīf." However,

---

[1] Brindāvana in *Safīnā-i Khushgo*, *Op. Cit.*

[2] SSRS, p. 1165, also SSMC, p. 41.

[3] The word *Firaṅga* does not necessarily mean "Europe." In the present context it should be interpreted as "the land overseas under the control of Europeans."

[4] The latitude measurements taken at "Mahailā" indicate it to be some island in the Seychelles archipelago in the Indian Ocean. It could not be the island of Mahe, however, as the island of Mahe was named much later, sometime in 1742-43, by a French explorer after the Christian name of the French governor of Mauritius at the time.

there are several entries of gifts given away in cash and in kind to one "Sheikh Muḥammad Shafī." It is possible that Shafī and Sharīf are the same person. The scribes of medieval India were not always careful with people's names. Different scribes entered the same name differently depending on how it sounded to them. Besides, the possibility of error in copying from one record to other always exists. If this is the case, then Sheikh Muḥammad Shafī (Sharīf) left on his overseas journey shortly after 1729.

As the involvement of the Muslim astronomers slackened, participation by the Europeans increased, indicating the Raja's growing appreciation of contemporary astronomy in Europe. It is interesting to note that as the number of gifts to the Muslim *nujūmīs* decreased, gifts to the Europeans (*Firaṅgīs*) increased, reaching a similar peak in 1733.[1]

EUROPEAN ASTRONOMERS

Prior to erecting his observatories, Jai Singh had been in contact with several Europeans. A letter from the Mughal emperor Muḥammad Shāh quoted in the preface to *Zīj-i Muḥammad Shāhī* substantiates this fact. Muḥammad Shāh wrote, ". . . you have already assembled the astronomers and geometricians of the faith of Islam and the Brahmins and Pundits and the astronomers of Europe."[2] However, the Europeans began to participate more actively in the program of the Raja after 1727, and they performed essentially the same tasks as the Muslims did except that they were not involved in erecting the observatories and in collecting data.[3] The replacement of the Muslim astronomers by Europeans indicates the Raja's growing appreciation of the contemporary astronomy of Europe.

Considering the involvement of Europeans in his astronomical program, modern scholars have often raised this question: why did Jai Singh, an enlightened scholar, a man far ahead of his times, remain unaware or choose to ignore the brilliant theoretical and experimental progress of European astronomy initiated by the Copernican revolution and instead attempt "to revive the spirit of Ulugh Beg at a time that seems—in retrospect—to have been a century too late?"[4]

Jai Singh's ignorance or repudiation of the Copernican revolution becomes all the more baffling when one discovers that before beginning work on his observatories he is said to have familiarized himself with the different systems

---

[1] DK, Vol. 18 and 20. See under *Kaum Musalmān* and *Kaum Firaṅgī*. Also see *Tozī* bundles of *Daftar Nuskhā Puṇya*, R.S.A.

[2] ZMS, f. 0.

[3] A number of authors suggest that Europeans participated in erecting Jai Singh's observatories. See Noti (1906), p. 403, *Op. Cit.*, Ch. 6. In the Rajasthan State Archives, we have found no evidence that Europeans were involved in erecting observatories or in data-taking on a regular basis.

[4] Blanpied (1975), p. 1033, *Op. Cit.*, Ch. 3.

of astronomy, including the European.[1] To understand the impact of contemporary European astronomy, or the lack of it, on the program of the king, it is essential to explore the religious and political background of the Europeans the king came in contact with.

When Jai Singh contemplated his ambitious program in astronomy, Europeans were everywhere in the country. European travelers, soldiers of fortune, entrepreneurs, quack-doctors, missionaries, and traders were a common sight. European factories, trading posts, and settlements were scattered all over the country. The Dutch had a factory at Tranquebar, on the east coast near Madras that had been operating since 1620. The French had their settlement at Pondicherry since 1674. The British, who later succeeded in taking over the entire country, had been trading since 1615 and had established themselves at Masulipatam, and, soon thereafter, they founded a fortified post that later became the modern city of Madras. They also had fortified trading posts at Surat and Bombay on the west coast. However, among the Europeans, the Portuguese were the most prominent and conspicuous.[2]

The Portuguese

By the time of Jai Singh, the Portuguese, having come to India first in 1498, had been in the country for almost 200 years.[3] In fact, at that time they controlled a 60-mile-wide strip on the western coast between Damao and Bombay, which had the town of Goa as its capital. A viceroy sent from Lisbon and stationed at Goa ruled over their eastern empire, which extended all the way from East Africa to modern Sri Lanka. Until 1640, when they were defeated by the Dutch at sea, the Portuguese were the leading power in the Indian Ocean. Because of their power at sea and also because of their interest in trade and commerce, the rulers of North India and the Portuguese viceroys at Goa frequently exchanged emissaries of rank lower than ambassador.

During the days of Jai Singh, thousands of Portuguese mercenaries, traders, and adventurers were in India from the north to the south. The soldiers sought employment mostly in the artillery units of the local rulers, and the sailors in the navy of the Mughals. At one time the Mughal fleet in the Bay of Bengal had 923 Portuguese sailors.[4] After replacing the Arab merchants at sea, Portuguese traders directly or indirectly controlled the coastal trade of India for a long time. Their currency, minted at their possessions in India, was accepted by local business people. By the late 17th century, because of their maritime enterprises and trade, the Portuguese language had become the means of international communication. The Dutch, the English, or any other Europeans for that

---

[1] ZMS, f. 0.

[2] *The Oxford History of India*, pp. 327 ff., Oxford, 1958.

[3] Marques de Oliveira, *History of Portugal*, p. 335 ff., New York, 1972.

[4] *Ibid*, p. 340.

matter, had to learn Portuguese in order to be understood by one another or by an interpreter in Asia.[1]

It appears natural, therefore, that desiring to learn about European astronomy, Jai Singh should have turned to the Portuguese. His relations with them, once established, remained unbroken until his death in 1743, and through them he acquired his principal European helpers. The Portuguese authorities at Goa and to some extent at Lisbon, were his political contacts for obtaining men and materials for his astronomical pursuits.

The Jesuits

Although at the political level Jai Singh's contacts were the Portuguese, the Europeans working directly under him as astronomers were Jesuit missionaries, belonging to a priestly religious order—"The Society of Jesus."[2] The society, with its head office at Rome, was formed in 1539 on a military pattern by a Spanish soldier, Ignatius of Loyola, who demanded absolute discipline from its members. Members of the society, commonly known as Jesuits, took a special vow of obedience to the Pope. The society selected its members carefully and put its recruits through a long period of rigorous training and apprenticeship—often lasting for a decade or more, during which the ones with shaky faith were weeded out—before ordaining them as full-fledged priests. The careful selection of its members, coupled with strict discipline, led to rapid growth of the society and made the Jesuits one of the most influential and powerful orders in the Roman Catholic Church. The Jesuits cultivated scholarship among their members from the very beginning and by the 17th century had quite a few eminent mathematicians and astronomers within their ranks. But as Geymonat has pointed out, this cultivation of learning did not imply any open-mindedness toward new ideas.[3] Rather, Jesuit scholarship constituted an intelligent effort to explain new researches according to the established orthodox pattern of their Church in order to enhance the prestige and the power of the Church. Moreover, during the 17th and 18th centuries, the society actively opposed any assimilation of new scientific ideas.

By the time Jai Singh planned his observations, the Jesuits were no longer new arrivals on the Indian soil. Arriving first in 1542, they had been in the country as long as the Portuguese.[4] Scholars have pointed out that the Jesuit order sent its members—at least in the beginning—after careful selection, and "the average Jesuit missionary in India was a man of culture, observation and judgement, and possessing some knowledge of the languages of the land."[5]

---

[1] C.R. Boxer, *Four Centuries of Portuguese Expansion*, p. 92, Berkeley, 1969.

[2] *Encyclopedia of Science and Religion*, Vol. 7, pp. 500-05, J. Hastings (ed.), New York.

[3] Ludovico Geymonat, *Galileo Galilei*, p. 74, New York, 1965.

[4] Boxer, *Op. Cit.* p. 37.

[5] John Correia-Afonso, *Jesuit Letters and Indian History*, p. 59, Bombay, 1955.

An invitation from the emperor Akbar introduced them to the Mughal court, where, enjoying favorable treatment for a while, they furthered Portuguese national interests while carrying out their priestly duties with their congregations.[1] However, in the reign of Aurangzeb, and during the years following his death, the patronage of the Delhi court ceased, and they looked for its substitute elsewhere. The interest shown by Jai Singh in European astronomy, therefore, was viewed by the Jesuits as a golden opportunity to regain some of their former privileges, and they participated in his pursuits wholeheartedly.

Jesuits' Scientific Beliefs

The participation of Jesuits in Jai Singh's astronomical endeavors, once initiated, continued on and off for several decades until his death in 1743. In this participation they were unable to go against the dictates of beliefs mandated by the hierarchy of the Catholic Church.

The Catholic Church and the Copernican Revolution

Behind the Jesuit reluctance to accept the Copernican revolution, which had swept the intellectual circles of Europe, lay the attitude of the Catholic Church. The Church from the very outset had opposed the heliocentric theory of Copernicus, and this opposition to the theory is now a well-documented fact. The Catholic Church had opposed this theory because it contradicted the established doctrine of the immovability of the earth and therefore was in disagreement with Catholic dogma. The Church's opposition to the theory, beginning at first on a mild note, became loud and bitter when its spread appeared to get out of hand. The Church sought out adherents of the theory and pronounced harsh punishments on them with the intention of sounding a clear warning to others. Giorduno Bruno was imprisoned in a dungeon and then burned alive in 1600 for proclaiming that the sun, and not the earth, was the center of the heaven.[2] About three decades later, Galileo was forced to stand trial before the Roman Inquisition for supporting the heliocentric theory with his telescope and upholding it as truth.[3] He was publicly humiliated, forced to recant his beliefs, and sentenced to spend the rest of his days under virtual house arrest.[4] It is noteworthy that the Jesuits of Italy played an important part in bringing the 70-year-old ailing scientist to trial before the courts of the Roman

---

[1] E. Dennison Ross and Eileen Power, *Akbar and the Jesuits*, New York, 1926.

[2] Andrew Dickson White, *A History of the Warfare of Science with Theology*, p. 116 ff., reprint, New York, 1965.

[3] For a detailed account of the institution of Inquisition see H.C. Lea, *History of Inquisition of Spain*, 4 Vols., London, 1907. The Inquisition was abolished in Europe during the early decades of the 19th century. See Vol. 4, p. 436.

[4] Georgeo De Santillana, *Crime of Galileo*, pp. 237 ff., Chicago, 1955.

Inquisition.[1]

The Catholic Church did not stop to rest after punishing Bruno and Galileo, however. The recantation of Galileo was read throughout Italy at universities and colleges before professors of mathematics and astronomy, and concerted efforts were made to ensure that these professors did not hold or defend Galileo's views regarding the mobility of earth.[2] Further, the works of Galileo were added to the "Index," the list of prohibited books, which already banned those of Copernicus and Kepler, and the Inquisitors were ordered to deny publication of any writings upholding the views of these three authors. The commissioners of the Inquisition, accordingly, searched public and private libraries, book shops, and foreign vessels at all ports for prohibited and smuggled books, and the smugglers were punished by death.[3] As a consequence, the Copernican revolution withered away completely from countries where the Catholic Church was strong.

The Jesuit attitude toward the Copernican revolution, reflecting the views of their Church, had three distinct phases to it, punctuated by the two condemnations of Galileo in 1616 and 1633. Prior to 1616, the Jesuits in Italy and everywhere else, dazzled by the telescopic discoveries of Galileo, and also in the absence of any clear guidelines from their superiors, were undecided and divided on the question of the Copernican scheme. Some favored the Copernican theory, whereas others opposed it. For example, Christopher Grienberger, a professor of Roman College, favored it.[4] He even defended Galileo from the attacks of a fellow Jesuit. Wenceslaus Kirwitzer, a young Jesuit missionary to China, was also a Copernican.[5] Kirwitzer criticized Ptolemy in one letter and then supported Copernicus in another letter.[6]

After 1616, however, the situation changed. The period between 1616 and 1633 may be called the duration of the second phase. In 1616, alarmed by Galileo's somewhat open advocacy of the heliocentric theory, the theologians of Rome issued their famous injunction:[7]

> "The first proposition that the sun is the center and does not revolve about the earth, is foolish, absurd, false in theology, and heretical, because expressly contrary to Holy Scripture; and the second proposition that the earth is not the center but revolves about the sun, is absurd, false in philosophy, and from a theological point of view, at

---

[1] Geymonat, *Op. Cit*,. pp. 151-52.

[2] White, *Op. Cit.*, p. 139.

[3] G.H. Putnam, *The Censorship of the Church of Rome*, Vol. 1, pp. 128-29, p. 309 ff, Vol. 2, p. 323, pp. 455-59, New York, 1967; also White, *Op. Cit.*, p. 193.

[4] Pasquale m. D'Elia, *Galileo in China*, p. 13., tr. Suter R. and Sciascia M., Cambridge, 1960.

[5] *Ibid.*, pp. 15-16.

[6] *Ibid.*

[7] White, *Op. Cit.* p. 132.

least, opposed to the true faith."

As faithful servants of the Church, the Jesuits thus had no choice but to mold their views accordingly. Consequently, after the injunction of 1616 Jesuit astronomers completely disassociated themselves from Copernicanism. Having disassociated themselves from Copernicanism, however, a few of the Jesuits adopted the Tychonic System.[1] This system was not only modern but it preserved the geocentric aspect of Ptolemy.

After the final condemnation of Galileo in 1633, orders were sent to archbishops, bishops, nuncios, and Inquisitors of Italy requiring them to inform their parish priests of the sentence and recantation of Galileo. The order read:

> "That you and all professors of philosophy and mathematics may have knowledge of it, that they may know why we proceeded against the said Galileo, and recognize the gravity of his error, in order that they may avoid it, and thus not incur the penalties which they would have to suffer in case they fell into the same."[2]

The Papal order precipitated the final phase during which even the Tychonic system was given up, and strict adherence to Ptolemy prevailed.[3] Books written by the Jesuits after this period are all Ptolemaic. In the Encyclopedia of Mathematics of 1645, published in China under Jewish supervision, the author Adam von Bell advocates the Ptolemaic system.[4] Similarly, a treatise written in China by the Jesuit, Kögler, in 1738, a century after Galileo's final condemnation, was Ptolemaic. [5]

Strictly speaking, the papal injunctions did not apply against the telescope; the telescope could still be used. This explains why Jesuits such as Kögler and

---

[1] In the Tychonic world system, the sun revolves around a stationary earth, and the planets revolve around the sun.

[2] White, *Op. Cit.*, p. 139.

[3] For example, when the news of Galileo's trial reached in China, "a curtain descended and a return to the Ptolemaic view took place." Needham, pp. 445-46, *Op. Cit.*, Ch. 1.

[4] Bolesław Szcześniak, "Notes on the Penetration of the Copernican Theory into China," *J. Roy. Asiatic Soc.*, pp. 30-38, (1945).

[5] It will be misleading to assume, however, that the Catholic Church was the sole and the only organized religious group active in condemnation of the new astronomy. White writes, "Doubtless many will exclaim against the Roman Catholic Church for this, but the simple truth is that Protestantism was no less zealous against the new scientific doctrine. All branches of the Protestant Church—Lutheran, Calvinist, Anglican—vied with each other in denouncing the Copernican doctrine as contrary to the Scripture; and at a later period the Puritans showed the same tendency." Martin Luther denounced Copernicus in these words: "People gave ear to an upstart astrologer who strove to show that the earth revolves, not the heavens of the firmament, the sun and moon. Whoever wishes to appear clever must devise some new system, which of all systems is, of course, the very best. This fool wishes to reverse the entire science of astronomy; but sacred scripture tells us that Joshua commanded the sun to stand still and not the earth." See White, pp. 122-123, *Op. Cit.*

Antoine Gaubil in China, Boudier and Duchamp in India, and a number of others elsewhere, occupied themselves primarily with the observational aspects of neo-astronomy and shied away from its theoretical advancements.

Furthermore, according to Cardinal Bellermine, the foremost theologian of the 17th century, a person could use the Copernican theory as a mathematical device to solve astronomical problems, provided he did not uphold the theory as truth. This interpretation of the injunction explains why Kögler used the Newtonian method, a development of the Copernican theory, in his calculations and yet avoided the Keplerian system in his compendium written in 1738.[1]

One might suppose that in a land far removed from Catholic Europe, such as Portuguese India, the vigilance of the Inquisition might not have been so intense, and that it would have been safer to advance the discoveries of Copernicus, Kepler, Galileo, and Newton in India. But the facts show otherwise. The authorities of the Inquisition at Goa were even more vigilant and active than their counterparts in Europe.[2,3] When stories of the Goan Inquisition trickled back to Europe, they startled even those who were well aware of the horrors committed in the Iberian peninsula.[4]

As faithful servants of the Church, the Jesuits believed in the dictates of the Church with all their hearts. Their training, absolute discipline, and obedience to their superiors had left them incapable of thinking otherwise.[5] It is not surprising therefore that in the letters of the Jesuit fathers, from their missions in India, there is not even an isolated reference or comment to reflect that there was even a single Copernican among them.[6] It is noteworthy that Jesuits such as Moduit and de Lane, while arguing with the local Brahmins, criticize Hindu

---

[1] Gaubil, p. 646, *Op. Cit.*, Ch. 1. For the Jesuits' attempts in China to circumvent the heliocentric theory, see N. Sivin, "Copernicus in China," *Colloquia Copernicana II (Studia Copernicana VI)*, Warsaw, 1973.

[2] An Italian traveler, Gemelli Carreri writes, "Inquisition is much respected and dreaded by the Christians at Goa and about it." Gemelli Carreri, *The Collection of Voyages and Travels*, IV, p. 221, London, 1704, as quoted by Henry Heras, *The Conversion Policy of the Jesuits in India*, p. 11, Bombay, 1933.

[3] The Inquisition in Goa was finally abolished in 1812. See William Wilson Hunter, *The Indian Empire: Its People, History, and Products*, p. 304 ff, London, 1893; also de Oliveira, *Op. Cit.*, p. 352, and Lea, *Op. Cit.* Vol. 3, p. 310.

[4] In some of the *autos da fe* (trials) of the Goan Inquisition for which the figures are available, it is revealed that 4,046 persons were sentenced to various kinds of punishment. These punishments included 105 men and 16 women condemned to flames, of whom 57 were actually burned alive and 64 burned in effigy. See W. W. Hunter (1893), *Op.Cit.*, p. 306.

[5] Ignatius, founder of The Society of Jesus, desired "blind" obedience from his members. He wrote: "(members) must endeavor to be resigned interiorly conforming their will and judgement wholly to the superior's will and judgement in which no sin is perceptible." *Encyclopedia of Science and Religion*, *Op. Cit.*, pp. 500 ff.

[6] For example see: *Lettres*, Also Stöcklein, Nos. 643, 644, 645, *Op. Cit.*, Ch. 6. Also see Goubil, *Correspondence de Pekin*, *Op. Cit.*, Ch. 1.

astronomy but do not assail its geocentric feature.[1]

Jesuits and Jai Singh

From the correspondence of the Jesuit missionaries in India at that time, and from the Jaipur State Archives, a number of European names associated with Jai Singh's astronomical enterprises, have come to light. Table 13-1 gives a brief summary of these people.

Year	Name	Nationality	Vocation
1727–	Manuel de Figueredo[2,3]	Portuguese Jesuit	Priest, led a delegation to Europe
1730–	Joseph du Bois[4]	French Jesuit	Physician
1730–	Pedro de Silva[5]	Portuguese	Physician, Astronomer (settled in Jaipur)
1734–36	Fr. Pons[6]	French Jesuit	Priest, Astronomer
1734–36	Claude Boudier[7]	French Jesuit	Priest, Astronomer
1740–43	Anton Gabelsberger[8,9]	Bavarian Jesuit	Priest, Astronomer
1740–43	Andreas Strobl[10]	Bavarian Jesuit	Priest, Astronomer

Table 13-1 Europeans in the service of Jai Singh

As the note of Muḥammad Shāh, quoted earlier, indicates, there were

[1] Lettre du P. Mauduit au P. Le Gobien, Carouvepondi, 1st Jan. 1702, in *Lettres*, pp. 307-308; and Lettre du P. de La Lane au P. Mourgues, Pondichery, 30 Jan. 1709, also in *Lettres*, pp. 401-02.

[2] "Observations . . . ," *Lettres*, pp. 772-78.

[3] Figueredo appears to have been primarily a liaison between Jai Singh and the Portuguese authorities of Goa and Lisbon.

[4] Du Bois, *Op. Cit.*, Ch. 11.

[5] Hunter reports that de Silva died in the 1780's and that he was survived by a son living at Jaipur while Hunter was visiting the city. See Hunter, p. 210. De Silva's descendants still live in Jaipur.

[6] Lettre du P. Pons au P. du Halde, 23, November, 1740, *Lettres*, pp. 645, *Op. Cit.*; and *Lettres*, pp. 772-778, *Op. Cit.*

[7] *Ibid.*

[8] Stöcklein, No. 643, *Op. Cit.*,

[9] Jean Bernoulli, *Description Historique et Geographique de l'Inde*, Tome I, p. 5, Berlin, 1786.

[10] *Ibid*, Stöcklein, Nos. 643-646, *Op. Cit.*,

Europeans knowledgeable in astronomy in the service of Jai Singh, but little is known about them. Their names may very well be buried in the archives of Jaipur or Goa, or lost forever, along with the other Jesuit documents which disappeared during a shipment from Goa to Portugal.[1]

The *Dastūr Kaumvār* records of Rajasthan State Archives, list a number of gifts and presents given away to Europeans or *Firaṅgīs*.[2] Some of the recipients of these gifts, can be readily identified with the names in Table 13-1, and the gifts they received were clearly for services related to the king's astronomical undertakings. In the same records, quite a few recipients are identified merely as *Padrīs* or Fathers.

The following is a sample from the records: "Padre Vaidiyā (?) was asked to come over from Agra and was paid Rs.100 for expenses, . . ., via Kevalarāma." Kevalarāma, as discussed earlier, was one of the prominent astronomers of the Raja. According to another entry, Father Vaidiyā was called from Delhi and paid Rs.1,100 for expenses. He was kept on the payroll even when he was away from Jaipur and visiting his country, and he was paid in 1733, a sum of Rs.2,095 as wages and for various expenses he incurred.[3] Another entry in the Archives says, "Father Francis Crūsa (?) came from Delhi, and (thus) was paid Rs.100 for his expenses."

In addition, Father Joseph Tieffenthaler (d. 1785), a Jesuit from the Austrian Tyrol,[4] arrived from Europe in 1743, supposedly to partake in the astronomical program of the Raja. Owing to the untimely death of the Raja a few months earlier, however, he was unable to do so. However, there is no definite proof that Tieffenthaler's arrival in India was connected with the Raja's astronomical program.

Delegation To Europe

In 1727 Jai Singh dispatched a scientific delegation to Europe. In the preface to the *Zīj-i Muḥammad Shāhī*, he says, "After seven years had been spent in this effort (observing the stars), information was received that observatories had been built in Europe, . . ., and that the business of the observatory was still being carried on there."[5]

The delegation, the first of its kind from the East, left Amber in 1727. The

---

[1] Correia-Afonso, p. 134, *Op. Cit*.

[2] In the *Dastūr Kaumvār* records of the R.S.A., the Europeans are listed under the caste *Firaṅgī* or under the caste *Musalamān*. Another source of information is the file of *Daftar Nuskhā Puṇya* of the Jaipur State at the R.S.A.

[3] R.S.A., *Daftar Nuskhā Puṇya*. Father Vaidiyā may not be identified with Boudier, because Boudier did not reach Jaipur before July 1731. However, he could be du Bois.

[4] Edward Maclagan, *The Jesuit and the Great Mogul*, p. 137, London, 1932. Also see, Noti, *Op. Cit*. p. 146.

[5] ZMS, f. 0.

Raja saw to it that the delegation was taken good care of on its way to the seaport and wrote letters to this effect to his friends. The Mahārāṇā of Udaipur confirms Jai Singh's request for assuring safe passage of the delegation through his domain.[1] The delegation traveled to Goa, paid a courtesy visit to the Portuguese viceroy, delivered presents to him, and then finally reached Portugal in January 1729.[2] A dispatch from Lisbon dated January 20, 1729, appearing in a Paris newspaper, reads:

> "Father Manuel de Figueredo of the Company of Jesus has arrived on the vessel Vaisseau . . . from East Indies with two gentlemen from the court of the great Mughal with presents to the king (of Portugal). They have visited the Viceroy of the Indes at Goa and presented him with a *siropā* or a habit in the mode of their country, and a handful of diamonds and rubies."[3]

The delegation was led by Father Figueredo, the rector of the college of Agra,[4] and it had two additional members. The *Gazeta De Lisboa Occidental*, in its issue of March 10, 1729, reports:

> "Around the end of the month His Majesty gave private audience to Father Manuel de Figueredo of the Society of Jesus, a missionary to the court of Agra. (He) turned over to the King the letters and the gifts from the king of Amber, Sawai Jai Singh. . . . He also brought along with him Pedro Jī, a Catholic and a Mughal by birth; (and) Sheikh Jī, a Mohammeden." [5]

The full name of the "Sheikh Jī" mentioned above, it appears, was Sheikh Asad'ullah Nujūmī.[6] According to the *Dastūr Kaumvār*, the Sheikh was given a variety of gifts in 1726 by the Raja, a few months before the delegation left

---

[1] R.S.A., A letter from Mahārāṇa Saṅgrāma Singh to Jai Singh, *Pauṣa Śukla* 9, V.S. 1784 (1727 A.D.)

[2] *Gazeta de Lisboa Occidental*, p. 24, Jan. 20, 1729.

[3] *Recveil De Touves Les Gazettes Nouvelles, Ordinaires and Extraordinaires & Autres Relations*, Paris, p. 88, 1729. The parentheses in the quotation are by the author. Also *Ibid*.

[4] Sommervogel as quoted by Hosten: Sommervogel, *Bibl. Cottoniana, Vespasianus F. 7, XII, Catalogo dos. MSS. Portuguezes existentes no Museu Britannico*, per Ferd. Fr. de La Figaniere. Lisbon, 1853. Vol. IX, Col. 337. See H. Hosten, *Jesuit Missions in North India and Inscription on Their Tombs*, p. 37, Calcutta, 1927. The so-called Jesuit college at Agra run for the education of Catholic children was more like a junior high school of modern times. Figueredo might have come in contact with Jai Singh when Jai Singh was the governor of the province of Agra around 1725.

[5] *Gazeta de Lisboa*, p. 80, March 10, 1729.

[6] DK, Vol. 18, p. 540.

for Europe.[1]

The *Gazeta de Lisboa* of March 10 adds that the delegation had come to resolve questions regarding the astronomical tables used in Portugal and in India and to acquire knowledge about old and new instruments of astronomical observation.

Although news of the delegation's arrival in Lisbon was duly published in the Paris Gazette, there is no mention in the Gazette of the delegation ever going to Paris. The London papers are also silent about the delegation visiting their city. Apparently, the delegation did not visit these two places where the most advanced work in astronomy had been done and was still continuing. The author also requested a check of the records of Coimbra University of Portugal, but found no references regarding the delegation's visit there either. Coimbra University was the only seat of higher learning in Portugal at the time. It appears the delegation did not consult the scholars of astronomy there either.

The delegation stayed on in Portugal for a few months and finally returned to India with instruments, books, and astronomical tables, including the tables of de La Hire published in 1702. The delegation landed at Goa in October or November 1730 and finally reached Jaipur in July 1731.[2,3] Figueredo was reinstated on the payroll beginning July 14, 1731 at daily wages of Rs.2.50, and most likely that is the date when he and the other members of the delegation reached Jaipur.[4] Pedro de Silva, a physician and an amateur astronomer, and his son Xavier, also came with the delegation with the intention of settling down in the country.[5] According to the Jaipur State Archives, Figueredo was profusely rewarded for his labors after his return with a purse of Rs.1,000.[6]

The delegation brought back with it European tables and instruments, but it did not bring back any information regarding the theoretical advances of astronomy in Europe. For example, it did not bring any books of Newton, Galileo, Kepler, or Copernicus.[7] At the Sawai Man Singh II Museum of Jaipur, where most of the books of Jai Singh are preserved, there is not a single book by these

---

[1] *Ibid*.

[2] A. B. de Bragança Pereira, *Archivo Portugues Oriental,* Tome I, Vol. III, Pt. IV, No. 83, p. 181, Bastora: Rangel, 1940. The letter is dated Nov. 4, 1730.

[3] From the state treasury of Jaipur a demand-draft for Rs.500 was sent to Surat for the expenses of Figueredo. The date on the order is March 20, 1731. DK, Vol. 20.

[4] In the R.S.A.'s *Daftar Nuskhā Puṇya Tozis*, Manuel Figueredo has been written as "Menvela Pirerā."

[5] One of de Silva's descendants, perhaps his grandson, became the head of the department of artillery during the reign of Jai Singh's grandson, Pratāpa Singh, and he was involved in casting cannons for the army. Records of the years 1792-1794 bear his seal. R.S.A., Jaipur State, *Kārkhānā Topān*.

[6] Entry dated *Bhādrapada Kṛṣṇa*, 7, 1788 V.S. (Sept. 14, 1731), *Tozīs*, *Daftar Nuskhā Puṇya*, Jaipur State, R.S.A.

[7] A close scrutiny of the inventory of Jai Singh's library suggests that the delegation did not bring any advanced scientific instruments such as a quadrant or a sextant of the type used at the Paris or Greenwich observatories. See Appendix VI.

authors.

The reason is not difficult to understand. The relentless war waged by the Church in the aftermath of Galileo's trial had taken its frightful toll in Catholic Europe. By the time Jai Singh's mission under Fr. Figueredo arrived in Europe, there was not a single scholar left in the Iberian Peninsula, particularly in Portugal, who could speak about Kepler or Newton. In Portugal, where King John V reigned, the Holy Office of Inquisition had cleansed the institutions of higher learning of all Copernicans decades before. The king himself was one of the most intolerant monarchs in Portugal's history. As a matter of fact, the "Most Faithful King" had presided over one of the most vigorous periods of the Inquisition in the history of his country.[1,2]

Because the mission under Fr. Figueredo never stepped outside the Catholic world, so far as neo-astronomy was concerned, it produced very little of substance.[3] Unfortunately, the Raja seemed to have had no alternative but to rely on the Jesuits to lead his fact-finding mission to Europe because no high-born Hindu in those days would dare "to cross the ocean" and thereby risk "losing his caste."

Joseph du Bois

Joseph du Bois, a French Jesuit with some knowledge of astronomy who was in the service of the Raja, writes that the tables brought from Portugal were received with great jubilation at Jaipur and that he was the one entrusted with their translations and was paid a sum of Rs.500 for his labors.[4] He further

---

[1] The title of the "Most Faithful King" was conferred upon John V, the king of Portugal, by Pope Benedict XIV.

[2] Lea, p. 310, *Op. Cit.*

[3] The tables of Flamsteed in the Jaipur palace library are later acquisitions.

[4] See du Bois, *Op. Cit.*, Ch. 3. The following is an extract from du Bois's introduction to the manuscript:

> ". . . A certain noble (Gentile) ruler called *Saway Yassang* [Sawai Jai Singh], master of astronomers, like Alphonsus Castella (*sic*) in the deserts of Africa. [Jai Singh] spent for the maintenance of astronomers 400,000 gold pieces including a monthly sum, over a period of six years, of Rs.4,000 [per month], valued at 1000 gold Venuses, or Dutch pieces (Ungeris). As time passed, under his clever supervision, he increased facilities in different cities of his empire with magnificent observatories furnished with big machines, and if I may say so, made with his own hands, as God is a witness, and as I have seen, not once, nor twice, but many times. . . . As a very graphic example, he used the form of a right angle in a triangle, divided in the middle with an equinoctial line [arc] and twice divided, giving three parts, and further divisible (indefinitely). Once the sketches were complete, he himself made the prototype with his own hands, from modeling wax, then [constructed them] as monuments (even for posterity). All these structures were carefully supervised by him personally, one to the approximate height of 73 Roman feet, another was made as an astrolabe of some 12 yards or 36 Roman feet, another was a plain sphere, and I know as I write these things that he is just beginning another 108 feet in similar fashion and many others to follow."

adds that the Raja advised his astronomers to use European tables from then on. The jubilations were short-lived, however, as discrepancies were noted between the translated text and the observed facts, particularly regarding the moon, as Jai Singh elaborates in the preface to the *Zīj-i Muḥammad Shāhī*. In the *Zīj*, Jai Singh writes:

> "On examining and comparing the calculations of these tables (European tables), with actual observation, it appeared there was an error in the former, in assigning the moon's place, of half a degree: although the error in the other planets was not so great, yet the times of solar and lunar eclipses he found to come out later or earlier than the truth, by the fourth part of a *ghaṭikā* or 15 *palas*."[1]

Because du Bois and the other Europeans at Jaipur were unable to account for the discrepancies, Claude Boudier, an astronomer priest stationed at Chandernagore in West Bengal, was recommended to the Raja.

Claude Boudier

At Chandernagore, Fr. Claude Boudier, a French Jesuit, headed a small mission. The Jesuits at Chandernagore under the guidance of Boudier had been collecting astronomical data for some time.[2] Around 1731-32, Jai Singh addressed a set of five questions to the Jesuits of Chandernagore which were reproduced by Fr. Calmette in a letter sent to Europe in January 1733.[3] We

---

Next, du Bois talks about Jai Singh acquiring tables from Europe. His narration is rearranged in the following for clarity.

> "A certain priest of the Society of Jesus, by birth a Portuguese, the Rector of the College of Agra in that same empire, had been commissioned by (Jai Singh) to go to Europe in search of an expert astronomer. That priest went and returned to Sawai (Jaipur) with those tables mentioned [already], and, along with other mathematical instruments, he brought a symbolic gift from the king of Portugal: a certain honorable man, educated by (Jesuit) priests, a very talented man, called Pedro de Silva, nicknamed Praditay. Born in Portugal and under the care of Father John Baptist Carvone at Rio Cleris. He [Jai Singh] gave a major boost to astronomy in India. Meanwhile the Ruler Jai Singh, with great jubilation at obtaining the tables, ordered that those tables be transcribed in his registers at Sawai [Jaipur] and also ordered that henceforth all his astronomers should use them in future calculations."

[1] ZMS, f. 3.

[2] Boudier and his colleagues began collecting astronomical data sometime after 1730. Although they were quite busy in the beginning, their activities tapered off within a decade due to lack of satisfactory equipment and funds. Boudier relayed his data to Europe and thus kept his superiors at Paris informed. He also sent copies of his data to Gaubil in China. See *Fonds Brotier*, Vol. 78, f. 1 ff., *Op. Cit.*, Ch. 5; Gaubil, p. 655, *Op. Cit.*, Ch. 1. For Boudier's observations at Chandernagore see V.N. Sharma, L. Huberty (1984), *Op. Cit.*, Ch. 5.

[3] Lettre du P. Calmette, à M. de Cartigny, Vencatiguiry, 24 January, 1733, *Lettres*, p. 610.

reproduce the questions as follows:

1. What is the reason for the difference in the observed-longitude of the moon and the calculations made for it according to the tables of de La Hire, which (Jai Singh) has had translated. This difference is nearly one degree; however, the instruments which were employed (by Jai Singh) are large and exact, and the observations were made with the necessary care. Is this difference also found at the meridian of Paris?
2. Are there any tables that give the moon (its coordinates) exactly, in agreement with observations? If there are, who is their author, and what hypothesis does he use?
3. What hypotheses did M. de La Hire follow, and according to what geometrical model did he construct his tables of lunar movements?
4. In Europe, what method is used to observe the moon in its off-meridian position, and with what instruments?
5. On what basis did M. de La Hire establish his third equation of the moon; further, how would one reduce this to a hypothesis so as to calculate it geometrically?

The Raja and Boudier

Realizing an excellent opportunity for gaining favor with a powerful ruler, Boudier and his associate Pons set out for Jaipur on January 6, 1734, carrying with them a 5 m long telescope, a 0.6 m quadrant, and possibly some other equipment. Availing themselves of the opportunity their long journey provided them with, they measured longitudes and latitudes of the towns on the way.[1] They also recorded the occultation of Jovian satellites, the meridian altitude of a number of stars in the same general period, and observed the solar eclipse of May 3, 1734, at Delhi. Because the observed hour of this eclipse did not agree with de La Hire's tables, Boudier was able to confirm the errors in the tables the Raja had found earlier. He also took some data at the observatory of Jaipur.

Boudier had several discourses with the Raja, whom he calls "the most capable astronomer among the Indians." He states that the Raja, among other things, was interested in the mathematical predictions of eclipse phenomena and the occultation of stars by the moon.[2] Boudier returned to his mission two years later in 1736.[3]

---

[1] "Observations ... ," *Lettres*, p. 772 ff.

[2] *Fonds Brotier*, Vol. 88, ff. 126-135, *Op. Cit.*, Ch. 5.

[3] At Amber, Boudier met Fr. du Halde. It is not sure if Halde had also been in the service of the Raja. *Fonds Brotier*, Vol. 88, f. 126, *Op. Cit.*, Ch. 5.

Anton Gabelsberger and Andreas Strobl

Because Jai Singh's interest at the time had been in mathematical astronomy, he arranged for two Jesuit mathematicians, Anton Gabelsberger (1704-1741) and Andreas Strobl (1703-1752),[1] to come to his court from Europe. The authorities of Portuguese Goa paid the role of intermediary.[2] Strobl and Gabelsberger landed at Goa on September 30, 1737. Jai Singh sent a Brahmin to escort them to Jaipur,[3] and after a delay of more than two years, due to the invasion of India by Nādir Shāh, the two Jesuits finally reached Jaipur in 1740.[4] Strobl claims to have satisfied the Raja with his computations but does not elaborate as to what the computations were for. One may reasonably assume, however, that they were related to lunar astronomy because Jai Singh's main interest at the time was lunar astronomy.

Strobl apparently impressed Jai Singh with his calculations to the extent that the Raja decided to house European mathematicians in each of his observatories. Strobl writes, "Shortly before his death, (Jai Singh) had decided to request from Rome, from our reverend General Father, European priests knowledgeable in mathematics whom he would then appoint head of the observatories that were to be established throughout the realm."[5]

In the same letter, Strobl makes a curious observation. He writes that Jai Singh had worked on a perpetual motion machine, and had spent a sizable sum, equivalent to 50,000 guilders, on it. Strobl in another letter writes that Jai Singh was contemplating sending a second mission to Europe under Strobl, this time to the "Roman Emperor" and to the Pope to establish friendly relations.[6] Preparations were being made and presents being readied, but due to a war having broken out and due to the untimely death of Strobl's colleague Gabelsberger in 1741, the journey was at first postponed and then never accomplished.

Other Contributions of the Jesuits

The other contributions of the Jesuits include helping with the translations of European texts. Du Bois translated portions of the *Tabulae Astronomicae* of de La Hire. He also copied the Latin text of the *Spherics* of Theodosius in 1732 at Jaipur.[7] Kevalarāma's *Dṛkpakṣa Sāraṇī*, which is a Sanskrit translation of

---

[1] For Strobl's date of birth refer to Anton Huonder S.J., *Deutsche Jesuitenmissionare des 17 und 18 Jahrhunderts*, Frieburg, 1899.

[2] De Bragança Pereira, *Op. Cit.*, Ref. 100.

[3] De Bragança Pereira, Tome I, Vol. III, pt. V, No. 105, p. 213, *Op. Cit.*

[4] George M. Moraes, "Astronomical Missions to the Court of Jaipur, 1730-1743," *J. Bombay Roy. Asiatic Soc.*, Vol. 27, pp. 61-65, 85, (1951).

[5] Stöcklein, No. 644, *Op. Cit.*, Ch. 6.

[6] Stöcklein, No. 643, *Op. Cit*, Ch. 6.

[7] Bahura II, p. 55.

the *Tabulae Astronomicae* of de La Hire, must have also been accomplished with the help of du Bois or some other Jesuit.[1] Du Bois also copied the manuscript of the *Tabulae Astronomicae*, now in the Sawai Man Singh II Museum, Jaipur. Similarly, *Vibhāga Sāraṇī*, *Layyara Vedhapatrāṇī*, *Jīvāchāyā Sāraṇī* and *Pratibimba Siddhānta* were also translated with the help of the Europeans. Hunter saw several of these works with a grandson of Kevalarāma, whom he met at Ujjain around 1790.[2]

The Jesuits and Europeans also procured many books for the royal library. In addition to the books and instruments brought by the delegation to Europe, a number of books were added to the royal library in later years. According to the Rajasthan State Archives, a three-volume set, most likely the tables of Flamsteed,[3] was purchased at a cost of Rs.200 in 1733. Similarly, globes, a Ptolemaic model of the world system, and maps of Europeans received from Surat might also have been acquired with the Jesuits' help.[4]

Assessment

Despite close contact with European astronomers lasting for decades and with enormous sums of money invested on a delegation to Europe, Jai Singh's endeavors do not reflect any influence of the European advances of the preceding century, particularly the influence of the Copernican revolution. When Jai Singh consulted the Jesuit scholars about the status of astronomy in Europe, it is unlikely that they ever discussed the Copernican revolution in its true perspective, if at all. To them it was heresy, and thus to be shunned at all cost.[5]

In retrospect, it appears that the Raja relied too much upon and was clearly influenced by his Jesuit assistants who erected an invisible shield between him and the so-called heretical neo-astronomy they avoided. To the end of his days, Jai Singh trusted the expertise of these faithful servants of the Church and made no attempts to consult anyone else from a country such as England or Holland, where the Copernican revolution had gained a solid footing. European involvement with the astronomical program of Jai Singh produced mixed results. On one hand, the Europeans introduced Indian astronomers to improved methods of calculating astronomical problems, particularly in the use of the logarithm.

---

[1] See Ch. 12.

[2] Hunter, p. 209.

[3] Flamsteed collected data on the moon, the sun, the planets, and on some 3000 stars with an accuracy of 10". His data was published first in 1712 and finally posthumously in 1725 in three volumes under the title of *Historiae Coelestis Britannica*.

[4] SSRS, p. 1165, and SSMC, p. 41. A model of the Ptolemaic world system is still preserved at the Jaipur Jantar Mantar.

[5] Moraes's contention that Father Figueredo referred Jai Singh to the discoveries of Copernicus, Kepler, and Newton is not plausible considering the background of the Jesuits. See Moraes, *Op. Cit.*

On the other hand, these Europeans held back from Indian astronomers the theoretical advances made by Copernicus, Galileo, Kepler, and Newton.

CHAPTER XIV

CONCLUSION

Jai Singh initiated his program in astronomy with the object of revising the existing astronomical tables. Toward this objective, he constructed observatories in different cities of north India. He made the observatories of Delhi and Jaipur his show cases and equipped them with his highest precision instruments. He built the largest number of instruments for the observatory of Jaipur, which was near his principal residence.

Jai Singh relied primarily on his masonry instruments. His instruments were graceful in design and aesthetically pleasing to the eye. They were functional and without adornments and unnecessary decorations. In this regard they were surprisingly modern. The surviving specimens of his instruments are excellent examples of the art of constructing large instruments in masonry and stone that became obsolete with the arrival of the telescope.

Instruments

Jai Singh constructed 15 different types of instruments of masonry for his observatories. Of these 15, he himself invented seven instruments: the Jaya Prakāśa, two Kapāla yantras, Rāma yantra, Digaṁśa, Samrāṭ yantra, and Rāśivalayas.[1] Jai Singh made prototypes of these instruments from modeling wax with his own hands and had them constructed under his own personal supervision.[2]

The design of Jai Singh's instruments varies from the very simple to the highly complex. The Dhruvadarśaka Paṭṭikā is his simplest instrument and Jaya Prakāśa the most complex. He designed his Dhruvadarśaka Paṭṭikā for the layman and the Jaya Prakāśa for the professional astronomer. Just as Jai Singh's instruments vary in complexity, they also vary in precision of their scales. In the next three tables, we compare the precision of the instruments.

---

[1] Jai Singh did not invent the Palabhā, Agrā, Śaṅku, Nāḍīvalaya, Dakṣinottara Bhitti, and Ṣaṣṭhāṁśa. These instruments had been known for some time.

[2] Joseph du Bois, *Op. Cit.*, Ch. 13, footnote.

	Instrument	Number	Precision in Time	Precision in Angle
1	Dhruvadarśaka Paṭṭikā (North Star indicator)	1	none	none
2	Nāḍivalaya (Equinoctial dial)	5	1 min	15′
3	Palabhā (Horizontal dial)	2	5 min	–
4	Agrā (Amplitude instrument)	5	–	1°
5	Śaṅku (Horizontal dial)	1	–	6°
6	Śara yantra (Celestial latitude dial)[1]	1	–	unknown
7	Unknown Instrument[2]	1	–	–

Table 14-1 Low-precision instruments of Jai Singh

	Instrument	Number	Precision in Time	Precision in Angle
1	Jaya Prakāśa (Hemispherical inst.)	2	1/2 min (at best)	3′ (at best)
2	Rāma yantra (Cylindrical inst.)	2	–	3′ (at best)
3	Rāśivalaya (Ecliptic dial)	12	–	10′
4	Digaṁśa (Azimuth circle)	3	–	6′
5	Kapāla (Hemispherical dial)	2	–	3′ (at best)

Table 14-2 Medium-precision instruments of Jai Singh

---

[1] Little is known about the Śara yantra of Jai Singh. The incomplete Krāntivṛtta of Jaipur could be remains of this instrument. See Ch. 7.

[2] The unknown instrument has been described by Bāpūdeva Śāstrī, pp. 21-28, *Op. Cit.*, Ch. 8.

	Instrument	Number	Precision in Time	Precision in Angle
1	Samrāṭ yantra (Equinoctial sundial)	6	2 sec - 1 min	3″ - 10′
2	Ṣaṣṭhāṁśa (60 deg meridian chamber)	5	–	1′
3	Dakṣinottara Bhitti (Meridian dial)	6	–	2′ - 10′

Table 14-3 High-precision instruments of Jai Singh

In the early stages of his data gathering, Jai Singh relied on the Jaya Prakāśa and the Rāma yantra, as Jagannātha writes.[1] But the Jaya Prakāśa and Rāma yantra are not his highest-precision instruments, as Tables 14-1,2,3 indicate. Besides, their precision depends on the location of the object in the sky. Apparently, Jai Singh realized these limitations and did not build these instruments at places other than Delhi and Jaipur. The Great Samrāṭ of Jaipur, the Samrāṭ of Delhi, the Ṣaṣṭhāṁśas, and Dakṣintottara Bhittis of Jaipur, Delhi, and Ujjain were Jai Singh's highest-precision instruments. With these instruments, Jai Singh extended the precision of measuring angles to the very limit of naked-eye observations, i.e., 1′ of arc. Lime plaster can be etched with divisions as small as 2.0 mm wide and marble with divisions of 2 to 3 mm wide. Keeping this fact in mind, Jai Singh apparently planned the size of his high-precision instruments.[2]

Samrāṭ Yantra

The division scheme of the gnomon scale of the Great Samrāṭ of Jaipur suggests that Jai Singh intended to measure declination to the nearest arc-minute. As our measurements show, the Great Samrāṭ of Jaipur can still achieve this precision. The small divisions on the quadrant scales of the Great Samrāṭ suggest that Jai Singh intended to measure time down to the nearest 5 *vipalas* (2 sec). The accuracy of time measurements at a Samrāṭ is affected by the penumbra of the gnomon's shadow. It is possible that Jai Singh recognized this limitation and applied a correction. With the correction applied, the Great Samrāṭ and its sister unit at Delhi, which has similar dimensions, can measure time within 2 to 3 seconds of its true value. The Great Samrāṭ of Jaipur and its

---

[1] SSRS, p. 1163; and SSMC, pp. 38, 39.

[2] For example, a 90° arc with 2 mm wide small divisions reading 1 arc-minute would have a radius of 6.88 m and would be 10.8 m long.

sister unit of Delhi were, therefore, optimally designed for both time and declination measurements.[1]

Although Jai Singh designed the Great Samrāṭ of Jaipur and its sister unit to achieve the highest precision possible for naked-eye observations, he might not have realized this precision in practice. We have analyzed the data collected with the Samrāṭ of Delhi. This data is given in the *Yantraprakāra*. The *Yantraprakāra* reports that on the night of April 18, 1729, the observed right ascension of Jupiter was 91;13. This result is 6 arc-minutes higher than our computer-generated results. Similarly, the observed declination of Jupiter, according to the *Yantraprakāra*, at the same time was 23;36. This result is higher by 3 arc-minutes according to our calculated results.[2] Evidently, the Samrāṭ of Delhi suffered from one or more defects elaborated by us in Ch. 3 and did not achieve the precision inherent in its design.

Ṣaṣṭhāṁśa

Jai Singh's other high-precision instrument, the Ṣaṣṭhāṁśa, was both well-designed and well-constructed. Jai Singh built his Ṣaṣṭhāṁśas only at Delhi and Jaipur. That he built four units of this instrument at Jaipur shows his high regard for it. The Ṣaṣṭhāṁśas of both Delhi and Jaipur had the same degree of precision. They could measure to the nearest arc-minute. We believe that in these instruments, the two main sources of uncertainties, namely, the aperture broadening and the diffraction broadening, were less than 1 arc-minute.[3] As a result, the data collected with the Ṣaṣṭhāṁśa of Delhi had an accuracy of less than 1 arc-minute. Jai Singh's Ṣaṣṭhāṁśa at Jaipur still maintains the accuracy of 1 arc-minute, as our measurements show.

Dakṣinottara Bhitti

Dakṣinottara Bhitti is another of Jai Singh's precision instruments. With the Dakṣinottara Bhitti, Jai Singh measured the obliquity of the ecliptic within 28 arc-seconds of the true value calculated by us.[4] Therefore, his Dakṣinottara Bhitti was also well-constructed. With the Samrāṭs, Ṣaṣṭhāṁśas and Dakṣinottara Bhittis, the accuracy of Jai Singh's measurements for the sun approached that of de La Hire's, who had the benefit of a quadrant fitted with a telescopic-sight for his observations. Jai Singh measured the mean motion of

---

[1] A Samrāṭ larger than the Great Samrāṭ of Jaipur will not improve the accuracy of declination measurements any further, because this accuracy is limited by the resolving power of the eye. A smaller instrument, on the other hand, cannot read the declination angle down to 1′ of arc.

[2] See Ch. 5.

[3] *Ibid.*

[4] *Ibid.*

the sun as 349;16,49,46 per Arabic year. This parameter, when calculated according to de La Hire, has a value of 349;16,49,51 per Arabic year.[1] Jai Singh's parameters for the other planets are not so impressive, as Table 11-4 of Chapter 11 suggests. His results are considerably off for the planets Jupiter and Saturn, which require long-term observations. Jai Singh based his parameters on less than 10 years of observing, which is insufficient for calculating accurately the parameters of planets such as Jupiter and Saturn, whose sidereal periods are nearly 12 and 30 years.

Jai Singh's Attempts to Rejuvenate Astronomy in India

As the years went by, Jai Singh considerably widened his initial aim of revising the existing planetary tables. He tried to make the astronomical instruments of observation readily available to "every person devoted to these [astronomical] studies, whenever he wished to ascertain the place of a star . . ."[2] He built five observatories. He probably would have made many more, one in every major city of his realm, according to Strobl, had he not passed away at an early age.[3] He advised future kings to facilitate astronomical observations in their realm by constructing astronomical instruments.[4]

Jai Singh had texts on mathematics and astronomy translated into Sanskrit from Latin, Portuguese, Arabic, and Persian. Sanskrit was the language of the Hindu intellectuals at the time. He compiled an excellent library on subjects of astronomy and mathematics by acquiring books from far and wide. His library had books published in France, England, Portugal, and Germany. He patronized scholars of different schools of astronomy such as the Hindu, the Persian-Arab, and the European. To sum up, he tried to rejuvenate astronomy in his country.

Jai Singh sent a scientific fact-finding mission to Europe. The fact-finding mission was the very first of its kind from the East to Europe. He wanted to establish a scientific and cultural bridge between Europe and India and to import the scientific knowledge of the West. Strobl writes, ". . . the king had in mind to send me [Strobl] . . ., as he himself expressed it several times, to Europe, to the Roman emperor, and to the Pope with presents so as to establish friendship with these leaders of the world . . ."[5]

---

[1] See Ch. 11.

[2] ZMS, f. 3r.

[3] Stöcklein, *Neuve Weltbott*, No. 644, *Op. Cit.*, Ch. 6.

[4] SSRS., p. 1165; and SSMC, p. 41.

[5] Stöcklein, *Neuve Weltbott*, No. 643, *Op. Cit.*, Ch. 13.

Jai Singh Enigma

Jai Singh's career has been described as an enigma.In spite of his close contacts with Europeans, Jai Singh's endeavors reflect little or no influence of the astronomy of contemporary Europe. His instruments do not exploit refinements such as the telescopic-sight, the micrometer, or the vernier. On the theoretical side, Jai Singh seems unaware of Kepler and Newton. A question is often raised: why did Jai Singh, an enlightened scholar, a man far ahead of his times, remain ignorant or ignore the neo-astronomy of Europe, and instead attempt "to revive the spirit of Ulugh Beg at a time that seems – in retrospect – to have been a century too late?"[1] Did he deliberately ignore the expertise of his European advisors and go ahead with the outmoded instruments of his own design? Did he feel uncomfortable with the heliocentric concept as the contemporaries of Galileo had felt before and thus fail to appreciate the true nature of the theoretical advances led by Copernicus, Kepler, and Newton? Was it because he failed to appreciate the theoretical advances that he requested a Sanskrit translation of an already discredited text, the *Almagest* of Ptolemy, and not the *Revolutionibus* of Copernicus nor the *Principia* of Newton, both of which had been in existence for some time?

Jai Singh and the Telescope

Jai Singh's instruments are non-telescopic, although he had at least one telescope in his possession. He bought this telescope at a cost of Rs. 100 as the inventory of his library shows.[2] When Jai Singh embarked upon his astronomical program, telescopes were being built in the country as he himself points out. Jai Singh was aware of the capabilities of the telescope to some extent, and in his *Zīj-i Muḥammad Shāhī* he mentions using it.[3,4]

Then why are his instruments non-telescopic in nature? In order to explore this question a bit deeper, a distinction ought to be made between a general purpose telescope and a telescopic-sight. A telescopic-sight is a modified telescope with a cross hair at its prime focus. The cross hair enables an observer to align his instrument toward the object with precision and thus helps

---

[1] Blanpied, *Op. Cit.*, Ch. 3.

[2] R.S.A., *Tozīs* of Pothikhana, V.S. 1798-1800. The entry says, "Telescope, cost Rs.100, obtained from Asadu'llah, *Bisātī*."

[3] See Ch. 11.

[4] A number of authors in addition to the present author have discussed the subject of Jai Singh and the Telescope. For example, see:

1. S.A. Khan Ghori, pp. 50-57, *Op. Cit.*, Ch. 13.
2. Eric G. Forbes, "The European Astronomical Tradition: Its Transmission into India, and its Reception by Sawai Jai Singh II," *Indian J. Hist. Sci.*, **17**, pp. 234-243, (1982).
3. S.M.R. Ansari, *Introduction of Modern Western Astronomy in India During 18-19th Centuries*, pp. 3-6, New Delhi, 1985.

measure an angle with greater accuracy. Although the telescope had become increasingly common with European astronomers following Galileo, the telescopic-sight came in much later. Jean Picard introduced it first into astronomy in 1667.[1] However, even with telescopic-sights the accuracy of the instruments of the 1670's were no better than 1′ of arc, a limit easily matched by accomplished observers such as Hevelius with their non-telescopic instruments.[2] Consequently, acceptance of the telescopic-sight was slow. Eventually, when Flamsteed achieved an unprecedented accuracy of 10″ in his star catalogs of 1712 and 1725, the telescopic-sight became a regular tool in the hands of European astronomers.[3]

In the astronomical endeavors of Jai Singh, therefore, it is not so much the telescope, but the telescopic-sight that is missing. His instruments do not incorporate the telescopic-sight. His disregard of the telescopic-sight, however, is not intentional but rooted in his general ignorance of European instruments and observational methods at the time. In a letter written around 1731, long after his Delhi observatory had been in operation, he asks Boudier how the longitude of the moon is observed in Europe in its off-meridian positions.[4]

The invention of the telescopic-sight with improvements such as the micrometer and vernier, which had come into vogue with European astronomers only a few decades earlier, did not reach him in time. He did not have the opportunity to examine and evaluate European instruments such as the sextant and the quadrant before he settled for the masonry structures. His delegation to Europe did not bring back any sextants or quadrants either. The instruments that Boudier had brought along with him to Jaipur in 1731, also came too late.[5] Besides, they were not of sufficiently superior quality to have convinced the Raja that his own had become outdated.

Jai Singh made the best use of the technology available to him, the technology of building large masonry and stone structures, which was highly developed in the country. He found in the masonry structures an added advantage of "perfect stability."[6] He might have begun to realize the superiority of the European observing techniques later in his career when better tables of Flamsteed had been delivered to him. But by then the observatories had been built.

Jai Singh did engage in some sporadic observing with the telescope. In his *Zīj-i Muḥammad Shāhī*, he writes about seeing the sunspots and the four moons

---

[1] Pannekoek, p. 259.

[2] *Ibid*, p. 260.

[3] John Flamsteed collected data on planets, the moon, the sun, and some 3000 stars with an accuracy of $\pm 10''$. His data was published first in 1712 and finally posthumously in 1725 in three volumes under the title of *Historiae Coelestis Britannica*.

[4] See Ch. 13.

[5] These instruments included a 5 m long telescope and a 0.6 m quadrant. *Lettres*, pp. 772-778.

[6] ZMS, f. 1v.

of Jupiter.[1] However, the telescopes available to him were of inferior quality. With these telescopes he could not make out the rings of the planet Saturn. He states that the shape of the Saturn is oval, which suggests that his telescopes were of the same quality or only slightly better than those of Galileo. His erroneous statement that the sun takes one year to rotate once on its axis indicates that his observatories did not have any sustained program of observing the disc of the sun with a telescope. He was apparently more interested in measuring the coordinates of the sun than in observing sunspots. His fellow astronomers such as Jagannātha show no interest in telescopic observations. Jagannātha does not mention any telescopic observations in his *Samrāṭ Siddhānta*. He also does not include the telescope among the yantras for an observatory.

Jai Singh and the Copernican revolution

Because Jai Singh's astronomy program shows no influences of the Copernican revolution that had swept the intellectual circles of Europe, it has been suggested that Jai Singh deliberately avoided the heliocentric discoveries. Some have argued that Jai Singh, learning from the Jesuits of the unsettling social upheaval the Copernican revolution caused in Europe, decided to play it safe and suppressed it from the populace. This argument is weak when one considers that in India the geocentric world view never held the same grip on the religious beliefs of the masses as it did in Catholic Europe. Besides, the argument presupposes that Jai Singh's Jesuit employees made him aware of the discoveries of Kepler, Galileo, and Newton.

As we have shown in Chapter 13, Jai Singh's European assistants, primarily the Jesuits or Catholic laymen, were prohibited under the threat of the Inquisition, "to hold or defend the views supported by the discoveries of Kepler or Galileo.[2] It is unlikely, therefore, that they ever presented to the Raja the neo-astronomy of Europe in its true perspective. To them it was heresy. The Jesuits had come to the East to seek converts to their faith. They were not ready to engage in an activity which contradicted their own theological beliefs and which their superiors had branded as heresy, infidelity, and atheism.

Jai Singh might have had some idea regarding the mobility of the earth because the concept was not new to the Hindu astronomers of India. Āryabhaṭa (b. 476) in his *Āryabhaṭīya* explains the cause of day and night based on the rotation of the earth on its axis.[3] Jai Singh might have learned about the heliocentric

---

[1] See Ch. 11., The Bombay copy of the ZMS has diagrams of Jupiter and its four moons, and phases of the planets Venus and Mercury. S.M.R. Ansari, "The Observatories Movement in India during the 17-18th Centuries," *Vistas in Astronomy*, Vol. 28, pp. 379-385, (1985).

[2] See Ch. 13.

[3] *Āryabhaṭīya* of Āryabhaṭa, ed., K.S. Shukla and K.V. Sarma, Ch. 4, verse 9, New Delhi, 1976.

system of Copernicus as one hypothesis of the planetary arrangement from the Grosser Atlas published in 1725 and acquired by him from Germany. The atlas shows a large drawing of the Copernican system.[1] The atlas also shows the geocentric systems of Riccioli, of the Egyptians, and of Tycho. The Atlas does not show the Keplerian system, which had been the most successful system to that date, however. As the Jesuits brought most of the European books for the royal library, it is not a mere coincidence that these European books now preserved at the museum of Jaipur are all Ptolemaic. The Raja never saw the *De Revolutionibus* of Copernicus and the *Principia* of Newton. Evidently, Jai Singh never became aware of Copernican thought in its fullest scope. It is clear, therefore, that he did not deliberately choose to ignore the theoretical advances of the Copernican revolution. There is little merit in de Solla Price's criticism that ". . . it was out of a conservatism of purpose than any ignorance of the new astronomy of Europe" that made Jai Singh adhere to the medieval astronomy.[2] Jai Singh was simply unaware of this new astronomy.

Jai Singh's Primary Interests

Jai Singh's stated purpose had been to determine new planetary constants, but his primary interests in astronomy centered on the moon. He was more interested in observing and mathematically predicting the positions of this heavenly body. The questions he addressed to Boudier around 1731-32 relate exclusively to the moon.[3] He wanted to learn from Boudier the observational techniques, and the hypotheses on which the European astronomers based their calculations of lunar positions in the sky. Boudier traveled to Jaipur at Jai Singh's invitation and had several conversations with him there. Boudier, impressed with Jai Singh's knowledge of astronomy, calls him the most capable astronomer in India. Reporting these conversations in a letter to Europe, Boudier writes that Jai Singh was interested then in the prediction of solar eclipses, and in calculation of the occultation of stars and planets by the moon.[4]

Jai Singh clearly grasped the advantages of mathematical tools such as the logarithm, developed in the West, and had the logarithmic tables translated into Sanskrit.[5] Jai Singh appreciated the mathematical skills of the Europeans he

---

[1] Johann Baptist Homanns, *Grosser Atlas*, Nuremberg, 1725. The Sawai Man Singh II preserves a copy of it.

[2] Derek J. de Solla Price, "Astronomy's past preserved at Jaipur," *Natural History*, Vol. 73, No. 6, pp. 48-53, June-July 1964.

[3] Jai Singh's letter to Boudier, *Op. Cit.*, Ch. 13.

[4] Fonds Brotier, Vol. 88, *Op. Cit.*, Ch. 13.

[5] See Ch. 12.

met. Strobl writes that he pleased the Raja with his mathematical calculations.[1] Strobl further states, "Shortly before his death he [Jai Singh] had determined to request to our reverend General Father [head of the Jesuits] European priests knowledgeable in mathematics. He would appoint them as the heads of the observatories that were to be established throughout the realm."[2,3]

Jai Singh and the Perpetual Motion Machine

Jai Singh carried out his scientific activities in a day and age when people still thought that perpetual motion machines were possible. He was interested in such machines. According to Strobl, Jai Singh experimented with perpetual motion machines and spent more than 50,000 *Guilders* on his experiments.[4,5] It is not certain what kind of perpetual motion machines Jai Singh experimented with. Perhaps his initial inspiration came from Bhāskarācārya's *Siddhāntaśiromaṇi* where such a machine based on mercury is described.[6]

Jai Singh's Impact on India

Jai Singh's major contribution to India was the compilation of the *Zīj-i Muḥammad Shāhī*, a set of astronomical tables based on his own observations. The scholars of India who were trained according to the Islamic school of astronomy readily adopted the *Zīj-i Muḥammad Shāhī*. They wrote commentaries on it. To the world at large the *Zīj* was of little value, but to traditional Islamic scholars of India, to whom Western science was out of reach, the *Zīj* served a valuable need. These scholars prepared almanacs with its aid for more than 100 years.[7] Hindu astronomers also might have embraced the parameters of the *Zīj* and prepared their *pañcāṅgas* with it, however, the extent of their acceptance is still to be ascertained.

---

[1] Strobl writes, "I am being of service to him [Jai Singh] in matters of mathematics which is a favorite subject of his, as I have pointed out at greater length earlier. Besides he seems to be pleased with my services to the extent that he gave the emperor in MOGOL, who invited me to Delhi for the sake of mathematics, a negative answer to two written requests and refused to allow me to leave." Stöcklein, *Neuve Weltbott*, No. 643, *Op. Cit.*

[2] Stöcklein, *Neuve Weltbott*, No. 644, *Op. Cit.*

[3] We disagree with Kaye's conclusion that Jai Singh was indifferent to European achievements. Kaye, p. 89.

[4] The Guilder was a basic monetary unit of the Netherlands. According to the exchange rate of 1991, two guilders equalled approximately one U.S. dollar.

[5] Stöcklein, *Neuve Weltbott*, No. 644, *Op. Cit.*, Ch. 13.

[6] *Siddhāntaśiromaṇi* of Bhāskarācārya, p. 223, *Op. Cit.*, Ch. 1.

[7] See Ch. 11.

After Jai Singh

Jai Singh passed away in 1743, and his eldest son and successor, Īśvarī Singh, forced by the demands of imminent war, diverted all his resources to raising an army, and thus put aside the astronomical pursuits of his father. Strobl, an eyewitness of the events, writes: "In order to protect it [his inheritance] the present ruler is beginning to assemble a sizable army and he is expending tremendous sums of money for the procurement of all necessary war material. This money was formerly spent on splendid buildings and other splendors. Because of these worries, he [Īśvarī Singh] would easily forget the calculation of eclipses, the observations of the orbits of heavenly bodies and things of that nature."[1] The astronomical pursuits at Jai Singh's observatories came to a halt after his death. As Hunter puts it, "Urania fled before the brazen fronted Mars."[2]

In 1750, seven years after Jai Singh's death, his second son, Madho Singh, ascended the throne. Madho Singh revived the astronomical interests of his father. He built the Miśra yantra of Delhi and had some brass instruments fabricated, which now are in storage at Jaipur. However, he added or incorporated little that could be called genuinely new in astronomy. He merely kept alive the traditions of his father. Apparently, he also had no opportunity or desire to learn about the neo-astronomy that was developing by leaps and bounds in Europe. In Madho Singh's own lifetime the Delhi observatory ceased to operate, and its instruments were vandalized for their material.[3] After Madho Singh, astronomical activities at Jai Singh's other observatories declined. The decline was so severe that the astronomers employed there had no choice but to look for work elsewhere.[4] For example, a grandson of Kevalarāma, finding that astronomy was no longer "held in estimation, undertook a journey to *Deccan* (the South), in hopes that his talents might there meet with better encouragement."[5] At about this time the observatory compound at Jaipur was turned into a gun factory.[6,7] After Madho Singh, his successors did little that was worthwhile except now and then repairing an instrument or two that had become dilapidated.

---

[1] Strobl's Letter, No. 645, Stöcklein, *Neuve Weltbott, Op. Cit.*, Ch. 6.

[2] Hunter p. 210.

[3] See Ch. 5.

[4] Hunter, p. 209.

[5] *Ibid.*

[6] See Ch. 6.

[7] Hunter, p. 210. Hunter himself did not visit the Jaipur observatory.

Why Jai Singh's efforts failed to make a lasting impact

Jai Singh failed to initiate a new age of astronomy in India. He did not usher in the Copernican revolution that had swept the intellectual circles of Europe. It would be unfair to conclude, however, that he is solely to blame for this fact. Lack of good communication systems, and a complex interaction of intellectual stagnation, religious taboos, theological beliefs, national rivalries, and the simple human failings of his associates also share the blame to some extent.

Communications

The meagerness of the communications facilities played a larger role in Jai Singh's ignorance of contemporary European astronomy than is generally realized. Communications between Asia and Europe during the 17th-18th centuries were pitifully bad. There was no postal service in those days. Letters were hand carried to Europe by sailing ships which took almost six months to reach their destinations. Often, the sender would dispatch a second copy of his letter by another carrier to ensure that at least one message would reach its destination. Under these conditions, an Eastern scholar such as Jai Singh, thousands of miles from the scientific circles of Europe, could easily have remained ignorant of the contemporary European research in astronomy. He did receive some knowledge of this research, but that knowledge had to pass through the filters of theological beliefs of its carriers, the Jesuits.

Tradition-bound Scholars of India

The Copernican revolution could have come to India with Jai Singh's efforts had India's own scholars been a little less biased or less tradition-bound. The Indian scholars, particularly the Brahmins, had an inflated view of their own science and culture. They were unwilling to concede anything significant coming out of the West. For them their ancient texts of astronomy, such as the *Sūryasiddhānta*, were divinely inspired and contained the complete astronomical knowledge within. They believed that there was little useful information, if any, which was not already in their texts. Consequently, no one felt compelled by scientific curiosity to take a voyage to Europe to get first-hand information about European science.

Taboos and Prejudices

Jai Singh's Brahmin pundits would not undertake a journey to Europe because of a second reason. They faced a religious taboo against "crossing the ocean." If they crossed the ocean they could lose their caste. The Europeans, primarily the Jesuits, had their own taboos and prejudices. They had rejected Copernicus and Kepler because of their theological beliefs. These Jesuit astronomers such as Boudier and Strobl, as a result, also missed out on the unprecedented opportunity of being instrumental in initiating the Copernican revolution in a land where a scholar-prince had his purse strings wide open for such an adventure.

National Rivalries of European Powers

When Jai Singh sent his fact-finding mission to Europe, the European powers were fiercely competing with each other for territorial gains, colonial expansion, trade monopolies, and other petty national interests. They were engaged in cloak and dagger games and had every intention of preventing their rivals from gaining advantages on the soil of India or elsewhere. The national rivalries of Europe thus prompted the Portuguese to keep the valuable contacts they had established with a powerful prince of the East all to themselves. Portuguese rulers could have dissuaded the fact-finding delegation of Jai Singh from venturing outside Portugal in search of gaining first-hand information for their mission. The delegation did not visit England or France, where the most advanced observatories were operating at the time. Besides, England was a Protestant country, possibly another reason for the Portuguese authorities to discourage the mission from traveling there. The Indian members of the delegation also did not show any initiative of their own by insisting on visits to other countries.

Jai Singh, a Medieval Astronomer

Despite his best efforts, Jai Singh remained to the very end of his days a medieval astronomer, the last link in the tradition of Naṣīr al-Dīn al-Ṭūsī and Ulugh Beg. However, his instruments are the finest examples of a by-gone era that had come to a close with the advent of the telescope. Jai Singh shared the interests and concerns of the medieval astronomers of the Islamic period and opted for large masonry structures whenever high precision in measurements was called for. With his measurements, he concluded that the orbit of the sun was of an oval shape, but there is no evidence to suggest that he doubted the geocentric model of the heavens. Finally, to complete the picture, he wrote a *Zīj* on the pattern of the *Zīj-i Ulugh Begī*.

It is a tragic fact that the science of astronomy, despite Jai Singh's multi-faceted efforts, did not revive in India. Jai Singh failed to usher in the modern age of this science in his country. The neo-astronomy was to come from an entirely different route, i.e., the colonial British rule in the North.[1] Its arrival had to await a few decades more until the colonial British established themselves

---

[1] For an introduction to the Western astronomy at that time, see, S.M.R. Ansari, (1985), *Introduction of Modern . . . 18-19 Centuries, Op. Cit.*

firmly in Bengal. Then, a number of scholars of Persian, having acquainted themselves with the new science, wrote about it for the benefit of their fellow countrymen. For example: Abū Ṭālib al-Ḥusaynī wrote a monograph on modern astronomy in Persian in 1798. In 1826, Abu'l Khayr ibn Mawlwī Ghiath al-Dīn wrote *Majmūᶜa Shamsī*, a monograph on the Copernican system and also in Persian. Its Urdu translation appeared in 1843. An anonymous author wrote *Miftāh al-aflāk* (Key to Heavens) in Urdu in 1833.[1]

Jai Singh approached his astronomical researches with an open mind. Before embarking upon his project, he studied what was available to him. Moreover, during his investigations, he kept this attitude alive. His search for better instruments continued on even after his observatories had been built.[2] According to du Bois he was ready to put aside his own tables if better ones were available.[3] Jai Singh's accomplishments were medieval in retrospect, but his outlook was quite modern. His efforts were truly secular. For him scientific knowledge had no religion or nationality. Astronomers of all faiths participated in them—a fact that alone is a great compliment to a ruler born in an environment and age of intolerance and bigotry.

---

[1] S.M.R. Ansari in *History of Astronomy in India*, by Sen and Shukla, p. 401, *Op. Cit.*, Ch. 1. Also Ansari, (1985), *Introduction of Modern . . . Centuries,"* pp. 50-51, *Op. Cit.*

[2] He entrusted his delegation to Europe to learn about the new instruments of observing. In the series of questions addressed to Boudier he inquires about the instruments with which the moon is observed in its off meridian position. See Ch. 13.

[3] Joseph du Bois, *Op. Cit.*, footnote, Ch. 13.

APPENDIX I

THE EQUATION OF TIME FOR THE YEAR 1981[1]

Date	Equation of Time	Date	Equation of Time	Date	Equation of Time
	min. sec.		min. sec.		min. sec.
Jan. 0	−2 55.37	March 1	−12 28.64	May 1	2 53.12
3	−4 20.27	3	−12 04.73	3	3 06.92
6	−5 41.95	6	−11 25.43	6	3 23.47
9	−6 59.53	9	−10 42.36	9	3 35.04
12	−8 12.18	12	− 9 55.94	12	3 41.63
15	−9 19.20	15	−9 06.64	15	3 43.24
18	−10 20.04	18	−8 15.03	18	3 39.84
21	−11 14.35	21	−7 21.74	21	3 31.48
24	−12 01.87	24	−6 27.41	24	3 18.24
27	−12 42.44	27	−5 32.66	27	3 00.28
30	−13 15.91	30	−4 38.08	30	2 37.88
Feb. 1	−13 34.24	Apr. 1	−4 02.07	June 1	2 20.65
3	−13 49.35	3	−3 26.53	3	2 01.78
6	−14 05.96	6	−2 34.35	6	1 30.78
9	−14 15.28	9	−1 43.88	9	0 57.10
12	−14 17.37	12	−0 55.51	12	0 21.30
15	−14 12.44	15	−0 09.66	15	−0 16.05
18	−14 00.84	18	+0 33.17	18	−0 54.42
21	−13 43.00	21	1 12.51	21	−1 33.27
24	−13 19.42	24	1 47.91	24	−2 12.06
27	−12 50.56	27	2 18.98	27	−2 50.22

[1] As a first approximation, the equation of time for any other year would be similar.

Appendix I, The Equation of Time cont.

Date	Equation of Time	Date	Equation of Time	Date	Equation of Time
	min. sec.	Apr. 30	min. sec. 2 45.39	June 30	min. sec. 3 27.13
July 1	−3 39.04	Sept. 1	−0 07.35	Nov. 1	16 22.00
3	−4 02.09	3	+0 30.77	3	16 24.03
6	−4 34.37	6	1 30.02	6	16 21.08
9	−5 03.34	9	2 31.35	9	16 10.81
12	−5 28.49	12	3 34.25	12	15 53.06
15	−5 49.41	15	4 38.14	15	15 27.66
18	−6 05.79	18	5 42.38	18	14 54.54
21	−6 17.40	21	6 46.33	21	14 13.79
24	−6 24.08	24	7 49.41	24	13 25.63
27	−6 25.71	27	8 51.13	27	12 30.45
30	−6 22.10	30	9 51.02	30	11 28.75
Aug. 1	−6 16.71	Oct. 1	10 10.49	Dec. 1	11 06.82
3	−6 08.88	3	10 48.60	3	10 21.09
6	−5 52.50	6	11 43.41	6	9 08.14
9	−5 30.61	9	12 34.92	9	7 50.57
12	−5 03.34	12	13 22.60	12	6 29.05
15	−4 30.93	15	14 05.84	15	5 04.24
18	−3 53.69	18	14 44.05	18	3 36.87
21	−3 12.02	21	15 16.65	21	2 07.82
24	−2 26.32	24	15 43.20	24	0 38.01
27	−1 36.94	27	16 03.36	27	−0 51.54
30	−0 44.21	30	16 16.84	30	−2 19.82

APPENDIX II

The Nāḍīvalaya Plaque

The south plate of Nāḍīvalaya has a plaque with seven verses in Sanskrit. Bhavan re-engraved them in 1901-02, as he had found them, without making any corrections.[1] The verses are as follows:

1. dharmaglānimadharmavṛddhimavalokyātmā jagattasthuṣo
   rājendro jayasiṁha ityabhidhayāvirbhūyavaṁśeraghoh
   luptvā dharmavirodhino dhvaramukhai-
   ścācīrṇavedādhvabhirdharmaṁnyasya dharātale racitavān
   yantrān subodhānvavhan.
2. golapravṛttergagane caraṇāṁ jijñāsayā śrījayasiṁhadevah
   Ājñaptavān yantravidah punaste cakrurhi yāmyottara bhitti
   saṁjñam.
3. savajralepaṁṣuviśuddha pārśva dvayastha nāḍīvalayaika
   kendram
   dhruvābhi kendra śrutimārgakīlaṁ kīlāgrabhā sūcita
   naḍikādyam.
4. pitāmahocchiṣta mayaṁśca bhārkā rohāvarohā navananda vṛttān
   pratāpasiṁhascavivudhya vidbhyastān kārayāmās supārśva
   yugme.
5. bhāropama mlecchagaṇasya vṛddhi bhūbhāraśāntyai
   punarādidevah
   ikṣvākuvaṁśepyavatīrya pūrvāvatāritān devagaṇānayuṅkta.
6. dharmādhikārī vidhi devakṛṣṇah prāmukti saṁrohita
   dharmapadah
   yantreṣu vedaṅga vibhūṣaṇeṣu dvitīyayantroddharaṇam cakāra.
7. yasminnanhi caturṣu pakṣa tithi vārarkṣesu
   pakṣogatrighnaścānyai
   stribhiranvitah smṛtilavah syātsāṣti śākasyasah.
   nandaghnastithiranyayuk sacalavo viśvaghnavāronyayuk
   vātatvaghnabhamanyayukta mathavaiṣasyodhdṛtau syānmitih.

---

[1] Bhavan, pp. 46,47.

Appendix II, Nāḍīvalaya plaque cont.

Interpretation

The first five verses in the above are about Sawai Jai Singh and about his grandson, Pratāpa Singh. The next two relate to "the second restoration" of the instrument. David Pingree of the Brown Univ., Providence, R.I., U.S.A., has kindly interpreted the last verse for the author.[1] He writes:

> ". . ., I assume *sāṣṭiśāka* means 1600 *Śāka* {literally, it is *Śāka* (year) with 16} and the *smṛtilava* is 'portion (of the *Śāka* century) for memorizing.' The date contains of course a *pakṣa* (1 or 2), a *tithi* (1 to 16), a weekday (1 to 7), and a *nakṣatra* (1 to 27). The puzzle is that each of these numbers ($p,t,w,$and n) when multiplied by a given number and increased by three other numbers equals the year to the century to memorialized. Thus, if the year to be memorialized (0 to 99) is denoted s,

$$37\,p + t + w + n = s$$
$$9\,t + p + w + n = s$$
$$13\,w + p + t + n = s$$
$$25\,n + p + t + w = s$$

Solving these four equations, with the above conditions, we find $s = 92$, $n = 3$, $w = 6$ (Friday), $t = 9$, and $p = 2$. The date then is weekday 6, (Friday), *tithi* 9 in *pakṣa* 2, with the moon in *nakṣatra* 3 (*Kṛttikā*) in *Śāka* 1692. This corresponds to Friday 9, *Śuklapakṣa* of *Pauṣa* in *Śāka* 1692 = 25 January 1771."

---

[1] David Pingree, Brown University, private communication.

APPENDIX III

The *Abjad* Notation of Numbers in the *Zīj-i Muḥammad Shāhī*[1]

Number	Number in ZMS	Number	Number in ZMS	Number	Number in ZMS
1	ا	21	كا	41	ما
2	ب	22	كب	42	مب
3	ج	23	كج	43	مج
4	د	24	كد	44	مد
5	ه	25	كه	45	مه
6	و	26	كو	46	مو
7	ز	27	كز	47	مز
8	ح	28	كح	48	مح
9	ط	29	كط	49	مط
10	ى	30	ل	50	ن
11	يا	31	لا	51	نا
12	يب	32	لب	52	نب
13	يج	~~34~~ 33	لج	53	نج
14	يد	34	لد	54	ند
15	يه	35	له	55	نه
16	يو	36	لو	56	نو
17	يز	37	لز	57	نز
18	يح	38	لح	58	نح
19	يط	39	لط	59	نط
20	ک	40	م	60	س

[1] A similar notation of numbers is also seen on some Persian astrolabes engraved in India.

Appendix III, the *Abjad* notation of numbers cont.

Number	Number in ZMS	Number	Number in ZMS	Number	Number in ZMS
61	سا	81	فا	105	قه
62	سب	82	فب	110	قی
63	سج	83	فج	115	قیه
64	سد	84	فد	120	قک
65	سه	85	فه	125	قکه
66	سو	86	فو	130	قل
67	سز	87	فز	135	قله
68	سح	88	فح	140	قم
69	سط	89	فط	145	قمه
70	ع	90	ص	150	قن
71	عا	91	صا	155	قنه
72	عب	92	صب	160	قس
73	عج	93	صج	165	قسه
74	عد	94	صد	170	قع
75	عه	95	صه	175	قعه
76	عو	96	صو	180	قف
77	عز	97	صز	185	قفه
78	عح	98	صح	190	قص
79	عط	99	صط	195	قصه
80	ف	100	ق	200	ر

Appendix III, the *Abjad* notation of numbers cont.

Number	Number in ZMS	Number	Number in ZMS	Number	Number in ZMS
205	ره	270	رع	335	شله
210	ری	275	رعه	340	شم
215	ریه	280	رف	345	شمه
220	رک	285	رفه	350	شن
225	رکه	290	رص	355	شنه
230	رل	295	رصه	360	شس
235	رله	300	ش	400	ت
240	رم	305	شه	500	ث
245	رمه	310	شی	600	خ
250	رن	315	شیه	700	ذ
255	رنه	320	شک	800	ض
260	رس	325	شکه	900	ظ
265	رسه	330	شل	1000	غ
				0	[illegible]

APPENDIX IV

Table of Contents of the Zīj-i Muḥammad Shāhī[1]

[1] Mercier, *Op. Cit.*, Ch. 5.

Appendix IV, Contents of the *Zīj-i Muḥammad Shāhī* cont.

Appendix IV, Contents of the *Zīj-i Muḥammad Shāhī* cont.

Appendix IV, Contents of the *Zīj-i Muḥammad Shāhī* cont.

APPENDIX V

Astronomy-Astrology Books at Jai Singh's Personal Library,

The Pothīkhānā[1]

1. *Anūpavyavahārasāgara* (1)
2. *Āryabhaṭīya golādhyāya* (1)
3. *Āryasiddhānta* (1)
4. *Bālabodha* (1)
5. *Bālaviveka* (1)
6. *Bālavivekinī* (1)
7. *Bṛhaspatisiddhānta* (1)
8. *Bhāsvatī* (3)
9. *Bhūpālavallabha* (1)
10. *Bhuvanadīpikā* (3)
11. *Bījagaṇita* (4); original (2), one by Nārāyaṇa *Paṇḍita*
12. *Brahmatulā* (*Brahmatulya Sāraṇī*) (1)
13. *Bṛhatjātaka* (4); *ṭīkā* (2), original (2)
14. *Candrasūryagrahaṇādhikāra* (1)
15. *Devajñānamanohara* (1)
16. *Dvādaśabhāvagrahakalpa* (1)
17. *Gaṇakamaṇḍana* (1)
18. *Gajanirayana* (1)
19. *Garustavayantra* (?) (1)
20. *Grhalāghava* (6); *ṭīka* (1), *udāharaṇa* (1)
21. *Horājñāna* (1)
22. *Horātra* (?) (1)
23. *Jaiminasutravṛtti* by Bālakṛṣṇānanda Sarasvatī (1)
24. *Jātakajñānamuktāvali* (1)
25. *Jñānamañjarī* (1)
26. *Jyotiṣabīja* (2); *ṭīkā* (1)
27. *Jyotiṣyonnatisāra* (1)
28. *Kālajñāna* (1)
29. *Karaṇakutūhala* (1)
30. *Karaṇaprakāśikā* (1)
31. *Keśavīyapaddhati* (2)
32. *Laghujātaka* (3); original (1), *ṭīkā* (1)
33. *Līlāvatī* (9); original (6); *ṭīkā* (3) including two named as *buddhivilāsa*
34. *Mahādevīsāraṇī* (1)
35. *Makaranda* (1)
36. *Makarandaviveka* (1)
37. *Māsaratnamālā* (1)
38. *Mitākṣara ṭīkā* (1)
39. *Muhūrtacintāmaṇi* (1)
40. *Muhūrtacintāmaṇi ṭīkā* by Pīyūṣadhara (1)
41. *Muhūrtadarpaṇa* (1)
42. *Muhūrtakalpa* (1)
43. *Muhurtamālā* (1)
44. *Muhūrtamālā* (1)
45. *Muhūrtaśiromaṇi* (1)
46. *Nalikābandhanakarma paddhati* (1)
47. *Nāsatapatrikā* (1)
48. *Nīlakaṇṭhī* (1)
49. *Pañcapakṣī* (2)
50. *Padmakośa* (1)
51. *Pārasīprakāśa* (1)
52. *Pātasāraṇī* (3)
53. *Pitāmahasiddhānta* (1)
54. *Rājamṛgānka* (1)
55. *Rājapaddhati* (1)
56. *Ramala* (2); *sārapatra* (1)
57. *Rāmavilāsa* (2); *paṭīgaṇita* (1)
58. *Ratnamālā* (5); *ṭīkā* (2)
59. *Sajjanavallabha* (1)
60. *Sajñana tantra* (?) (1)

---

[1] Pothikhana records of the Jaipur State, R.S.A.

Appendix V, Library cont.

61. *Samassar* (?) (5); original (1), *ṭīkā* (1)
62. *Sāmudrika* (4)
63. *Sāmudrikatilaka* (1)
64. *Sarvatobhadra* (2)
65. *Siddhāntakaustubha majastī* (1)
66. *Siddhāntarahasya ṭīkā* (1)
67. *Siddhāntasaṁhitāsāra* (1)
68. *Siddhāntasindhu Nityānandī* (2)
69. *Siddhāntaśiromaṇi* (6)
70. *Siddhāntaviveka* (1)
71. *Śivalikhyatam* (4)
72. *Śrīpatiṭīkā* (1)
73. *Śrīyantracintāmaṇi* (1)
74. *Stotrahs* related to *tantra* in Bengali script (1)
75. *Sudhāsāraṇī* (1)
76. *Śukajātaka* (1)
77. *Sundarasiddhānta* (1)
78. *Surasiddhānta* (1)
79. *Sūryasiddhānta* (7); with *ṭīkā* (1)
80. *Sūryatulāvidhāna* (1)
81. *Svapnacintāmaṇi* (1)
82. *Svapnādhyāya* (1)
83. *Svarodaya* (1)
84. *Svarodaya* (3)
85. *Tajikasāra* (3); *ṭīkā* (1)
86. *Tatvaviveka* (1)
87. *Tithicintāmaṇi sāraṇī* (1)
88. *Tithivinodasāraṇī* (1)
89. *Toḍarānanda* (1)
90. *Trilokaprakāśa* (1)
91. *Tripraśnādhikāra* (1)
92. *Varṣagaṇitapaddhatibhūṣaṇa* (1)
93. *Vārāhīsaṁhitā* (6); *Bhatotpalaviricita* (1) *Bhatotpalaviricita-ṭīkā* (1), *ṭīkā* (1)
94. *Varṣaphalapaddhati* I (1)
95. *Varṣaphalapaddhati* II (1)
96. *Varṣatantra* (1)
97. *Vāsanābhāṣya adhyāyas' ṭīkās: pātādhyāya* (1), *golādhyāya* (1)
98. *Vasanābhāṣyamitākṣara* (1)
99. *Vāsanāyantrādhyāya* (1)
100. *Vasantarāja* (5); original (2), with *ṭīkā* (2)
101. *Vastupaddhati* (1)
102. *Vastuprakaranajñāna* (1)
103. *Vijaya* (?) *kalpalatā* (?) (1)
104. *Vijayavāsanābhāva* (1)
105. *Vivāhapatala* (1)
106. *Vrataśata* (1)
107. *Vyāsasiddhānta* (1)
108. *Vyavahāradīpikā* (1)
109. *Yantrarāja* (5); *ṭīkā* (1)
110. *Yogārṇava* (1)
111. *Yoginīdaśā* (2)
112. (*Zīj-i*) *Muḥammad Shāhī* in *Hindagī* (Devanagari) characters (1)

Appendix V, Jai Singh's Library cont.

Persian Books at Jai Singh's Library

The inventory of the library gives the total number of the Persian Books as 7, and then lists the following 6 titles.

1. *Suratuabdulrahaman* (1)
2. *Khamakehana Jayan* (1)
3. *Taharir Ukledas* (2)
4. *[Zīj-i] Muḥammad Shāhī* in Persian characters (1)
5. *Sharḥ Zīj Mirza* (1)

Books of European Origin at the Library

The inventory lists 24 books, and then counts them as follows:

1. *Zījes* of *Siddhānta* (3)
2. *Jyotiṣa* (17)
3. European *Zīj* with tables (1)
4. *Celestial Atlas* (1)
5. *Atlas of the World* (1)
6. A Book on battle tactics (1)

In addition *Rekhāgaṇita Ukalīdasa* (Euclid's Geometry) in *Phiraṅgī* (European) characters is listed separately.

APPENDIX VI

Astronomical Instruments at the Royal Library

(Total no. 36)

1. Astrolabes (6); including one with a cost of Rs.400.
2. Miscellaneous (11), including:
   a. *Phiraṅgī* (European) instrument, for measuring altitude of land objects.
   b. Zarqālī astrolabe (*Jarakālī yantra Sarvadeśī*).
   c. Cūḍā yantra
   d. Turahī yantra
   e. Yantra related to a Nāḍīvalaya *ghaḍī*
   f. *Bhaupatra* yantra for all four directions (?)
   g. Terrestrial globes (*gola*) (2)
3. Krāntivṛtta (2)
4. Celestial globes (*Khagola*) (7)
5. Wooden instruments (2), including
   a. spherical instrument (*gola yantra*)
   b. wooden box for a Palabhā yantra
6. Hemispherical (*Katorā*) yantras (4); including:
   Jaya Prakāśa (1), complete
7. *Phiraṅgī* (European) instrument of ivory
8. Telescope of wooden body, cost Rs.100
9. Nāḍīvalaya plates (2)
10. Set of drawing instruments, (European)
11. Silver astrolabe

The last two instruments are missing from the inventory of 1741-43. They are included, however, in the inventory of 1804-1805 taken during the reign of Īśvarī Singh.

APPENDIX VII

The Arabic and Persian books at the Sawai Man Singh II Museum, Jaipur

1. *Jāmī-i Shāhī*, Persian, (astrology) No.2 (AG).
2. *Zīj-i Sulṭānī* of Ulugh Beg with commentary by Mullah Chānd, Persian, (acquired 1725), No. 6 (AG).
3. *Zīj-i Sulṭānī* of Ulugh Beg with commentary by ᶜAlī al-Birjandī Persian, No. 5 (AG).
4. *Zīj-i Sulṭānī* of Ulugh Beg, Persian, (acquired 1727), No. 11 (AG).
5. *Zīj-i Khāqānī* of Ghiyāth al-Dīn al-Kāshī, Persian, (acquired 1728), No. 9 (AG).
6. *Zīj-i Shāhjahānī* by Farīd al-Dīn Masᶜūd ibn Ibrāhīm al-Dihlawī, Persian, No. 12 (AG).
7. ------, second copy, (acquired 1725), No. 14 (AG).
8. *Al-Tafhīm li-awā'il ṣināᶜat al-tanjīm* by Abu'l-Rayḥān al-Bīrūnī, Persian, (acquired 1725), No. 7 (AG).
9. *Almagest*, Arabic, (acquired 1725), two copies, Nos. 19 and 20 (AG).
10. *Kitāb al-Manāzir* of Ibn al-Haytham, as contained in *Tanqīḥ al-Manāzir* by Kamāl al-Dīn al-Fārisī, Arabic, No. 17,1 (AG).
11. The Arabic treatise on the rainbow and lunar halo by Ibn al-Haytham, No. 17,2 (AG).
12. *Lawā'iḥ al qamar* by Ḥusayn ibn ᶜAli al-Bayhaqī al-Kāshifī, Persian, (astrology, acquired 1725), No. 91 (AG).
13. *Al-Mulakhkhaṣ fi'l-hay'a* by Maḥmūd ibn ᶜUmar al-Jaghamīnī, with commentary by Qāḍīzāde al-Rūmī, Arabic, (acquired 1725), No. 18 (AG).
14. *Sharḥ Tadhkira* by Nizāmu'd-dīn al-Nīshāpurī, Arabic, (acquired 1725), No.21 (AG).
15. -----, second copy, No. 22 (AG).
16. *Sharḥ-Shamshīya-Ḥisāb of al-Birjandī* with commentary, Nizāmu'd-dīn al-Nīshāpurī, Arabic, (acquired 1725), No. 10 (AG).
17. *Risālāh-hai'at al-Kursī*(?), Arabic, (acquired 1725), No. 90 (AG).

APPENDIX VIII

European Books at the Sawai Man Singh II Museum

1. *Historia Coelestis Britannica* by Flamsteedious, Vols. I, II and III (Nos. 84-85); printed in 1725 A. D.
2. *Observations Chinoise*, Vol. II, by P. Gaubil (1732 A. D.); No. 47 J.M.
3. Ad Astrum (Astronomy) (1557 A.D.); bears a seal No. 48 J.M.
4. *Dictionarium Latinum* (16th cent.); No. 49 J. M.
5. *Description and use of sector and other instruments* (Geometry), (1636 A.D.); bears a seal; No. 51 J. M.
6. *The pathway to knowledge*; printed at Paules Churchyard, London; (1551 A. D.); bears a seal; No. 52 J. M.
7. *Seamen's Calendar* (English); 17th cent.; No. 53 J. M.
8. *Common accidence examined* (English Grammar) by Charles Hoole; printed at Mercus Chappel, London; (1663 A. D.); No. 55 J. M.
9. *Traite de physiques* (1675 A. D.); No. 57 J. M.
10. *Sphera* (Latin), by Aliay Joseph du-Buoy (du Bois?), written in 1732 A. D. at Jaipur; (Reserved collection) No. 36.

BIBLIOGRAPHY

History of Sawai Jai Singh and of Jaipur State

1. Virendra Swarup Bhatnagar, *Life and Times of Sawai Jai Singh 1688-1743*, Impex India, Delhi, 1974.
2. Harish Chandra Tikkiwal, *Jaipur and the Later Mughals*, Jaipur, 1974.
3. K.A. Nilakanta Sastri and G. Srinivasachari, *Advanced History of India*, Allied Publishers, reprint, New Delhi, 1980.
4. Jadunath Sarkar, *The Fall of the Mughal Empire*, 4 Vols., Calcutta, 1932-1950.

History of Astronomy

1. A. Pannekoek, *A History of Astronomy*, Interscience Publishers, New York, 1961.
2. Charles Singer, E.J. Holmyard, A.R. Hall and Trevor I. Williams, ed., *History of Technology*, Vol. 3, Oxford Univ. Press, Oxford, 1964.
3. Joseph Needham, *Science and Civilization in China*, Vol. 3, pp. 171 ff., Cambridge Univ. Press, 1959.

Instruments in Europe

1. Edmund Stone, *The Construction and Principal Uses of Mathematical Instruments. Translated from the French of M. Bion, Chief Instrument-Maker to the French King . . .*, First published in 1758, reprint, The Holland Press, London, 1972.
2. Derek J. Price, "Precision Instruments: To 1550," and "The Manufacture of Scientific Instruments from *c* 1500 to *c* 1700," in *A History of Technology*, ed., Charles Singer et al, *Op. Cit.*

Astronomy in India

1. S. N. Sen and K. S. Shukla, ed., *History of Astronomy in India*, Indian National Science Academy, New Delhi, 1985.
2. David Pingree, *Jyotiḥśāstra, A History of Indian Literature*, ed. Jan Gonda, Vol. 6, pp. 17-40, Otto Harrassowitz, Wiesbaden, 1981.
3. Shankar Balkrishna Dikshit, *History of Indian Astronomy*, tr., R. V. Vaidya, Delhi, 1969.

The Persian-Arabic or Islamic School of Astronomy

1. W. M. O'Neil, *Early Astronomy from Babylonia to Copernicus*, pp. 117-145, Sydney University Press, Sydney, 1986.
2. Aydin Sayili, *The Observatory in Islam*, Ankara, 1960.
3. J. L. E. Dreyer, *A History of Astronomy from Thales to Kepler*, Dover, New York, 1953.
4. E. S. Kennedy, "A Survey of Islamic Astronomic Tables," *Transactions of the American Philosophical Society*, Vol. 46, part 2, pp. 123-177, Philadelphia, 1956.
5. E. S. Kennedy, "Al-Kāshī's Treatise on Astronomical Observational Instruments," *J. Near Eastern Studies*, Vol. 20, pp. 98-107, (1961).
6. Hugo J. Seemann, "Die Instrumente der Sternwarte zu Maragha nach den Mitteilungen von al-ᶜUrḍī," *Sitzungsberichte der physikalischmedizunischen Sozietat zu Erlangen*, Vol. 60, pp. 15-126, (1928).
7. Louis P. E. A. Sedillot, *Materiaux pour Servir a L'Histoire Comparee des Sciences Mathematiques Chez Les Grecs et Les Orientaux*, Tome Premier, pp. 289-364, Librarie de Firmin Didot Freres, Paris, 1845-1849.
8. *Yantraprakāra of Sawai Jai Singh*, ed. and tr., Sreeramula Rajeswara Sarma, *Supplement to Studies in History of Medicine and Science*, Vols. X and XI, New Delhi, 1986, 1987. The Supplement has been published in a book form.
9. For Instruments at Maragha and Samarkand, See ᶜAbd al-Munᶜim Āmīlī, Ms. The British Museum, O.M.P., Add. 7702, Title "Tract, Astronom. Al-Monaem."

Sawai Jai Singh's Instruments and Astronomy

1. *Yantraprakāra of Sawai Jai Singh*, ed. and tr., Sreeramula Rajeswara Sarma, Supplement to Studies in History of Medicine and Science, Vols. X and XI, New Delhi, 1986, 1987. Available in book form from the publishers. At the end, the booklet has an excellent bibliography of essays on Jai Singh's astronomy.
2. A. ff. Garrett and Chandradhar Guleri, *The Jaipur Observatory and Its Builder*, Pioneer Press, Allahabad, 1902.
3. Gokula Candra Bhāvana, *A Guide to the Observatories in India* (in Hindi), Varanasi, 1911.
4. G.R. Kaye, *The Astronomical Observatories of Jai Singh*, Archaeological Survey of India, reprint, New Delhi, 1982.

INDEX